THE LEAN TOOLBOX

THE ESSENTIAL GUIDE TO LEAN TRANSFORMATION

FOURTH EDITION

THE LEAN TOOLBOX: THE ESSENTIAL GUIDE TO LEAN TRANSFORMATION

FOURTH EDITION

Published by:

PICSIE Books
Box 622
Buckingham, MK18 7YE
United Kingdom

How to order:

PICSIE Books
Telephone & Fax: +44 (0) 1280 815 023
Web site: www.picsie.co.uk
E-mail: picsiebook@btinternet.com

Copyright © PICSIE Books, 2009
Publication date: January 2009
ISBN: 978-0-9541244-5-8
British Library Cataloguing-in-Publication Data
A catalogue record for this book is available from the British Library

'The credit belongs to the man who is actually in the area, whose face is marred by dust and sweat and blood, who knows the great enthusiasms, the great devotions, and spends himself in a worthy cause; who, at best, if he wins, knows the thrills of high achievement, and, if he fails, at least fails daring greatly, so that his place shall never be with those cold and timid souls who know neither victory nor defeat.'

John F Kennedy on Theodore Roosevelt

New York City, December 5th 1961

To those that work day by day at the Gemba.

FOREWORD TO THE 4TH EDITION

By
Julie Madigan

That the book you are about to read opens with a chronology of Lean, is a curious and revealing thing. Lean has been described as a philosophy, a theory, a method, a methodology, even a toolbox, but none of these things obviously lend themselves to a chronology. Most theories, methods, philosophies enjoy a heyday, are accepted for a brief span, and are then superseded, becoming obsolete.

They do not evolve.

Lean, however, it appears, has evolved in the last half century. And I say evolved rather than developed, as it has done so only as part of an external environment, in reaction to, and as a prompt for, a changing landscape of business and production. It has travelled from Japan to America and back again, from Australia to the Urals. Every mile has taken Lean closer to the purism of Ohno and yet spawned a host of variants, each potentially more suited to its individual time and place.

There is now a schism opening between the disciples of Lean who cleave to the traditional canon of the Toyota Production System and those who favour an approach of adaptation, rather than adoption, of Lean.

Whether you feel Lean is converging towards perfection or expanding towards local specialisms, this book will give you the knowledge to judge how best Lean can help you.

The text has certainly played a central role in the educational programmes that have been delivered by The Manufacturing Institute since 1995. The Institute has, in fact, ordered over 4,000 copies of the previous editions, over the years, and it remains a core text on introductory courses through to the MSc, and at all levels in between. The Lean Toolbox's broad coverage, combined with depth of knowledge, make it a key choice for all students of Lean, whatever their level of experience.

Dr Julie Madigan

Chief Executive,
The Manufacturing Institute

FOREWORD TO THE 3RD EDITION

By
Peter Hines

I am frequently asked by firms on their Lean conversion journey what should they read. Another question often asked involves finding out what tools should be used. A third involves how apparently competing subjects such as Lean, theory of constraints and Six Sigma fit together. My usual response refers people to this publication and its forerunners.

The New Lean Toolbox sets out in a clear and reader friendly way how Lean fits together. It is an invaluable resource for staff right across an organisation. It provides a ready reckoner for directors wanting to know what is going on in their business, a guide to the improvement agent as well as an insight to those involved as part of a project team.

Whilst no organisation will ever use all of the tools illustrated here, understanding what is available and how these can be pulled to the needs of the organisation is an excellent starting point. Those who are likely to be most successful will integrate their approach into a systematic change programme taking into account the needs of the customer, business strategy as well as the needs of the people in the business. In order to start from the right place I would suggest reading this publication right through at an overview level before developing your own implementation plan. After that it will be a handy reference text towards your own desired future state.

Good luck on your Lean journey.

Professor Peter Hines

Director,
Lean Enterprise Research Centre

FOREWORD TO THE 2ND EDITION

By
Dan Jones

There is a growing realisation that there is much more to Lean than we thought. In its purest form it grows out of a genericised version of the Toyota Production System, pioneered by Taiichi Ohno and his colleagues in the early post war years. However they in turn rediscovered the path originally followed by Henry Ford in his first large plant in Highland Park. There every operation was arranged in a continuous flow sequence.

This in turn was inspired by Colt's rifle factory in Hertford back in 1855. So the idea of continuous flow has a long history.

But no one was able to follow the example of Highland Park. As soon as they needed to make a range of products they chose to organise by process, activity and department and to schedule work through each step, using ever more complicated scheduling systems. The River Rouge plant was a classic process village organisation, with big batched and long lead times. Even Henry Ford lost the plot and followed a different fork in the way! But in truth as long as markets were growing fast the hidden wastes in such a system did not matter and the error of his ways did not become apparent.

Except to Eiji Toyoda and Taiichi Ohno – who were determined to find a way of successively overcoming all the obstacles to linking every step into a continuous flow sequence, precisely synchronised with the demand of the end customer. Twenty five years later the world woke up to the power of this alternative path – and many have been struggling to follow their example.

As they do so they discover not only the power of this alternative path, but how profoundly different it is to the traditional batch and queue path. Tools and techniques for moving from batch to flow are just the first step – the next is to combine them in the right way using Lean principles, and then to design an implementation path to transform the organisation and its relationships with its suppliers and customers down each value stream. We are at the beginning of a long journey to explore the true power of Lean.

This book will be a great help on your journey.

Best wishes

Dan Jones

Founder and Chairman, The Lean Enterprise Academy UK
Co-author of 'The Machine that Changed the World', 'Lean Thinking' and 'Lean Solutions'

Table of Contents

A Lean Chronology – The Road to (and away from) Lean

1780s – 1790s Development of what is today called standard ops and quick changeover by Royal Navy enabling them to deliver a broadside twice as fast as the French or Spanish Navies

1797 Maudslay builds the world's first precision metal screw cutting machine. This was the 'parent' machine of the machine tool industry.

1810 Maudslay and Brunel (father of Isambard) set up the first mechanised production line that produced 160k pulleys per year with 10 men, for the Royal Navy – previously done with inferior quality by 100 men

1859 Smiles publishes 'Self-Help' – a book that inspires Sakichi Toyoda. The only book on display at his birthplace.

1871 Denny, a Scottish shipbuilder, asked his workers to suggest methods for building ships at lower cost

1893 Taylor begins work as a 'consulting engineer'.

1896 Pareto publishes law of economic distribution

1898 Taylor begins his time studies of shovelling of iron

1904 Cadillac begins building cars using interchangeable parts

1906 Oldsmobile builds first car with multiple parts coming from external suppliers

1908 Ford Model T

1909 Frank and Lillian Gilbreth study bricklaying. Beginnings of motion study.

1911 Wilson Economic Order Quantity (EOQ) formula

1913 Ford establishes Highland Park plant using the moving assembly line

1922 Gantt 'The Gantt Chart: A Working Tool for Management'

1925 Stuart Chase, 'The Tragedy of Waste', Macmillian

1925 'Mass Production' phrase coined by Encyclopaedia Britannica

1926 Henry Ford 'Today and Tomorrow'

1927-1930 Mayo and Roethlisberger studies at Hawthorn Plant of Western Electric

1929 Sakichi Toyoda sells the rights for a quick-change shuttle to Platt Brothers UK for £100k, and establishes a car business.

1931 Shewhart. 'Economic Control of Quality of Manufactured Product' Van Nostrand. First book on SPC and PDCA

1934 Maynard coins the term 'Method Study'

1936 An engineer at General Motors coins the term 'automation'. Toyota sells first car. The name 'Toyota' adopted because it can be written in Japanese with 8 pen strokes (8 is a lucky number in Japan). Toyoda requires 12 strokes. Kiichiro Toyoda visits USA, especially Ford, and starts 'just in time'.

1937 Establishment of Toyota Motor (Toyoda Loom Works established 1922)

1940 TWI (Training within Industry) programme begun by US military sets out 3 key tasks for supervisors: Job Instruction, Job Improvement, Job Relations. Introduced into Japan in 1949.

1942 Juran: Reengineering procurement for Lend Lease (90 days to 53 hours)

1943 Toyoda hospital established. Now known as Kariya Toyota hospital.

1943-44 Flow production of bombers at Boeing Plant II and Ford Willow Run

1945 Shingo presents concept of production as a network to JMA. Also identifies batch production as the main source of delays

1948 Deming first sent to Japan. Lectures on waste as being the prime source of quality problems

1949 Juran first goes to Japan

1950 Eiji Toyoda visits Ford's River Rouge plant. Impressed by Ford's suggestion scheme. Ohno visits in 1956. Ohno learns about pull from supermarket chain Piggly Wiggly.

1950 Ohno begins work on the Toyota Production System following strikes. First use of a U shaped cell at Toyota. TWI introduced into Toyota during early 1950s.

1951 Deming Award established in Japan

1951 Juran 'Handbook of Quality Control' (first edition). Includes cost of quality, Pareto analysis, SPC. (Fifth edition published 1999)

1955 First use of Andon lights.

1956 First sea container shipment

1961 Shingo devises and defines 'pokayoke', book published in 1985

1961 Ishikawa devises Quality Circles, and first are set up in 1962. Juran introduces the concept in Europe in 1966.

1961 Feigenbaum, 'Total Quality Control', McGraw Hill

1963 Toyota South Africa plant established

1969 First microchip designed at Intel by Ted Hoff.

1971 Mudge, 'Value Engineering: A Systematic Approach', McGraw Hill

1974 Skinner 'The Focused Factory', Harvard Business Review.

1974 First commercial bar code scan – with Wrigley's gum

1975 Orlicky, 'Material Requirements Planning', McGraw Hill

1975 Burbidge, 'The Introduction of Group Technology', Heinemann – lays down cell design principles

1978 APICS MRP Crusade

1978 First articles on 'Just in Time' appear in US magazines.

1980 NBC television screens 'If Japan Can, Why can't We'. Kawasaki opens factory in USA running 'Kawasaki Production System', based on TPS.

1981 Motorola begins a methodology called 'Six Sigma'

1982 Deming 'Quality, Productivity and Competitive Position' MIT Press and 'Out of the Crisis', MIT Press - contains his 14 point plan.

1982 Schonberger 'Japanese Manufacturing Techniques', Free Press

1982: Hewlett Packard video on 'Stockless Production' shown at APICS conference and widely in USA

1983 Hall, 'Zero Inventories', Dow Jones Irwin APICS

1983 Monden, 'Toyota Production System', Ind Eng and Management Press

1984 Goldratt 'The Goal'

1984 Hayes and Wheelwright, 'Restoring our Competitive Edge', Free Press (Second edition. 'Pursuing the Competitive Edge', 2005)

1984 Kaplan 'Yesterday's Accounting Undermines Production' HBR and 1987 Kaplan & Johnson 'Relevance Lost: The Rise and Fall of Management Accounting'

1985 Shingo: 'SMED', Productivity (But note that quick change methods and technology were in use at Ford River Rouge in the 1930s – now on display in Ford museum.)

1985 Skinner, 'Manufacturing: The Formidable Competitive Weapon', Wiley

1986 Hill, 'Manufacturing Strategy', Macmillan

1986 Imai 'Kaizen – The Key to Japan's Competitive Success'

1986 Goldratt and Fox, 'The Race'

1987 Baldridge award established.

1987 Davis in 'Future Perfect' makes first mention of Mass Customisation.

1987 Boothroyd and Dewhurst 'Design for Assembly'

1988 Nakajima 'Introduction to Total Productive Maintenance'

1988 Ohno, 'Toyota Production System', Productivity

1988 Motorola wins Baldridge Award. Winners are required to share their knowledge so Six Sigma becomes more widely known.

1988 Akao introduces QFD into manufacturing

1988 Cooper and Kaplan, 'Measure Costs Right: Make the Right Decision', Harvard Business Review (First paper on ABC)

1989 Ohno and Japanese Management Association, 'Kanban – Just in Time at Toyota', Productivity Press.

1989 Shingo Prize established

1989 Camp 'Benchmarking: The Search for Industry Best Practices', ASQ Quality Press

1990 Taiichi Ohno dies

1990 Stalk and Hout: 'Competing against Time', Free Press – sets out 'time based competition'.

1990 Hammer 'Reengineering Work: Don't Automate, Obliterate' Harvard Business Review, and 1994 Hammer and Champy 'Reengineering the Corporation'

1990 Pugh 'Total Design', and 1981 Concept Selection

1990 Womack and Jones, 'The Machine that Changed the World', Rawson

1990 Schonberger, 'Building a Chain of Customers', Free Press

1990 Quick Response initiative started by WalMart

1990 Osborn, Moran, Musselwhite, Zenger, 'Self Directed Work Teams', Business One

1992 Jack Stack, 'The Great Game of Business', Currency Doubleday (start of 'open book management')

1992 EFQM award established

1993 Hajime Ohba becomes general manager of Toyota Supplier Support Centre, and begins teaching TPS to US companies, many of them outside automotive.

1993 Pine, 'Mass Customisation', Harvard

1994 AME begins promotion of 'Kaizen Blitz' (Book by Laraia, Moody and Hall published in 1999)

1994 Altshuller, First English translation of book about TRIZ

1995 Clayton Christensen, 'Disruptive Technology', Harvard

1996 Womack and Jones 'Lean Thinking', Simon and Schuster

1997 Christensen, 'The Innovator's Dilemma', (and 2003, 'The Innovator's Solution').

1996 Hopp and Spearman, 'Factory Physics', Irwin (Second edition, 2000)

1998 Suri 'Quick Response Manufacturing', Productivity Press

1999 Rother and Shook, 'Learning to See', Lean Enterprise Institute

1999 Spear and Bowen, 'Decoding the DNA of the Toyota Production System', Harvard Business Review

1999 Cardiff Business School establishes the first MSc degree entirely devoted to Lean.

2000 Internet supply platform established by Big Three, called 'COVISINT'.

2000 Johnson and Bröms, 'Profit Beyond Measure', Nicholas Brearley

2001 Hinckley, 'Make No Mistake!', Productivity

2001 Schonberger, 'Let's Fix It!', Free Press

2001 Mackintosh et al, 'Improving Changeover Performance', B-H

2002 Jones and Womack, 'Seeing the Whole', LEI

2003 Seddon writes of 'failure demand and value demand' in 'Freedom from Command and Control'

2003 Mackle – Theory of Flow

2003 Schmenner, 'Swift Even Flow' as a service differentiator

2004 Liker, 'The Toyota Way', McGraw Hill

2004 Lee, 'The Triple A Supply Chain', Harvard Business Review

2004 Holweg and Pil, 'The Second Century', MIT Press

2004 Maskell and Baggaley, 'Practical Lean Accounting', Productivity

2005 Gershenfeld, 'FAB: The Coming Revolution on your Desktop', Basic Books, explains personal 'fabrication laboratories'

2005 Dinero, 'Training Within Industry', Productivity – a rediscovery of TWI principles that were the foundation of TPS

2005 Dassault builds the first aircraft (the Falcon) to be designed entirely in a virtual environment, cutting tooling and manufacturing time in half.

2005 Womack and Jones, 'Lean Solutions', Simon and Schuster

2006 Toyota overtakes Ford in cars sold. Honda car sales in USA approach Ford levels.

2006 Morgan and Liker, 'The Toyota Product Development System', Productivity

2006 Anderson, 'The Long Tail', RH Business – shows how the long-ignored tail of the Pareto can be a vital resource in a Lean business.

2007 Ward, 'Lean Product and Process Development', LEI, published posthumously.

2008 Joseph Juran dies.

2008 Toyota overtakes GM in vehicle sales to become the world's largest car company

2008 Schonberger, 'Best Practices in Lean Six Sigma Process Improvement: A Deeper Look', Wiley, points out the winners and losers in long term inventory turn trends – Toyota shown to be a poor performer on this measure

2030 Toyota aims for most of its vehicles to be non-fossil fuelled.

1   The Fourth Edition of the Lean Toolbox

Much has happened since 2004 when the previous edition was published – both in terms of new developments in Lean, and what the authors have learned. This book has been tailored for practitioners and graduate and postgraduate students in manufacturing, logistics, and hopefully construction as well. It discusses topics from concept to delivery, including support functions such as planning, measurement and Lean accounting. It also to some extent addresses people, leadership and cultural aspects. In short this book attempts to cover all operations aspects of Lean that you will need for a successful Lean transformation. Lean service is discussed, in as far as it is part of the Lean Enterprise, but more detail is to be found in the companion book, *Lean Toolbox for Service Systems*.

This new edition adds and revises material throughout, but particularly in the areas of:

- Implementation frameworks
- Variation and overload (mura and muri)
- New aspects of Value Steam Mapping (VSM)
- Supply chain aspects
- Lean accounting and performance measures
- Lean scheduling and heijunka
- People and cultural aspects
- Sustainability of improvement programmes
- Lean design and Lean product development
- Lean layout
- Total Preventive Maintenance (TPM)
- History of Lean
- Academic research on Lean

The core issues are discussed: people, leadership, and cultural issues to some extent. If these are neglected, everything fails, irrespective of the tools. The book does not explore these aspects at length because it is primarily a book about the tools and the system, within which they fit. The book is certainly lacking in respect of system – but even if we wrote a more comprehensive book that explored beyond the tools – on Lean people, leadership and culture – it would never be enough. Lean has to be developed and adapted within a **real world** context - these tools are only an aid for a Lean framework!

Three quotes are relevant:

'There are several keys to Lean Manufacturing, all of which relate to the people actually doing the work. The TPS is an interlocking set of three underlying elements: Philosophical Underpinnings, Managerial Culture, and Technical Tools - a triangle, where human development is at the core. It is very often overlooked as people tend to focus on the tangible aspects of TPS. All three must be in place and in practice for TPS to flourish.'

(G.S. Vasilash – VP of Manufacturing, Toyota Kentucky)

'Understanding the theory of (Lean Production) in the head is not the problem. The problem is to remember it in the body.'

(Taiichi Ohno, Father of the Toyota Production System)

'To learn anything other than the stuff you find in books, you need to be able to experiment, to make mistakes, to accept feedback and to try again. It doesn't matter whether you are learning to ride a bike or starting a new career, the cycle of experiment, feedback and new experiment is always there.'

(Charles Handy, Management Guru)

1.1   Going Back…

It is now 25 years since Richard Schonberger and Robert Hall wrote the two books that effectively launched (or re-launched) Lean in the West. It is over 15 years since Womack, Jones and Roos wrote their seminal book naming the approach 'Lean'. Huge changes have taken place, yet it is also true that for the majority of operations organisations the Lean potential has hardly been tapped.

The name 'The Lean Toolbox' - as pointed out in the last edition - may cause three reactions:

1. Reject it outright, thinking that Lean is not an eclectic selection of tools, but a system. You would be correct. But hopefully the book goes beyond tools. We aim to provide not just the tools, but frame them in the context of how you would apply them in your Lean transformation.

2. As the Lean-*Toolbox.* Lean, as an extension of Toyota Production System (TPS), continues to evolve so this book attempts to consolidate and explore traditional Lean areas. However, thinking that TPS is the answer to all is also naïve. Out of automotive TPS needs considerable adjustment. In fact this is one reason for disappointing Lean results. Of course, Toyota would be the first to admit that they do not have all the answers. Schonberger has pointed out that Toyota's inventory turns have steadily got worse over the past 20 years – even though they have become No 1 in the world. And they are battling in Formula 1 where continuous innovation is paramount.

3. Third, as the *Lean* toolbox. Lean has now expanded out well beyond the original TPS. We have begun to realise how Lean can become an even more powerful concept as it integrates with the theory of constraints, with 'factory physics', with service, with much of Six Sigma, and with ERP. This surely constitutes *Lean.*

A central theme of the book is the amalgamation of traditional Lean, Theory of Constraints, Six Sigma, and a range of relatively new concepts for measurement, analysis, and transformation.

1.2   Lean, Sustainability and Change

Sustainability has become one of the big themes in Lean. Sustainability is certainly important from an environmental perspective and the ideas of wasting fewer materials and energy and avoiding polluting emissions fit extremely well with wider Lean ideas. But sustainability may not be enough. The word carries connotations of freezing-in new ways of working. In change management we have the notion of 'unfreeze, change, refreeze'. What is now needed in Lean is to be continually unfrozen so adaptation will be continuous. We will try to emphasise this throughout this book.

To quote Jack Welch, 'People always ask, 'Is the change over, can we stop now?' You've got to tell them, 'No, it's just begun!''

1.3   Lean Evolution

For many, Lean started with 'tools'. Often, these were not even a set of tools but completely independent: 5S here, SMED there, kanban here and there. For a while, this is not a bad thing. Then comes Lean through Principles – often the 5 Lean Principles of Womack and Jones, or principles of self-help, respect, responsibility towards staff, customers and society. Much better, and better still if systemically brought together. But now some have begun to realise that 'real' Lean is behaviour-driven. What everyone does every day without being told. But how to get to this state of nirvana? It relies on feedback about the tools and principles. Behaviour is built by establishing principles such as pulling the Andon chord when a problem occurs but **ALWAYS** doing this, always expecting this, and always supporting this. Not 'lip service'. And by **ALWAYS** using tools such as PDCA, standard procedures for meetings, root cause problem solving, and visual management to highlight problems. Always is expected, not optional. Over time, with persistence, this builds the 'world view' – the things we take to be self evident. The most important behaviour is that, at every level, leaders are teachers – continually reinforcing the correct usage of the principles and the tools. Not relying on a 10-day Lean course, or a book, or intranet for their staff to learn the principles and tools – but by self demonstration and coaching every day.

For some the word 'Lean' is has connotations 'mean-ness' or 'cutting back', and redundancy. generally in terms of headcount. But, on the contrary, Lean is about growth and opportunity. Toyota, for example, has grown not cut back. They have grown because they have capitalised on the huge advantages that Lean brings. It is better to grow into profitability rather than to shrink into profitability.

This leads to another important idea – that of 'Lean Enterprise'. Womack and Jones have tried to emphasise that Lean is concerned with enterprise not just with manufacturing. If you have already started on your Lean journey without involving design, marketing, accounting, HR, distribution, and field service, you will have to do so very soon or risk the whole programme. These functions have a vital role to play in answering what the organisation will do with the improved flexibility, extra capacity, enhanced quality, reduced lead-times, and the rest. If the answer is just 'reduce costs' management has missed the point. The Lean enterprise also needs appropriate people policies, measures, accounting, and service – without which it cannot work.

David Cochrane makes an excellent point: Lean is not what organisations need to do. Lean is what organisations *should become* by effective system design and implementation.

A simple definition of Lean is 'Doing More with Less'. This is of course directly in line with the definition of productivity (outputs / inputs) but could be interpreted more widely as doing good with less resources – materials, energy, pollution – to achieve ultimate sustainability.

Which leads us to the views of Robert 'Doc' Hall who talks about Lean being ultimately concerned with 'compression' – reducing lead time, space, energy, materials, stress and overburden, defects, pollution, changeover time, processing time, and work time (to free up more time for leisure). The reduction should not be a trade-off – we need to do all simultaneously. The good news is that all this can be done profitability – by good design.

So, upfront, Lean is not just about waste reduction, or even waste prevention. These viewpoints are fairly negative. It is far better to think about the future, and to emphasise value and growth.

George Davidson, retired manufacturing director of Toyota South Africa, says that the first principle of the Toyota Production System (TPS) is 'the customer first'. George first said this in 1982, and in 2007, Katsuaki Watanabe in an interview with Harvard Business Review used exactly the same words. That's consistency! And, how do you do that? 'By creating thinking people', said he. And how do you do that? 'By creating workplaces and organisations that are more human'. Note what Davidson is NOT saying. It is not primarily about waste; waste is removed because you want to improve benefits to the customer. Hence Toyota is not averse to adding inventory where necessary – as indeed they have done recently. It is not about 5S – 5S is a just a tool for consistency and quality. It is not about SMED – SMED is just a tool for improving response time and service to the customer. TPS developed from first principles, with the customer in mind. In fact non-Lean systems do just the opposite. For instance, through 'economic order quantities', 'mass production', long lead times, reduced variety, 'push' systems, and the location of plants in China purely for cost – all of which are designed with the producer in mind, not the customer!

Appropriately, and following Davidson and others, some have begun to say that TPS stands for *Thinking People System*, rather than *Toyota Production System*. Appropriate.

One way of understanding Lean is to view it as a (proven) approach to dispense with increasingly inappropriate 'economies of scale' and to adopt 'economies of time'. To conclude, take Ohno's Method:

1. Mentally force yourself into tight spots.

2. Think hard; systematically observe reality.

3. Generate ideas; find and implement simple, ingenious, low cost solutions.

4. Derive personal pleasure from accomplishing Kaizen

2   Philosophy

2.1   Lean Seeks the 'Ideal Way'

Perfection, as we shall see, is Womack and Jones' fifth Lean Principle. It could have been the first. So we need to ask, continually, 'Will that move us closer to the Ideal?'. And what is the ideal? It is perfect quality, zero waste, perfect customer satisfaction. Is it so ridiculous to talk about 'Free Perfect and Now' as Robert Rodin did in transforming his company, Marshall Industries, pointing out that all the trends are going in those directions? Indeed, think of Skype and Google.

Peter Hines talks about the 'Six Rights'. Right product, right place, right time, right quantity, right quality, and right cost. That is a good start.

The Ideal Way is a re-statement of Fred Taylor, a hero of Toyota, who spoke of 'The One Best Way'. There is a best way for everything, so why not seek it out? But finding the best way requires deep understanding of customers and the participation of all employees. Toyota Chairman Watanabe also has a dream for the ideal state: A car that can improve air quality rather than pollute, that cannot injure people, that prevents accidents from happening, that can excite and entertain, and drive around the world on one tank of gas.

Ohno had a vision too – of one at a time, completely flexible, no waste flow. In fact, that has been the driving force of Toyota for the past 50 years. Ohno did not have a Lean toolbox. He had in mind a vision of where he wanted to be. The vision first, **THEN** the necessary tools. So look at every job, every process, and every system. What is the ideal way to do it? What is preventing us from doing that? How can the barriers be removed?

This is a repetitive process. Reducing the batch size moves you closer to the ideal, but you will need to come back and reduce it further. After you make the engine more fuel efficient, try it again then again and again.

Likewise, Levitt maintained that Ford was not a production genius, but a marketing genius. He realised that if he could make and profitably sell a car for $500, millions of cars could be sold. That being the case, he had to find a way to make such a car.

'Here is Edward Bear, coming downstairs now, bump, bump, bump on the back of his head, behind Christopher Robin. It is, as far as he knows, the only way of coming downstairs, but sometimes he feels that there really is another way, if only he could stop bumping for a moment and think of it.'

(From 'Winnie the Pooh')

2.2   Lean is not tools – or even a set of integrated tools!

Maslow, famous for his hierarchy of motivation, said in 1966, 'It is tempting, if the only tool you have is a hammer, to treat everything as if it were a nail'. Just so with Lean tools.

To quote Kate Mackle, 'A quick survey of what's available tells us one thing: early interpreters of what made Toyota different concentrated on bringing to a new audience the models that Toyota had developed to describe the elements of their well-established system. So we can now find many different ways of describing the 'house of Lean' and its different components: Just-In-Time, Jidoka, Standard Work etc. Many companies adopt such models to represent their own improvement programme: the (*insert your company's name here*) Production System. There are many excellent sources of information to explain the tools and systems of Lean: however, knowing what tools we have at our disposal does not help us to design what we want to build.

Maybe you have found yourself in the position of many other Lean enthusiasts: trying very hard to use the Lean tools and getting some good localised results but not making a breakthrough in performance that would emulate what Toyota achieved? Here's a simple but powerful lesson: if you want to get the same result, follow the same process. Toyota did not start with the tools: they did not start with their system. They started with an unremitting focus on how to use their resources to produce a product that is defined to be as close as possible to what the customer wants to buy now, and how to align the flow of production as close as possible to the flow of cash into the business. As the goal of the business is to make money, that makes sense, doesn't it?'

In their excellent book *Nudge*, Thaler and Sunstein recount the story of the discovery of car windscreen pits (or minor damage) at a small American town in the 1950's. The discovery of windscreen pits in one region led to discoveries in adjoining regions. Investigations were launched. Possible causes, ranging from radioactivity to aliens, were postulated. But the pit phenomena continued to grow. Eventually an in-depth scientific study found that windscreen pits occur in almost all cars as a result of routine use. It was just that drivers were sensitised to notice pits that were always there.

So it may be with Lean implementation. What do you need to do and what is important? First we thought quality circles. Aha! That is the thing to do! Later came changeover, then kanban, then 5S, then kaizen blitz, then value stream mapping, then people issues, then policy deployment, and now leadership and sustainability.

Like windscreen pits, these issues were always there. So, stand back and try to look at the total system. Pfeffer and Sutton recommend that one should try to benchmark the thinking, rather than the technique. Is the company you admire achieving success because of the technique or approach, or in spite of the technique, they ask.

2.3   Muda, Muri, and Mura

Womack and Jones, began their book *Lean Thinking* with the words, 'Muda. It's the one Japanese word you really must know.' Today, there is widespread awareness of waste (Fujio Cho, former President of Toyota, defined waste as 'anything other than the minimum amount of equipment, material, parts, space, and worker's time, which are absolutely essential to add value to the product'.)

But Toyota also talks about three 'M's' – Muda (waste), Muri (overburden), and Mura (unevenness). Knowing about all three gives a more complete understanding of Lean. The three are interlinked.

Do you have a situation where orders arrive exactly evenly? Where capacity is always adequate? Where machines never break down? Or where all process times are exactly even? If you answer is 'No' to any of these questions, then you need to understand the wisdom of Muri and Mura.

You can play a dice game with a friend, one generating the orders, the other making the product – that is the capacity available. You play 'day by day' for 20 days, over 5 rounds, so you will need 5 columns of 20 days each. Each 'day' you each roll a dice. During round 1 the maker adds 3 to the dice roll to represent capacity available for the day. If the capacity exceeds or equals the orders, no jobs will be left over at the end of the day. If orders exceed capacity, there will be jobs uncompleted at the end of the day that need to be carried over to the next day. Record the jobs left at the end of each day. When you have completed 20 days, repeat the procedure but this time capacity is the dice roll plus 2. Order generation remains the dice roll for the day throughout all five games. Again record the jobs waiting at the end of the day. Repeat for when capacity is dice roll plus 1. Record the jobs waiting each day. Then again for dice plus 0. Now you have completed 4 x 20 = 80 days. Add each column and divide by 20 to get the average jobs waiting. Plot the 4 readings on a graph. It will look something like that shown below – marked 'high variability'. Now repeat the game one last time with capacity adding zero to the dice roll, but taking only the numbers 3 and 4 to be valid. (If you roll any other number you have to re-roll.) The order dice still rolls 1 to 6 as before. Very likely you are going to a find shorter average queue – as shown by the 'low variability.

By the way, it is even better if several pairs do this exercise together. The average results are more interesting and valid.

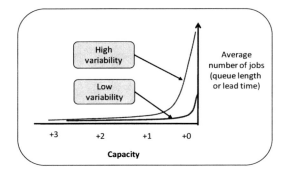

Notice, from the graph:

- Job queues are near zero where there is quite a bit of extra capacity.

- Job queues start to accumulate severely when the arrival rate of orders equals the capacity. (i.e. at +0).

- In fact the 'turn up' of the graph starts happening at about +1 – or about 80% to 90% of capacity

- Where variation in capacity is less, the average queue length is less

- And, if you played with several pairs, you would notice that variation in the average number of jobs is small when there is lots of capacity (i.e. +3 and +2), but as capacity tightens the spread of queue length – or uncertainty in lead time, increases.

- If you had no variation – both players had a dice with only the number 3, the queue would have been zero. Of course if the customer had just had one occasion with an excess demand (like 5) – perhaps representing a promotion – the queue would build instantly and remain at that level forever. The same would apply to just one breakdown.

Now, variation in the order arrival rate and variation in the capacity is unevenness (Mura). Capacity is directly linked with overburden (Muri).

So, the lessons are that unevenness and being overburdened are the big enemies. They are a major source of waste. You need to reduce the process variation and be very careful of order variations like promotional activity. And you need to run at less than 100% of capacity – because if you don't your lead times will be both long and unpredictable. Mura and Muri lead to Muda.

Now let us consider this more widely: Muri causes waste by 'overburdening' people and machines. First, people: if people are to be partners in the process, if they need to be willing participants in improvement. If they are to produce good quality work and be responsible for that quality, they cannot be expected to do so if they are stressed or overworked. They must enjoy good 'quality of work life' at the workplace, so the ergonomics (temperature, lighting, vision, comfort, lifting, and risk of repetitive strain) must be as friendly as possible. Safety must be paramount. This enables 'bring your brain to work' and 'hire the whole person' to be possible. Second, machines: machines also can be overburdened by working them beyond their limits – this is actually the basis of TPM. Third, people and machines together: this book frequently makes reference to the problem of queues and variation. We know that, unless there is no variation (a virtually impossible situation), if the ratio of the arrival rate of work to the service rate – or the 'utilisation', is above about 90% (or lower in the case of higher variation), queues will accumulate and missing targets will become inevitable. This is overburden. Note that overburden begins at less than 100% utilisation because of variation.

Mura is unevenness. Fast, uninterrupted flow is not possible with uneven demands. Queues and lead time will build up. Extra materials and inventory is required to meet peak demands. Of course, customer demands are not entirely even. First, do not amplify the unevenness by your own policies – such as end of month reporting, quantity discounts, and the like. Second, encourage both suppliers and customers to order and produce more evenly – often to mutual advantage. Does the supplier really want to deliver in bulk; does the customer really want six months supply of toothpaste? Can you move closer to your customers to understand their true longer term requirements – thereby enabling your own operations to be smoothed in terms of working hours, leave and so forth? Buffering will often be necessary, but working together with your customers and suppliers smoothes the overall flow! Also consider postponement as a way to reduce the risk of over- or understocking.

Perhaps Mura is the root problem. Unevenness causes people and machines to be overburdened, which in turn causes many other forms of waste. Periods of peak work cause stress to people and may cause lack of maintenance for machines, and high utilisation will result in deadlines being missed. But Muda can result in Mura due to lead times being long, quality being uncertain. So in fact it is circular: Mura causes Muri causes Muda causes Mura, and so on.

See also the section on Factory Physics for a powerful mathematical relationship between these factors.

2.4   A Formula for 'Lean'

A good way of representing Lean in a single formula is given in the figure below. 'Load' is the amount of work imposed on the system. 'Capacity' is the resources available to do the work. Load minus Capacity gives a gap. If Load is greater than Capacity there is overload or overburden or Muri. If Load is less than Capacity there is underload. But we now know that, because of variation, some underload is desirable – a plant loaded to 100% of capacity will frequently fail to meet requirements.

Load is made up of two types of demand – Value (or true) Demand, and Failure Demand which results from a failure to do it right the first time, or from a failure to take action – a powerful idea from John Seddon. Understanding the nature of demand is the beginning point of Lean. Value Demand in turn in made up of demands that are 'Runners, Repeaters, or Strangers'. It may be possible to shift Repeater and Stranger demands in time to level the demand – say by persuading customers to take smaller, more frequent batches. Shifting or levelling the demand frees up capacity, or at least makes better use of available capacity.

Ohno said that Capacity is Work plus Waste. So, of course, removing waste allows more efficient use of total capacity, or to further cut failure demand.

If there is a negative gap (and there should always be some at least), then any spare resources should be directed towards waste removal.

Reducing variation enables the gap to be narrowed. The result is improved customer service or more demand met at the same level of service.

Thus, there are several feedback loops – reducing failure demand, reducing wastes, spreading the load, reducing variation, and making use some slack capacity to improve efficiency.

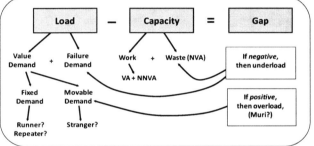

2.5 Lean is 'System'

The essence of Lean is the Systems Approach. Systems Thinking is holistic. As Peter Checkland, UK Systems guru – a person more widely recognised in Japan than in his home UK – says, 'the systems approach seeks not to be reductionist'. 'Seeks' because it is quite hard to keep the end-to-end system in view when almost the whole business world, and the whole academic world, is organised by function.

Ohno said, 'All we are trying to do is to reduce the time from order to cash'. Note the systemic nature of this statement. Lean is not about manufacturing or service but about the system that brings both of these together. Toyota is a 'systems' company rather than a manufacturing company. Toyota learned their systems craft from, amongst others, Deming. Ohno saw economies of flow rather than economies of scale.

Believing that optimising the individual parts will lead to optimising the whole represents possibly the greatest barrier to Lean. Thus, buying a faster machine, automating the warehouse, or outsourcing a process step, may well look good from a departmental or vertical silo perspective, but may be a disaster from a Lean system perspective.

The systems approach means the focus should be on the organisation or entity as a whole before attention is paid to the parts. This is a very hard thing to do for most managers and other workers who have been brought up and educated in vertical silos. If we don't remain systemic we quickly get into the 'push down, pop up principle' whereby we solve one problem then another emerges because we have sub-optimised.

The great, and hugely entertaining, systems thinker Russell Ackoff, Emeritus Professor at Wharton, talks about resolving problems (by discussion), better is solving problems (by fact-based tools approach), and best of all is dissolving problems (by understanding the purpose of the system and using innovative thinking). Non-Lean practitioners resolve 'inefficiencies', beginning

Lean practitioners solve problems to remove waste, but the experienced Lean practitioner improves the whole system. Ackoff also says that a system is more about INTERACTIONS than ACTIONS. In health care, for example, it is interactions between many professionals that count towards the success of a patient's recovery. In office systems it is the handoffs and rework loops that make the difference.

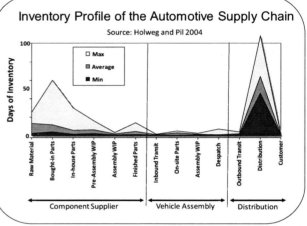

Inventory Profile of the Automotive Supply Chain
Source: Holweg and Pil 2004

Sub-optimisation is very clearly visible when you look at the inventory profile across the automotive supply chain. See the figure by Holweg and Pil. One can clearly see how JIT implementation has reduced inventories on site for the car manufacturers, yet this has only created an 'island of excellence' in the supply chain by pushing stocks into the component supply chain and the distribution system. As a result, stock is held at the most expensive level in the supply chain, accounting for three quarters of all capital employed in the supply chain. Remember to consider the Lean value stream, from raw materials to the end customer, to avoid such islands of so-called excellence.

Systems also have feedback loops. Success breeds success. Excess work can drive out the time for improvement effort. Most feedback loops have delays. So it takes time for success or failure to develop. Small things can set off chain reactions in connected loops, which may amplify. An example is the actions of a manager ignoring wastes or not participating in 5S. The loops may have several stages, which is one reason why root cause problem solving is important. Drawing out some of these causal loop diagrams can be thought provoking. Understanding that there are many loops at work at any time, and that they take time to work through is what is important for Lean transformation.

Water is a liquid at normal temperatures. Water constituents are the elements oxygen and hydrogen, which are gases. You can never understand the properties of water by studying oxygen and hydrogen. Likewise Lean and Lean tools. Lean is a system – more than the sum of its components. Systems are in constant interplay with their environment – it is not obvious where the boundary is, what should be outsourced, the extent to which customers and suppliers are involved. Systems adapt continuously but at a faster rate when threatened, like ant colonies (true kaizen culture). Systems evolve – like bugs combating insecticides, a question of how to recognise and kill off inappropriate tools whilst developing new and stronger ones.

Another analogy is the human body. Layout, supermarkets and buffers provide the skeleton; pull systems and information flows are the circulation system. Eyes and brain give vision and strategy; control, deployment and measurement come from the nervous system, quality and improvement from the muscular system, energy and getting rid of waste comes from the digestive system. The body needs them all for 'simple, slim and speedy' Lean (to quote Katsuaki Watanabe, Toyota President).

There is a danger that Lean is thought of as a box of tools to be cherry-picked. The problem is that it is an end-to-end value stream that delivers competitiveness. Tools are cause and effect – actions not interactions. No doubt some tools used individually give good results. A wonderful cell serving a morass of poorly controlled inventory is waste. A 5S programme without follow through into flow is largely waste. Kanban working in a situation of unlevelled demand can be waste. And so on. Even if all these were sorted out through good value stream mapping and a well-directed kaizen programme, Lean may still fail to deliver its true potential.

In Pfeffer's book 'What Were They Thinking?' numerous examples are given of top management from major corporations adopting practices that could be termed non-systems thinking. Pfeffer does not use the term systems thinking. Examples are top-down announcements of cutting staff or wages when a company gets into financial trouble, or believing in driving staff with forces such as threats and rewards. These practices ignore the impact on the system in favour of short-term 'solutions'. They prove ineffective, even disastrous, for the organisation sometimes immediately and almost invariably in the medium term – although, unfortunately, not always disastrous for the executive himself. Systems bite back.

2.6   Lean is Continuous Learning

One of the most common mistakes is to perceive Lean through its tools and concepts; these are only visible 'symptoms', but the real source of the power of Lean lies in its ability to learn from mistakes, and to continuously improve. In a Lean organisation: **mistakes are seen as 'opportunities to improve',** not as something that needs to be monitored and punished. There is no blame-game if something goes wrong. People are not rewarded for how few mistakes they make, but on how well they improve the process when mistakes have occurred. This ability to continuously learn is a 'dynamic learning capability' that provides Lean firms with their real competitive advantage. It rests upon two pillars: Kaizen, or continuous improvement, and Hansei, or honest self-reflection. Toyota does not commit fewer mistakes when it designs a new product or manufacturing plant, but it is much quicker at identifying them, and improving the process to make sure they do not occur again.

2.7 Lean is both Revolution and Evolution

The Toyota Production System (TPS) grew through revolution and evolution. Revolution rejected the concepts of mass production and economies of scale, and steered the organisation. Evolution developed the details and the tools. When TPS began there were few Lean tools - most developed from first principles over several decades, but fitting in with the top-level vision. Lean ideas developed from first principles; Taiichi Ohno believed in developing managers by the Socratic method – asking tough questions rather than providing answers. If you give the answer the person does not learn as much, and is less committed, compared with thinking it out him or herself. This is in line with the practice of Hoshin Kanri or Policy Deployment, whereby top management sets the strategic direction (the 'what' and the 'why') but evolves the detail level by level by a process of consultation (the how). In the reverse direction decisions are taken locally, only migrating upwards in exceptional circumstances. This is much like the human body's management system. So a top-level system that imposes the detailed (Lean) tools, and makes decisions that are in conflict with Lean, is precisely the wrong way around.

2.8 Lean is 'Distributed Decisions'

Ohno saw the TPS ideal as minimising the amount of information and control. Like the human body where most routine operations are decentralised and self-repair takes place locally. Excess information must be suppressed, said Ohno. Perhaps the greatest opportunity for Lean is not on the office or factory floor, but in simplification and decentralisation that enables whole swathes of overhead and administration to be eliminated.

Today large ERP systems, with data warehouses are 'in'. In the late 1990's and early 2000's many organisations moved in this direction by implementing large centralised systems – at great cost and often with mixed results. Today's need is for fast reacting distributed decision-making. There is a case for centralised strategic decisions, but not for operational decisions. The Lean way is to deal with resolving even dissolving problems with schedules, with quality, and even some aspects of design and supply at a localised level, but still keeping the end-to-end value stream in mind. This is not only more effective, but also more human. You may want ERP for planning but probably not for execution. Similarly you may use cost accounting for planning, but not for execution. Organisationally, the essence of the Hoshin process is Nemawashi (consensus building) and Ringi (shared decision making).

2.9 Two Analogies and the 'F's: The Orchestra and Fitness

Deming used the analogy of an orchestra to illustrate system and quality. Bob Emiliani has applied music and the orchestra to Lean. It is an appropriate analogy. The orchestra is more than the sum of the parts. Coordination from a conductor is required. An orchestra's star players are not necessarily the best. People cannot 'do their own thing'. There are many experts but they are subservient to the whole. Regular practice is required. Timing is vital. Standard operating procedures are used in the form of sheet music. The players know more about their instrument than the conductor, but the conductor knows abut integration. Symbols and notation have precise meanings. The note is the part number; pitch is the frequency; melody is the level schedule; harmony is the functions that support one another; tempo is tact time

Simon Elias has used the Fitness analogy. Fitness is built up over time; some every day. Self discipline is required, over a long term. Fat reduction is important. Reduce fat by exercise or improvement, but also don't add fat unnecessarily. There are different types of fitness – one may be fit for a sprint, another for a marathon – but you train for the specific event. It is difficult to remain at peak all the time. Periods of consolidation are required. Natural ability is important, but is useless unless developed. Coaches can help – but you need to be honest with the coach for him to really assist in areas of weakness.

Tony Hoy proposes the 'F's of Lean. They make a good list:

- Frugal
- Fast
- Flexible
- Flow
- Focused
- Flat
- Feasible
- Fit
- Forthright
- Factual
- Firm but Fair
- Forgiving
- Faithful
- Far-sighted

2.9.1   Lean, Six Sigma or Lean Six Sigma?

Lean and Six Sigma are no longer at odds, nor should they be. Lean is better at the big picture, at establishing the foundation, whilst Six Sigma offers a powerful problem solving methodology through DMAIC. The Lean service value stream mapping tools are generally superior but synergistic with those used in Six Sigma, and many concepts and tools are entirely synergistic. But, to be sure, there are also doubtful practices in Six Sigma, as pointed out by one of the originators, Keki Bhote, one-time of Motorola.

But much of Six Sigma itself is waste! What can be done to reduce or eliminate that variation before it even arises should be a prime question. Why is Toyota not falling over itself to get after Six Sigma or alternatively how has it been able to achieve Six Sigma levels of performance without legions of black belts and statistical software? See also sections 4.5.3 and 11.5.

2.9.2   Lean is 'Green'

Recently we have seen that not only is it possible to be both Lean and 'Green', but that by focusing on the wastes of materials and energy it can also be profitable and attractive to customers. Pamela Gordon in *Lean and Green* has researched many examples of the successful adoption of the three

'R's' (Reduce, Reuse, and Recycle). Hawken points out the scandalous fact that about 99%of the original material used in the production of goods in the US becomes waste within six weeks of sale. Presumably other countries are similar. Surely this is the ultimate waste. But Hawken, Lovins and Lovins in *Natural Capitalism* give hope, giving examples of a carpet made from recycled material that lasts four times longer, is stain proof and can be water cleaned, and of windows that remain cool at high temperatures and warm at freezing temperatures thereby drastically reducing the need for air conditioning and central heating. Surely, this is ultimate Lean.

Further Reading and References

Russell Ackoff, *Management in Small Doses*; *Ackoff's Fables*; *Beating the System*; *Management F laws*, *Creating the Corporate Future* (to name a few!)

Thomas Stewart and Anand Raman, 'Lessons from Toyota's Long Drive: Harvard Business Review Interview with Katsuaki Watanabe', *Harvard Business Review*, July August 2007

Louis Savary and Clare Crawford-Mason, *The Nun and The Bureaucrat*, CC-M Productions, 2006, an excellent book on systems and Lean healthcare

James Womack and Daniel Jones, *Lean Thinking*, revised edition, Free Press, 2003

Jinichiro Nakane and Robert Hall, *Ohno's Method*, Target, First Quarter, 2002

Richard Schonberger, *World Class Manufacturing: The Next Decade*, Free Press, 1996

Matthias Holweg and Frits K Pil, *The Second Century: Reconnecting Customer and Value Chain through Build-to-Order*, MIT Press, 2004

Thomas Johnson and Anders Bröms, *Profit Beyond Measure*, Nicholas Brealey, London, 2000

Jim Womack, *The Perfect Process*, Presentation, AME Conference, Chicago, 2002

Paul Hawken, Amory Lovins and L Hunter Lovins, *Natural Capitalism*, Little, Brown 1999

Phil Rosenzweig, *The Halo Effect*, Free Press, 2007

Richard Thaler and Cass Sunstein, *Nudge*, Yale Univ Press, 2008

Jeffrey Pfeffer and Robert Sutton, *Hard Facts Dangerous Half Truths and Total Nonsense*, Harvard Business School Press, 2006

Jeffret Pfeffer, *What Were They Thinking?*, Harvard Business School Press, 2007

Robert Hall, Seminar at Cardiff Business School, October, 2006

Bob Emiliani, Lecture at Cardiff Business School, July 2007, and *Real Lean*, Volume One, Centre for Lean Business Management, 2007

David Cochrane, 'The Need for a Systems Approach to Enhance and Sustain Lean', in Joe Stenzel (ed), *Lean Accounting: Best Practices for Sustainable Integration*, Wiley, 2007

2.10 The Five Lean Principles

In *Lean Thinking*, Womack and Jones renewed the message set out in *The Machine that Changed the World* (that Lean was, at least in automotive, literally Do or Die), but extended it out beyond automotive. These reflective authors have given manufacturing, but to an extent also service, a vision of a world transformed from mass production to Lean enterprise. The five principles set out are of fundamental importance. Reading the Introduction to *Lean Thinking* should be compulsory for every executive.

Throughout *Lean Thinking*, Womack and Jones emphasised *Lean Enterprise* rather than Lean Manufacturing. In other words it was emphasising systems. But unfortunately the book became thought of as a manufacturing book, and the system message was missed.

In this section, whilst using Womack and Jones' 5 principles, some liberties have been taken, particularly in relating them to service. Some managers are upset by the five principles, believing them not to be feasible within their industry. But this is to miss the point, which is vision: you may not get there within your lifetime, but try - others certainly will.

1. The starting point is to **specify value from the point of view of the customer**. This is an established marketing idea (that customers buy results, not products - a clean shirt, not a washing machine). Too often, however, manufacturers tend to give the customers what is convenient for the manufacturer, or deemed economic for the customer. Womack and Jones cite batch-and-queue airline travel, involving long trips to the airport to enable big batch flights that start where you aren't and take you where you don't want to go, via hubs, and numerous delays. How often are new product designs or service operations undertaken constrained by existing facilities rather than by customer requirements? Of course we have to know who the customer is: the final customer, next process, next company along the chain, or the customer's customer.

2. Then identify the **Value Stream**. This is the sequence of processes all the way from raw material to final customer, or from product concept to market launch. If possible look at the whole supply chain (or probably more accurately the 'demand network'). Again, an established idea from TQM/Juran/business processing. You are only as good as the weakest link; supply chains compete, not companies. Focus on the object (or product or customer), not the department, machine or process step. Think economies of time rather than economies of scale. Map and measure performance of the value stream end-to-end, not departmentally.

3. The third principle is **Flow.** Make value flow. If possible use one-piece or one-document flow. Keep it moving. Avoid batch and queue, or at least continuously reduce them and the obstacles in their way. Try to design according to Stalk and Hout's Golden Rule - never to delay a value adding step by a non value adding, although temporarily necessary, step - try to do such steps in parallel. Flow requires much preparation activity. But the important thing is vision: have in mind a guiding strategy that will

move you inexorably towards simple, slim and swift customer flow.

4. Then comes **Pull**. Having set up the framework for flow, only operate as needed. Pull means short-term response to the customer's rate of demand, and not over producing. Think about pull on two levels: on the macro level most organisations will have to push up to a certain point and respond to final customer pull signals thereafter. An example is the classic Bennetton 'jerseys in grey' that are stocked at an intermediate point in the supply chain in order to retain flexibility but also to give good customer service at low inventory levels. On the micro level, respond to pull signals as, for instance, when additional staff are needed at a supermarket checkout to avoid excessive queues. Attention to both levels is necessary. Of course, pull needs to take place along the whole demand flow network, not only within a company. So this ultimately implies sharing final customer demands right along the chain. Each extension of pull reduces forecast uncertainty.

5. Finally comes **Perfection**. Having worked through the previous principles, suddenly now 'perfection' seems more possible. Perfection does not mean only defect free - it means delivering exactly what the customer wants, exactly when (with no delay), at a fair price and with minimum waste. Beware of benchmarking - the real benchmark is zero waste, not what the competitors or best practices are doing. Remember that a human activity system (unlike a mechanical process) cannot be copied – because although the actions can be seen the interactions are invisible. You can learn good practice, but you cannot duplicate it and expect it to work the same way.

You quickly realise that these five principles are not a sequential, one off procedure, but rather a journey of continuous improvement. Start out today.

2.11 The 25 Characteristics of Lean

The literature on Lean contains several seminal books, amongst them by Womack and Jones, Schonberger, Hall, Goldratt, and Imai. These built on the 'greats': Deming, Juran, and Ohno. To distil them is a daunting task, but certainly there are common themes. These 25 seem to be at the core:

1. *Customer* The external customer is both the starting and ending point. Seek to maximise value to the customer. Optimise around the customer, not around internal operations. Understand the customer's true demand, in price, delivery and quality - not what can be supplied.

2. *Purpose* The 'big picture' question is 'What is the purpose?' This simple question is the way forward to reduce waste, complexity, and bureaucracy. And 'do the measures actually work for or against the purpose?'

3. *Simplicity* Lean is not simple, but simplicity pervades. Simplicity in operation, system, technology, control, and the goal. Simplicity is best achieved through avoidance of complexity, rather than by 'rationalisation' exercises. Think about ants that run a complex adaptive system without any management information system. Simplicity applies to product through part count reduction and commonality. Simplicity applies to suppliers through working closely with a few trusted partners. Simplicity applies in the plant, by creating focused factories-within-a-factory. Beware of complex computer systems, complex and large automation, complex product lines, and rewards. Select the smallest, simplest machine possible consistent with and without compromising quality requirements.

4. **Waste** Waste is endemic. Learn to recognise it, and seek to reduce it, always. Everyone from the chairman to the cleaners should wear 'muda spectacles' at all times. Seek to prevent waste by good design of products and processes.

5. **Process** Organise and think by end-to-end process. Think horizontal, not vertical. Concentrate on the way the product moves, not on the way the machines, people or customers move. Map to understand the process.

6. **Visibility** Seek to make all operations as visible and transparent as possible. Control this by sight. Adopt the visual factory. Make it quick and easy to identify when operations or schedules are diverging.

7. **Regularity** Regularity makes for 'no surprises' operations. We run our lives on regularity (sleep, breakfast, etc); we should run our plants on this basis too. Seek 'repeater' products and run them in the same time slots - this cuts inventory, improves quality, and allows simplicity of control. Regularity in new product introduction shortens the development cycle and makes innovation the norm.

8. **Flow** Seek 'keep it moving at the customer rate', 'one piece flow' manufacture. Synchronise operations so that the streams meet just in time. Flow should be the aim at cell level, in-company and along supply chains. Synchronise information and physical flows. If the process cannot flow, at least pulse one at a time or in small batches.

9. **Evenness** 'Heijunka' or levelling is the 'secret weapon' for flow and quality. Seek ways to level the schedule, to level sell, to level buy. Avoid those killer waves. Be proactive – ask both customers and suppliers if they would not prefer smaller, more frequent batches.

10. **Pull** Seek for operations to work at the customer's rate of demand. Avoid overproduction. Have pull based demand chains, not push-based supply chains. Pull should take place at the customer's rate of demand. In demand chains this should be the final customer, not distorted by the intermediate 'bullwhip' effect.

11. **Postponement** Delay activities and commit to product variety as late as possible so you retain flexibility and reduce waste and risk. This characteristic is closely associated with the concept of avoiding overproduction, but includes plant and equipment, information, and inventory. Note that this is not the same as simply starting work at the last possible moment, but is about retaining flexibility at the right levels.

12. **Prevention** Seek to prevent problems and waste, rather than to inspect and fix. Shift the emphasis from failure and appraisal to prevention. Inspecting the process, not the product, is prevention. Seek to prevent errors through pokayoke - Japanese for 'failsafing' or 'mistake-proofing'.

13. **Time** Seek to reduce overall time to make, deliver, and introduce new products. Use simultaneous, parallel, and overlapping processes in operations, design, and support services. Seek never to delay a value-adding step by a non value-adding step. Time is the best single overall measure. If time reduction is a priority you tend to do all the right things – waste, flow, pull, and perfection.

14. **Improvement** Improvement, continuous improvement in particular, is everyone's concern. Make improvement both 'enforced' and passive, both incremental and breakthrough. Improvement goes beyond waste reduction to include innovation.

15. **Partnership** Seek co-operative working both internally between functions, and externally with suppliers. Seek to use teams, not individuals, internally and externally. Employees are partners too. Seek to build trust. Another way of saying this is *Win-Win*, which is one of Stephen Covey's principles of highly effective people. You must find a win-win, never win-loose, solution and if you

can't you should walk away.

16. **Value Networks** The greatest opportunities for cost, quality, delivery and flexibility lie with cooperating networks. Supply chains compete, not companies. But each member of the chain also needs to add value. Expand the concept of the one-dimensional supply chain to a two-dimensional value network.

17. **Gemba** Go to where the action is happening and seek the facts. Manage by walking around. Implementation takes place on the floor, not in the office. Encourage the spirit of Gemba throughout.

18. **Questioning (and Listening)** Encourage a questioning culture. Ask why several times to try get to the root. Encourage questioning by everyone. As Bertrand Russell said, it is 'a healthy thing now and then to hang a question on things you have long taken for granted'. A manager who asks questions empowers. Listen actively, not passively. Restate the other's viewpoint. 'Seek first to understand, then to be understood', said Covey.

19. **Variation reduction** Variation in time and quantity is found in every process from supply chain demand amplification to dimensional variation. It is a great enemy of Lean. Seek continually to reduce it. Measure it, know the control limits, and learn to distinguish between natural variation and special deviation. Manage it. Build in appropriate flexibility.

20. **Avoiding Overburden** Overloading means less than full loading, and applies to people and machines — otherwise with just a little disturbance or variation you will miss the schedule. And bottlenecks need special attention. 'An hour lost at a bottleneck is an hour lost for the whole system'. So protect them.

21. **Participation** Give operators the first opportunity to solve problems. All employees should share responsibility for success and failure. True participation implies full information sharing.

22. **Thinking Small** Specify the smallest capable machine, and then build capacity in increments. Get best value out of existing machines before acquiring new ones. Break the 'economy of scale' concept by flexible labour and machines. Specify a maximum size of plant to retain 'family focus' and to develop thinking people. Locate small plants near to customer sites, and synchronise with their lines. Internally and externally make many small deliveries — runners or water spiders - rather than few big ones.

23. **Trust.** If we truly believe in participation and cutting waste, we have to build trust. Trust allows great swathes of bureaucracy and time to be removed internally and externally. In supply chains, Dyer has shown how trust has enabled Toyota to slash transaction costs which represent as much as 30% of costs in a company, and enable huge savings in time and people to be achieved. Building trust with suppliers gives them the confidence to make investments and share knowledge. Internally, trust allows a de-layered, streamlined, and more creative organisation.

24. **Knowledge** Since Peter Drucker's original work on knowledge workers being the engine of today's corporation, the importance not only of building knowledge but distributing it has become increasingly important. Spear and Bowen have shown how Toyota builds knowledge in a systemic, scientific way. Dyer has shown how Toyota cultivates both explicit knowledge (such as tools in this book), but also tacit knowledge, involving softer or stickier skills. It is tacit knowledge that is hard to copy and gives sustainable advantage. Knowledge is built through the scientific method, through PDCA.

25. **Humility** The more you strive for Lean, the more you realise how little you know, and how much there is yet to learn. Learning begins with humility.

2.12 The Toyota Way

The 'Toyota Way' was launched by Fujio Cho, then President of Toyota, in 2001. The 'Toyota Way' is not fundamentally different from Lean, but it aims to take Lean beyond its traditional applications in manufacturing and product development, into the entire organisation. For example, Cho said that it was now time for Toyota to apply its mastery in Just-in-Time [manufacturing], to its sales and marketing operations. The Toyota Way is an attempt to translate Lean into all business processes within Toyota. It is worth noting that even after 50 years, Lean is still a journey for Toyota, where they are learning and expanding their Lean efforts! The Toyota Way is based on five core values that employees at all levels are supposed to use in their daily work:

1. **Challenge:** to maintain a long-term vision and strive to meet all challenges with the courage and creativity needed to realise that vision.

2. **Kaizen:** to strive for continuous improvement. As no process can ever be declared perfect, there is always room for improvement.

3. **Genchi Genbutsu:** to go to the source to find the facts to make correct decisions, build consensus and achieve goals.

4. **Respect:** to make every effort to understand others, accept responsibility and build mutual trust

5. **Teamwork:** to share opportunities for development and maximise individual and team performance.

Jeffrey Liker of the University of Michigan has identified 14 principles, reproduced here with permission of The McGraw Hill companies, from Jeffrey Liker, *The Toyota Way*, McGraw Hill, 2004:

Long-Term Philosophy

1. Base your management decisions on a long-term philosophy, even at the expense of short-term financial goals.

The Right Process Will Produce the Right Results

2. Create a continuous process flow to bring problems to the surface.

3. Use 'pull' systems to avoid overproduction.

4. Level out the workload (heijunka). Work like the tortoise, not the hare.

5. Build a culture of stopping to fix problems, to get quality right the first time.

6. Standardised tasks and processes are the foundation for continuous improvement and employee empowerment.

7. Use visual control so no problems are hidden.

8. Use only reliable, thoroughly tested technology that serves your people and processes.

Add Value to the Organisation by Developing Your People

9. Grow leaders who thoroughly understand the work, live the philosophy, and teach it to others.

10. Develop exceptional people and teams who follow your company's philosophy.

11. Respect your extended network of partners and suppliers by challenging them and helping them improve.

Continuously Solving Root Problems Drives Organisational Learning

12. Go and see for yourself to thoroughly understand the situation ('genchi genbutsu').

13. Make decisions slowly by consensus, thoroughly considering all options; implement decisions rapidly ('nemawashi').

14. Become a learning organisation through relentless reflection ('hansei') and continuous improvement ('kaizen').

Further Reading

Jeffrey Liker, *Toyota Way*, MacGrawHill, 2004

2.13 The Lean Enterprise House

Toyota and TPS continue to evolve. Toyota, like many others, have recognised the limitations of too much emphasis on tools. They now use a Lean Enterprise house that differs from the 'tools' house. The enterprise house is a stage back from tools, and emphasises philosophy and approach. The 'whats', not the 'hows'. The Toyota *Production* System may be a house of tools, but the Toyota *Enterprise* system is far more broad.

The foundation is the ongoing challenge of continually adapting to the needs of customers, employees, and environment. There is kaizen or continuous change for the better. There is teamwork and emphasis on working together. And there is Gemba - the approach of hands-on, going to see oneself rather than management by remote control.

The pillars are now continuous improvement and respect for people. These two go back to the origins of Toyota in the 1930s to 1950s with Sakichi and Kachiro Toyoda. Perhaps they go back to a main source of their inspiration, Samuel Smiles' *Self Help*. These two support the Toyota Way – that hard to capture set of principles that Jeffrey Liker has attempted to capture. And finally, the roof – thinking people – the real root of sustained performance.

The concept of *enterprise* is important. Even today, some 25 years after the 'rediscovery' of Lean in the west, there is a still a tendency, as Schonberger has pointed out, to regard Japanese Manufacturing, or TPS, or indeed Lean as only one of these separate areas, depending on the observer's background. The three areas that Schonberger says are frequently treated separately, but are in fact integrated, are:

- Unique employment practices
- Extraordinary attention to Quality
- Production, especially just-in-time methods.

To these more recently have been added accounting, design, and supply chain practices. Of course, Lean is all of these, and more.

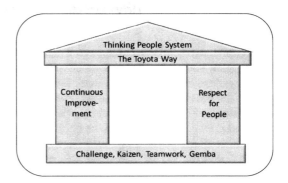

Further Reading

Jeffrey Liker, *The Toyota Way*, McGraw Hill and subsequent books, *Toyota Talent* (2007) and *Toyota Culture* (2008)

Samuel Smiles, *Self Help*, Oxford Classics, originally published 1856

David Magee, *How Toyota Became No 1*, Portfolio, 2007

Richard Schonberger, 'Japanese Production Management: An evolution – With mixed success', *Journal of Operations Management*, Vol 25:2, 2007

3 Value and Waste

3.1 Value

It is unsatisfactory to think that Lean is all about Muda, Muri, and Mura. It is also about value. The Toyota system is really about growth and profit. One of the most common misperceptions in Lean implementation is to walk around the factory shop-floor, and start eliminating obvious 'waste', such as inventory. What is and what is not waste is solely determined by the customer, not the Operations Manager on the shopfloor! The inventory that appears to be wasteful might well enable a short delivery lead-time that the customer values, and is prepared to pay for, so it does not constitute waste!

Hence Womack and Jones' first Lean principle is about understanding value. Activities that do not contribute to value are waste – either total waste or temporarily necessary non value adding. So we need to understand these concepts a little more deeply. This section considers value, the next section considers waste.

Gitlow has the useful concept that value is a function of Time, Place, and Form – at least one has to be improved, if not all three. Time is delivery lead time. Place is to do with customer convenience. Form is to do with design and utility.

Porter says, 'In competitive terms, value is the amount buyers are willing to pay for what a firm provides them. Value is measured by total revenue, a reflection of the price a firms' product commands and the units it can sell'.

From a Lean perspective, this is a little too simple. First, there is present value – what present customers are willing to pay for. This is the usual way of identifying waste. Then there is future value – what tomorrow's customers are willing to pay for, but today's may not be. This is relevant in research and development and design. These represent different value streams – so you should not judge a current manufacturing stream in the same way as an R&D stream.

Similarly there are today's customers and tomorrow's customers. And today's customers come in different categories – those that are very valuable, an intermediate set, and a third set that are just not worth having. Possibly your products or services are inappropriately focused. So waste may be different depending on the customer group, for instance the pensioner chatting to a checkout attendant and a time-hassled businessman.

One straightforward interpretation of value is Perceived Benefit/Perceived Sacrifices (Saliba and Fisher). TRIZ (Darrell Mann) says that value is the ratio of Benefits divided by Cost plus Harm. Benefits may accrue before, during, or after the event. Think of drink or sex! Sacrifices may be in terms of cost, convenience, time, or exchange. There are probably upper bounds on sacrifices that a customer is willing to pay, and lower bounds on benefits – like Kano basics. Both are dynamic. Harm includes all the possible 'victims' – of overburden, environment, energy, and safety.

Kano, speaking about quality, talks about 'Basics', 'Performance Factors', and 'Delighters'. See separate section. Much the same can be said about value. There are some activities that are basic to value – defect free has become a basic in some industries. There is 'performance' value – lead time for example in some businesses, and 'delighter' value. Moreover, as in the Kano model, value is dynamic.

Similarly, Terry Hill talks about 'order qualifiers' and 'order winners'. Qualifiers get you into the league, but winners win the match.

References

Michael Saliba and Caroline Fisher, *Managing Customer Value*, Quality Progress, June 2000, p.63-69

Howard Gitlow and David Levine, *Six Sigma for Green Belts and Champions,* FT Prentice Hall, 2005

Darrell Mann, *Hands-on Systematic Innovation*, CREAX, 2004

Terry Hill, *Manufacturing Strategy*, 2nd edn, MacMillan, 1993

3.2 Value and TRIZ

The innovation methodology TRIZ provides a useful way of looking at Value. TRIZ uses the notion of the 'Ideal Final Result' (or IFR) – that which the product or service is tending towards. The IFR would represent the ideal state of an attribute linked to the product or service. 'Ideal' may be free, or perfect, or now or some other desirable state. An example will illustrate. Consider a home dishwasher. See the table.

What we notice here is that ideal result for some attributes are the same for the customer and the manufacturer, but on others there is a contradiction. Of course, individual customers may well have differences, but the table reflects the general condition. The asterisk indicates that the attributes may be related – you may not want a totally silent dishwasher because you may want to be assured that it is working. Also provocative is to ask if there is any reason why a customer might not want an IFR of zero or infinity. For example, the ideal life of the dishwasher may be infinite for many customers, but only so long as the product stays within current acceptable appearance and technologies.

The related attributes are the first to look at, and are generally easier to overcome – for example by providing a visible light to indicate normal operation. But it is the contradictions that are of concern. The needs remain, but the means to meet the need varies according to current thinking and technology. This is the 'holes not drills' concept – is a power tool manufacturer in the holes business or in the drills business?

Sooner or later someone will find a way of solving each contradiction that has been uncovered. In this case a dishwasher that can be economically run with only one plate, but that can also handle a large dinner party. This may mean a different washing technology.

Conflicts may be resolved outside of the current value stream – for example by providing a service that removes dirty crockery and returns it clean, thereby removing the necessity for a home dishwasher completely. This is a trend that has been developing steadily – from window cleaning services, to ready meals and to 'power by the hour' in aircraft engine design, and aircraft sales. Such a trend can change the design parameters – from a solution that has limited life, and thereby the possibility of replacements, to one of much longer, more robust life where maintenance costs are minimised.

This in turn leads onto wider considerations – what is the system IFR? It is self-cleaning crockery, utensils, and pots! Non-feasible? Yes, but only for now. Perhaps the dishwasher company should be thinking about some R&D work in ceramics.

Attribute	Customer IFR	Manufacturer IFR
Noise *	0	0
Life	Long / life of technology	Warranty period only
Cost	0	0
Servicing	0	0 (yearly if profitable?)
Size	Variable	Standardized
Danger	0	0
Power Consumption	0	0
Environment impact	0	0
Ease of use *	0	0

The table helps identify areas of opportunity – but also brings out the real needs and competitive values and choices. (Note 0 = zero or nil.)

Further Reading

Darrell Mann, *Hands on Systematic Innovation for Business and Management*, IFR, 2004

3.3 Muda and the 7 Wastes

'Muda' is Japanese for waste. Waste is strongly linked to Lean. But consider:

- Waste elimination is a means to achieving the Lean ideal – it is not an end in itself.

- Waste prevention is at least as important as waste elimination.

- Value is the converse of waste. Any organisation needs continually to improve the ratio of value adding to non-value adding activities. But there are two ways to do this – by preventing and reducing waste, but also by going after value enhancement specifically.

Before getting carried away with waste reduction, pause. Waste reduction is not the same as Cost reduction. You can reduce waste with zero impact on cost. To translate waste reduction into cost reduction requires follow-up action – like improving flow, or increasing sales.

And, before we get to discuss the 7 Wastes, we need to be aware that there is almost always another level of resolution of waste. Within a 'value adding step' there are more detailed micro wastes, like in a robot cycle. Such as the production engineer shaving seconds off a machine cycle, when the end to end lead time is weeks. So it is important to home in on the right level of resolution. Go after big picture wastes first.

Taiichi Ohno, father of the Toyota Production System, of JIT, and patriarch of Lean Operations, originally assembled the 7 wastes, but it was Deming who emphasised waste reduction in Japan in the 1950s. Today, however, it is appropriate to add to Ohno's famous list, presumptuous though that may be. The section after next begins with Ohno's original seven, then adds 'new' wastes for manufacturing and service.

3.4 Type 1' and 'Type 2' Muda, Elimination and Prevention

Womack and Jones usefully talk about two types of waste. *Type 1 Muda* are activities that create no value but are currently necessary to maintain operations. These activities do not do anything for customers, but may well assist the managers or stakeholders other than customers or shareholders. Type 1 should be reduced through simplification. It may well prove to be greatest bottom-line benefit of Lean. Moreover, Type 1 muda is the easiest to add to but difficult to remove, so *prevention* of type 1 muda should be in the mind of every manager in every function. *Type 2 Muda* is pure waste. It creates no value, in fact destroys value, for any stakeholder, including customers, shareholders, and employees. Elimination should be a priority. Type 2 tends to grow by 'stealth', or carelessness.

Waste Elimination is achieved, by as Dan Jones would say, by 'wearing muda spectacles' (a skill that must be developed), and by kaizen (both 'point' and 'flow' varieties). Elimination is assisted by 5S activities, standard work, mapping, level scheduling and by amplification reduction. Ohno was said to require new managers to spend several hours in a chalk circle, standing in one place and observing waste. This should happen more frequently in offices, in warehouses, and in factories.

Waste Prevention is another matter. Womack and Jones talk about the eighth waste – making the wrong product perfectly – but it goes beyond that. Waste prevention cannot be done by wearing muda spectacles, but requires strong awareness in system, process, and product design. It is known that perhaps 80% of costs are fixed at the design stage. Of that 80%, a good proportion will be waste. System design waste prevention involves thinking through the movement of information, products and customers through the future system. For instance, questioning the necessity for ERP and the selection of far-removed suppliers - from a total perspective, and removing layers in a supply chain. Process design waste prevention involves the avoidance of 'monuments', the elimination of adjustments, and working with future customers and suppliers to ensure that future processes are as waste-free as possible. For instance, incorporating flexible service workers in a flexible layout. Prevention involves much more careful pre-design considerations. It also involves keeping options open via 'mass customisation' and recycling considerations. Of course, pokayoke is a prime technique for defect prevention in service and manufacturing.

In the opinion of the authors, waste *prevention* is likely to assume a far greater role than waste elimination in the Lean organisation of the future – in the same way that prevention in quality is now widely regarded as more effective than inspection and fault elimination.

3.5   Value Added, Non Value Added (Necessary and Avoidable)

In Lean manufacturing the terms 'value adding', 'avoidable non value adding' and 'necessary non value adding' are widespread, meaningful, and useful. Abbreviate these to VA, NVA, and NNVA. Value added activity is something that the customer is prepared to pay for. In some types of service, for example, health care and holidays, the customer is certainly prepared to pay for experience-enhancing activities so VA, NVA and NNVA designations need to be treated with care.

For other types of manufacturing and administration, for example many clerical procedures, one may argue that the customer is never happy to pay. Yet to call activities NNVA can be both unhelpful (since everything is NNVA) or de-motivating to employees. How would you like to spend most of your life doing necessary non value added work?

In such situations it may be more useful, and clearer, to talk about 'Not Avoidable' and Avoidable work. A problem here is negative connotation, and the abbreviation NA can be confused with not applicable!

One major maintenance organisation simply says that waste is anything other than the minimum activities and materials necessary to get the job done immediately, right first time to the satisfaction of customers. They don't get into Type 1 and Type 2, nor into NVA and NNVA semantics. Another definition of waste is anything that affects Form, Fit or Functionality.

Yet another alternative is to refer to Organisation Value Added (OVA) and Organisation Non Value Added (ONVA). Here it is probably not helpful to talk about Organisation Necessary Non Value Added – so this will not be used.

3.6   Ohno's 7 Wastes

Ohno's seven wastes were originally for manufacturing. However, they also have application in many types of service. There is the good news and the not-so-good news about Ohno's wastes. The good: they form a widely used set. The not-so-good: Ohno was reluctant to state them because they 'codified knowledge' - far better to derive your own set.

You can remember the seven wastes by asking, 'Who is TIM WOOD?' Answer: Transport, Inventory, Motion, Waiting, Overproduction, Over-processing, and Defects. This idea came from the Lean Office at Cooper Standard, Plymouth, UK. An alternative, used at Unipart, but source unknown, is WORMPIT (Waiting, Overproduction, Rework, Motion, Processing, Inventory, Transport).

In all these wastes, the priority is to avoid, only then to cut.

3.6.1   *The Waste of Overproduction*

Ohno believed that the waste of overproduction was the most serious of all the wastes because it was the root of so many problems and other wastes. Overproduction is making too much, too early or 'just-in-case'. The aim should be to make or do or serve exactly what is required, no more and no less, just in time and with perfect quality. Overproduction discourages a smooth flow of goods or services. 'Lumpiness' (i.e. making products or working in erratic bursts) is a force against quality and productivity. By contrast, regularity encourages a 'no surprises' atmosphere that may not be very exciting but is much better management. 'Too much, too early' often leads to 'too little, too late' because of the time knock-on effects.

Perhaps a better phrase for overproduction is 'over-activisation' – that is making batches too big, too early, just in case.

Overproduction leads directly to excessive lead time. For example, processing and passing on files in batches rather than one at a time. As a result defects may not be detected early, products may deteriorate, and artificial pressures on work rate may be generated. All these increase the chances of defects. Overproduction also impacts the waste of motion – making and moving things that are not immediately required.

Yet overproduction is often the natural state. People do not have to be encouraged to overproduce; they often do so 'just to be safe'. Sometimes this is reinforced by a bonus system that encourages output that is not needed. By contrast, a pull system helps prevent unplanned overproduction by allowing work to move forwards only when needed. Burger King makes Woppers this way. CONWIP is yet another form. Clerical operations are most effective when there is a uniform flow of work. The motto 'Sell daily? make daily!' is as relevant in an office as it is in a factory.

Overproduction should be related to a particular timeframe – first reduce overproduction (or early delivery) in a week, then a day, then an hour.

Some like to distinguish overproduction from 'over-purchasing' – buying too much, too early, just in case. For both overproduction and over-purchasing, why not label storage or inventory that is not due to be used today, with that tag.

3.6.2   The Waste of Waiting

The waste of waiting is probably the second most important waste. It is directly relevant to FLOW. In Lean we are more concerned with flow of service or customers than we are with keeping operators busy.

In early days of Toyota, waiting for a machine was considered an 'insult to humanity' (people should have far better things to do than to require them to wait for a machine). This is a useful in service because many service companies 'insult' their

customers by requiring them to wait – in effect saying 'your time is worth much less than mine'.

In a factory, any time that an item is seen to be not moving (or not having value added) is an indication of waste. Waiting is the enemy of smooth flow. Although it may be very difficult to reduce waiting to zero, the goal remains. Whether the waiting is for the delivery of a spare part or of customers in a bank there should always be an awareness of a non-ideal situation and a questioning of how the situation can be improved. Waiting is directly relevant to lead time – an important source of competitiveness and customer satisfaction.

A bottleneck operation that is waiting for work is a waste. As Goldratt has pointed out in his book 'The Goal', 'an hour lost at a bottleneck is an hour lost for the whole plant'. Effective use of bottleneck time is a key to regular work that in turn strongly influences productivity and quality.

3.6.3   The Waste of Unnecessary Motions

Next in importance is probably the waste of motion. Unnecessary motions refer to both human and layout. The human dimension relates to the importance of ergonomics for quality and productivity and the enormous proportion of time that is wasted at *every* workstation by non-optimal layout. A QWERTY keyboard for example is non-optimal. If operators have to stretch, bend, pick-up, move in order to see better, or in any way unduly exert themselves, the victim is immediately the operator but ultimately quality and the customer.

An awareness of the ergonomics of the workplace is not only ethically desirable, but economically sound. Toyota, famous for its quality, is known to place a high importance on 'quality of worklife'. Toyota encourages all its employees to be aware of working conditions that contribute to this form of waste. Today, of course, motion waste is also a health and safety issue.

The layout dimension involves poor workplace arrangement, leading to micro wastes of movement. These wastes are often repeated many, many times per day – sometimes without anyone noticing. In this regard 5S (see later section) can be seen as the way to attack motion waste.

3.6.4   The Waste of Transporting

Customers do not pay to have goods moved around unless they have hired a removal service! So any movement of materials is waste. It is a waste that can never be fully eliminated but it is also a waste that over time should be continually reduced. The number of transport and material handling operations is directly proportional to the likelihood of damage and deterioration. Double handling is a waste that affects productivity and quality.

Transporting is closely linked to communication. Where distances are long, communication is discouraged and quality may be the victim. Feedback on poor quality is inversely related to transportation length, whether in manufacturing or in services. There is increasingly the awareness that for improved quality in manufacturing or services, people from interacting groups need to be located physically closer together. For instance, the design office may be placed deliberately near to the production area.

As this waste gains recognition steps can be taken to reduce it. Measures include monitoring the flow lengths of paper through an office or the distance a customer needs to walk whilst catching a plane. The number of steps, and in particular the number of non-value adding steps should be monitored.

3.6.5   The Waste of Overprocessing (or Inappropriate Processing)

Overprocessing refers to the waste of 'using a hammer to crack a nut'. Think of a mainframe computer rather than distributed PCs, or a large central photocopier instead of distributed machines. But further, think of a large aircraft requiring passengers to travel large distances to and from a regional airport. Thinking in terms of one big machine instead of several smaller ones

discourages operator 'ownership', leads to pressure to run the machine as often as possible rather than only when needed, and discourages general purpose flexible machines. It also leads to poor layout, which as we have seen in the previous section, leads to extra transportation and poor communication. So the ideal is to use the smallest machine, capable of producing the required quality, distributed to the points of use.

How many have fallen into the trap of buying a 'monument' of a machine, that accountants then demand to be kept busy? The tail begins to wag the dog.

Inappropriate processing also refers to machines and processes that are not quality capable. An incapable process cannot help but make defects. In general, a capable process requires having the correct methods and training, as well as having the required standards, clearly known.

Note that it is important to take the longer-term view. Buying that large machine may just jeopardise the possibility of flow for many years to come, for both customers and employees. Think 'small is beautiful'. Smaller machines avoid bottlenecks, improve flow lengths, perhaps are simpler, can be maintained at different times (instead of affecting the whole plant), and may improve cash flow and keep up with technology (buying one small machine per year, instead of one big machine every five years).

3.6.6   The Waste of Unnecessary Inventory

Although having no inventory is a goal that can never be attained, inventory is the enemy of quality and productivity. This is so because inventory tends to increase lead-time, prevents rapid identification of problems, and increases space thereby discouraging communication. The true cost of extra inventory is very much in excess of the money tied up in it. 'Push' systems almost invariably lead to this waste.

Note the three types of inventory: raw material, work in process, and end items. The existence of any of these is waste, but their root causes and priorities for reduction are different. End items must sometimes be held to meet variation in customer demand, but excessive inventory is waste. It also represents risk of obsolescence. Raw material may be temporarily necessary due to supplier constraints – quality and reliability. WIP or distribution inventories need to move through the supply chain as fast as feasible. A section in this book is given to inventory considerations.

3.6.7 The Waste of Defects

The last, but not least, of Ohno's wastes is the waste of defects. Defects cost money, both immediate and longer term. In Quality Costing the failure or defect categories are internal failure (scrap, rework, delay) and external failure (including warranty, repairs, field service, but also possibly lost custom). Bear in mind that defect costs tend to escalate the longer they remain undetected. Thus a microchip defect discovered when made might cost just a few dollars to replace, but if it reaches the customer may cost hundreds, to say nothing of customer goodwill. So, central themes of total quality are 'prevention not detection', 'quality at source', and 'the chain of quality' (meaning that parts per million levels of defect can only be approached by concerted action all along the chain from marketing, to design, to supply, to manufacture, to distribution, to delivery, to field service). The Toyota philosophy is that a defect should be regarded as a challenge, as an opportunity to improve, rather than something to be traded off against what is ultimately poor management.

In service, 'zero defections' has become a powerful theme, recognising that the value of a retained customer increases with time.

3.7 The New Wastes

These may be added to Ohno's original list, and are appropriate in service and manufacturing:

3.7.1 The Waste of Making the Wrong Product Efficiently

This is Womack and Jones' eighth waste. It is really a restatement of the first Lean principle,

and closely related to the waste of overproduction.

3.7.2 The Waste of Untapped Human Potential

Ohno was reported to have said that the real objective of the Toyota Production System was 'to create thinking people'. So this 'new' waste is directly linked to Ohno. The 1980s were the decade of factory automation folly. GM and many others learnt the hard and expensive way that the automated factory and warehouse that does not benefit from continuous improvement and ongoing, innovative thought is doomed in the productivity race.

Today we have numerous examples, from total quality to self directed work teams, of the power of utilising the thoughts of all employees, not just managers. Human potential does not just need to be set free. It requires clear communication as to what is needed both from management and to management. It requires commitment and support because uncapping human potential is sometimes seen as a real threat to first line and middle managers. It requires a culture of trust and mutual respect which cannot be won by mere lofty words, but by example, interest and involvement at the workplace ('Gemba'). Basic education is also necessary. The answer to the question, 'What happens if I train them and they go?' should be 'What happens if you don't train them and they stay!'

Examples: Not using the creative brainpower of employees, not listening, thinking that only managers have ideas worth pursuing.

3.7.3 Excessive Information and Communication

Ohno himself spoke about the dangers of this waste when he said that 'excessive information must be suppressed'. Think e mails and, maybe, all those books about Lean! If we are not to be submerged we need to think carefully before copying e mails to all in the office, to have team briefings for the sake of having them, to send staff on Lean and Quality courses the material of which is never used. Today no one can read everything of relevance. Selection, discipline and trade-off is imperative on the part of receivers – for example, consider what value you get from reading a newspaper for 30 minutes each day. Senders have responsibilities also. But also, Stephen Covey says, you need time to 'sharpen the axe' every day – beware of having no time to sharpen the axe because you are cutting down the tree – so prioritisation is important. This is closely related to the next waste.

3.7.4   The Waste of Time

Everyone suffers from this. Stephen Covey, referring not to Lean but to the 7 Principles of Highly Effective People, has a useful 2 x 2 matrix. The axes are important and urgent. Most people spend excessive time in the urgent but not important activity category. The not urgent, not important category is OK for relaxation but is otherwise waste. Urgent and important work may be fine, but could also indicate out-of-control conditions or fire fighting. But everyone needs to prioritise time spent away from the urgent but not important to the important but not urgent category. This requires setting blocks of time aside. See also the section on OEE applied to people.

3.7.5   The Waste of Inappropriate Systems

How much software in your computer is never used – not the packages, but the actual code? The same goes for MRPII, now repackaged as ERP.

The Lean way is to remove waste before automating, or as Michael Hammer would say 'don't automate, obliterate!' The waste of inappropriate systems should not be confined to computers and automation. How much record keeping, checking, reconciling, is pure waste? Recall the categories of waste.

Recently, for example in the 3DayCar project, the waste of inappropriate systems has been highlighted. It is the order processing system, not the shopfloor that is the greatest barrier. Often, it's not the operations that consume the time and the money; it's the paperwork or systems. And we now understand a little more of the dangers of demand amplification, of inappropriate forecasting, and of measurement systems that make people do what is best for them but not best for the company. All this is waste.

3.7.6   Wasted Energy and Water

Energy here refers to sources of power: electricity, gas, oil, coal, and so on. The world's finite resources of most energy sources (except sun and wind) were highlighted in a famous report 'The Limits to Growth' written by the Club of Rome in 1970. Their dire predictions have not come to pass but the true impact of unwise energy use on the world's environments is growing. Wasting these resources is not only a significant source of cost for many companies, but there is also the moral obligation of using such resources wisely.

Although energy management systems in factory, office and home have grown in sophistication there still remains the human, common sense element: shutting down the machine, switching off the light, fixing the drip, insulating the roof, taking a full load, efficient routing, and the like. By the way, the JIT system of delivery does not waste energy when done correctly: use 'milkrounds', picking up small quantities from several suppliers in the same area, or rationalise suppliers to enable mixed loads daily rather than single products weekly.

Several companies that have 'institutionalised' waste reduction, Toyota included, believe that a good foundation for waste awareness begins with everyday wastes such as switching off lights and printers. You get into the habit.

3.7.7   Wasted Natural Resources

A most severe, and ever-more important waste is that of wasted natural resources. Hawken, Lovins and Lovins of the Rocky Mountain Institute estimate that 99% of the original materials used in production of goods in the USA becomes waste within 6 weeks of sale. Paper is a case in point –

the 'paperless office' is still a dream, in part due to it being low priority. According to Lycra Research 1.5 trillion pages were printed in US offices in 2006. In the US each person uses an average of 320 kg of paper annually. Xerox found that Dow Chemical had 16,000 printers, producing 480m printouts per year at a cost of $100m over 5 years. By contrast Easy Jet is a paperless company – almost. Conservation begins with awareness and measures.

Today conservation of materials is not only environmentally responsible, but is beginning to be profitable. To reduce the waste of materials a life cycle approach is needed, to conserve materials during design, during manufacture, during customer usage, and beyond customer use in recycling.

3.7.8   The Waste of Variation (Mura?)

Variation, amazingly, did not figure at all even in the revised edition of Womack and Jones' *Lean Thinking*!

We now recognise variation as one of the great enemies. It should enjoy prominence amongst the wastes – at least on a par with Overproduction.

3.7.9   The Waste of 'No Follow Through'

We began this section by saying that waste reduction is not the same as cost reduction. Further actions are generally required to reduce cost or to increase sales. If you don't do this, it is waste. So if you save walking distance, but don't do anything with the time saved, you have not really made a saving.

3.7.10   Waste of Knowledge

This waste results from simply letting knowledge disappear. It applies particularly in Design and Innovation, but also in many professional fields. So experience and knowledge that is gained when, for example, new products are designed, made, introduced, and marketed is not recorded and is simply forgotten about next time around. Such knowledge has to re-discovered all over again.

'Learning the hard way' is so silly when it already has been learned. Even if knowledge is re-used but not recorded, but instead is kept in the head of the person, there is the significant danger that it will be lost when that person leaves. This waste is similar to the waste of untapped human potential, but concerns knowledge and experience that the company has already used and paid for. So, have a procedure for recording lessons learned – even if this is as simple as a 'little black book'. Insist that it be done.

Allen Ward discusses Design Wastes in his book. Our own list of design wastes is given in the New Product Development section of the book.

3.7.11   The Seven Service Wastes

Although the wastes discussed above begin with the customer, they are nevertheless applied from the organisations perspective. What about the customer's perspective? Perhaps an improvement programme should begin with the service wastes:

1. **Delay** on the part of customers waiting for service, for delivery, in queues, for response, not arriving as promised. The customer's time may seem free to the provider, but when she takes custom elsewhere the pain begins.

2. **Duplication:** Having to re-enter data, repeat details on forms, copy information across, answer queries from several sources within the same organisation.

3. **Unnecessary Movement:** Queuing several times, lack of one-stop, poor ergonomics in the service encounter.

4. **Unclear Communication** and the wastes of seeking clarification, confusion over product or service use, wasting time finding a location that may result in misuse or duplication.

5. **Incorrect Inventory**: Out-of-stock, unable to get exactly what was required, substitute products or services.

6. **Opportunity Lost** to retain or win customers, failure to establish rapport, ignoring customers, unfriendliness, and rudeness.

7. **Errors** in the service transaction, product defects in the product-service bundle, lost or damaged goods.

Further Reading

Taiichi Ohno / Japan Management Association, *Kanban: Just-in-Time at Toyota*, Productivity Press, 1985

John Bicheno, *The Lean Toolbox for Service Systems* PICSIE Books, Buckingham, 2008

Allen Ward, *Lean Product and Process Development*, LEI, 2007

3.7.12   *A Final Thought: Can You Go too Far on Waste Reduction?*

The immediate response to this question from the Lean practitioner would be No! This is correct, but only if the bigger picture clearly is kept in mind. How so? Consider two stories:

- For over 30 years Henry Ford drove out waste on the Model T line. Cars have never been produced as efficiently since. But, of course, in the end, customer dissatisfaction forced the abandonment of the line to less efficient mass production.

- 40 years ago the best way to do the high jump was to use the 'Western Roll'. Then the Olympic rules changed. Dick Fosbury, a physics student, recognised the opportunities that were created by allowing competitors to land on air bags. He developed the 'Fosbury Flop' that has become the only way to improve the world record ever since. Now, you could go on 'kaizening' the western roll, driving up the maximum height achieved by that method. But you will never win the Olympic title.

Thus waste reduction needs to be seen as a partner to innovation. Both are necessary.

Further Reading

Taiichi Ohno / Japan Management Association, *Kanban: Just-in-Time at Toyota*, Productivity Press, 1985

George Stalk and Thomas Hout, *Competing Against Time*, The Free Press, New York, 1990

James Womack and Daniel Jones, *Lean Thinking*, revised edition, Free Press, 2003

Hawkin, Lovins and Lovins, *Natural Capitalism*, Little Brown and Co, 1999

Dominic Rushie, 'Business reports a loss on paper', *Sunday Times Business*, 19 Aug 2007

3.8   Gemba and 'Learning to See'

'Gemba' is the place of action – often but not necessarily the workplace. But this Japanese word has taken on significance far beyond its literal translation. Taiichi Ohno, legendary Toyota engineer and father of TPS, said, 'Management begins at the workplace'. This whole philosophy can best be captured by the single word: Gemba.

Contrast the Gemba way with the traditional (Western?) way. The Gemba way is to go to the place of action and collect the FACTS. The traditional way is to remain in the office and to discuss OPINIONS. Gemba can be thought of in terms of the 'four actuals': Go to the actual workplace, look at the actual process, observe what is actually happening, and collect the actual data.

Gemba is also a learning-to-see approach to daily work. As you walk around, question what you see. For instance:

- Is it really necessary to keep everything under lock and key? And at remote locations?

- Is it really necessary to fill in all those forms? Can they be pre-prepared? Why is it that at check-in or check-out there are such a lot keyboard strokes?

- Why is it that customers are asked to sign next to the X agreeing that they know all the conditions – when reading them though would take half an hour?

- Can customers (or patients) wait in more comfortable conditions? Why do they have to wait in line anyway – can they not be called when needed?

- Why are there interruptions in the middle of the process – necessitating start-up losses?
- If there was a 'near miss' don't just sigh with relief, try to put in place preventive measures.
- Is a one-stop procedure possible?
- What is the root cause of the problem? Don't just do a patch-up job, but ask if the cause can be eliminated.

None of these questions can be effectively asked or answered by managers sitting at their desks. A questioning culture is only possible if you are at Gemba.

So Gemba breaks away from the 'its not my problem' attitude and 'I only work here'. It is managers' fault if workers have such attitudes. What can be done to empower employees to make immediate improvements?

Under Gemba, if your organisation has a problem or a decision, go to Gemba first. Do not attempt to resolve problems away from the place of action. Do not let operators come to the manager, let the manager go to the workplace. Spend time at the service counter. This is the basis of so much Japanese management practice: that new Honda management recruits should spend time working in assembly and in stores, that marketers from Nikon should spend time working in camera shops, that Toyota sends its Lexus design team to live in California for three months, and so on. Ohno was famous for his 'chalk circle' approach - drawing a circle in chalk on the factory floor and requiring a manager to spend several hours inside it whilst observing operations, noticing variation, and taking note of wastes. The West too has its devotees. John Sainsbury who ran the supermarket chain in its heyday could pass a shelf and see at a glance if prices were wrong. His retirement may account for the decline of a once-great chain. An open plan office, with senior management sitting right there with 'the troops' is Gemba.

Gemba is, or should be, part of implementation. How often is the Western way based on 'change agents', on simulation, on computers or information systems, on classroom-based education? These have a place, of course, but Gemba emphasises implementation by everyone, at the workplace, face-to-face, based on in-depth knowledge. And low cost, no cost solutions rather than big-scale expensive information or technology solutions.

Gemba is often combined with other elements. The 5 Whys, Muda (or waste), Policy Deployment, Kaizen, 5 S, the 7 tools of quality. Gemba is the glue for all of these. So today one hears of 'Gemba Kanri', 'Gemba Kaizen', and 'Gemba TPM'. The word has already appeared in the English and American dictionary.

Finally let us remind ourselves that Gemba is not a 'Japanese thing'. They learned it from the Americans – specifically the famous Hawthorne experiments at General Electric in the 1930s. One study concerned the effects of lighting on productivity. Turn up the lights, productivity improves. Turn up the lights further, productivity improves further. Now turn down the lights. What happens? Productivity improves! What was happening is that workers were responding not to lighting levels but to the interest being taken in them by heavyweight researchers. It became known as the Hawthorn Effect. But the West promptly forgot the lesson. The Japanese took it on. So, don't sit in your office looking at your Excel spreadsheet and imagine that you are improving productivity – that is management by looking in the rear view mirror. Instead, 'get your butt to the gemba', and learn to anticipate problems.

Further Reading

Masaaki Imai, *Gemba Kaizen*, McGraw-Hill, New York, 1997

3.9 Time-Based Competition

The reduction of lead time – in manufacturing, in supply chains (including information flows), and in design is central in Lean. Stalk and Hout's classic work, *Competing Against Time*, was one of the first to identify the importance of time to the competitive edge. In the book, Stalk and Hout set out four 'rules of response' which are provocative rules of thumb, but apparently based on research by the Boston Consulting Group.

The 0.05 to 5 rule, states that, across many industries, value is actually being added for between 0.05% and 5% of total time. This is no longer a surprise; see for instance the earlier section on Lean Thinking.

The 3/3 rule, states that the wait time, during which no value is added, is split 3 ways, each accounting for approximately one third of time. The three ways are waiting for completion of batches, waiting for 'physical and intellectual rework', and waiting for management decisions to send the batch forward.

The 1/4-2-20 rule, which states that for every quartering of total completion time, there will be a doubling of productivity and a 20% cost reduction.

The 3 x 2 rule, states that time based competitors enjoy growth rates of three times the average, and twice the profit margin, for their industry.

Note that there may be good reasons for emphasising speed or reduction in lead-time even though customers may not be interested in reduced lead times. This was revealed in the 3DayCar project where it was found that some customers were not interested in having a car in three days, but this fact could allow those that require short lead times to be catered for.

Becoming a time-based competitor is really what this book is all about. Like Womack and Jones, Stalk and Hout recommend process mapping. They talk about the 'Golden Rule of Time Based Competitiveness' which is never to delay a customer value adding step by a non-value adding step. Instead, seek to do such steps in parallel.

Further Reading

George Stalk and Thomas Hout, *Competing Against Time*, The Free Press, New York, 1990

3.9.1　The DNA of TPS: Four Rules and Four Questions

In a now classic article in Harvard Business Review, Spear and Bowen proposed the '4 Rules of the Toyota Production System'. Included in the article were the Four Questions. The article has become immensely influential. It was written following extensive research at Toyota. Steven Spear claims that the four rules capture the essence or DNA of Toyota.

Rule 1: 'All work shall be highly specified as to content, sequence, timing, and outcome.' This simple sentence has enormous implications. It goes right back to the days of F W Taylor and his 'one best way'. There is a best way for almost everything. It is the basis of Plan, Do, Study, Act. Spear talks about 'highly' specified. We would rephrase as 'ALL work shall be appropriately specified'. In manufacturing, 'specification', or standard work, allows problems to be identified and reduced. In service, it requires tasks to be done in the current best known way to minimise error and maximise service. It also places the onus on management to see that specifications are developed. If there is a problem, it is not good enough to say 'you must try harder' or 'you must work more conscientiously', or 'we have a motivation problem'. Instead, the process needs to looked at to see why the problem arose in the first place and to prevent it happening again. If there is no standard method, this cannot be done. So 'why did the patient get the wrong medicine?' or 'why did was there a part missing?' puts the emphasis on the process, not the person. This is pure Deming: system, not person.

Rule 2: 'Every customer-supplier connection must be direct, and there must be an unambiguous yes-or-no way to send requests and receive responses.' If there is a problem or issue the single, shortest path of communication must be clearly known and used. In the west we like to reward problem solvers. This is OK as long as the problem is communicated and its solution built in to the new standard. But, too often, the problem is solved but hardly anyone apart from the problem solver and perhaps his immediate manager knows about it. The next shift does not know, so when the problem recurs it is 'solved' in another way. Both shifts get rewards for 'initiative', but the fundamental solution remains in the head of the solver. In service, when a customer complains to the front desk, does the front desk person communicate it; does it get to the source? And, more likely, management sits in a fools-paradise thinking everything is great. Sometimes the communication route is too long – most people have played the children's game of sending a message around a circle. Sometimes the problem is communicated to a 'CRM' or 'maintenance MIS' – leaving it up to someone else to follow up on. Sometimes, the problem is communicated through many informal networks – allowing distortion and mischief. And communication channels need to thought out the other way around – from top to bottom. 'Every-one's problem is no-one's problem'.

Rule 3: 'The pathway for every product and service must be simple and direct.' This is about clear value streams. And it is about the minimum steps in the stream. We do not want the spaghetti of the job shop in either manufacturing or service. Do not leap into Theory of Constraints scheduling before untangling the spaghetti. Simplify the streams, the routings, the priorities. If at all possible, don't have a complex of shared machines and conflicting priorities (and then add insult to injury by requiring a complex finite scheduling package). Paying for a few extra machines can be more than worthwhile. In service, have you ever experienced multi-stage automatic telephone answering – and finally got through to a person (or worse, a machine) that cannot deal with your 'unusual' request? Try a human. Better still, try a highly knowledgeable human at the first stage. So, value stream map it *from the customer's perspective*. And remember Stalk and Hout's Golden Rule – 'Never delay a value adding step by a non value adding step.'

Rule 4: 'Any improvement must be made in accordance with the scientific method, under the guidance of a teacher, at the lowest possible level in the organisation.' This is about all improvements being done under PDSA, even if it a

small improvement. Without PDSA there is no learning. If there is no plan, no hypothesis there can be no surprises in the outcome. So all changes must be tested and reflected upon. The 'lowest level' means both place and organisational level. So improvement must be done at *Gemba* by *direct observation* probably using the *Socratic Method*. Direct observation is needed for understanding and to anticipate problems. The Socratic Method asks 'why'; it does not show 'how'.

Spear and Bowen suggest that the rules are not learned by instruction, but by questioning. The rules are not stated. The rules are absorbed over time. The manager is a teacher, not a 'boss'. And Socratic teaching is highly effective. Challenging questions involve going to Gemba and asking:

- How do you do this work?
- How do you know that you are doing it correctly?
- How do you know that the outcome is defect free?
- What do you do if you have a problem?
- We would add a few:
- Who do you communicate with?
- How do you know what to do next?
- What signals cue your work?
- Do you do this in the same way as others?

In fact, it is learning by the ongoing use of (Kipling's) 'six honest serving men' – who taught me all I knew; their names are what and why and when; and where and how and who.'

This Socratic method encourages operators to think, question and learn. Persistent asking of the questions allows decentralisation to evolve.

Bear in mind that 'it is not the quality of the answers that distinguishes a Lean expert, but the quality of the questions' (source unknown), and as Yogi Berra said, 'Don't tell me the answer, just explain the question.'

To summarise: TPS is not rules or tools. It is not *instinct* (instinct can't be learned), but *instinctive* – like the unwritten rules of a society.

Further Reading

Spear and Bowen, 'Decoding the DNA of the Toyota Production System', *Harvard Business Review*, Sept-Oct 1999, pp 97-106

For a health care related perspective on the four rules see, Steven Spear, 'Fixing Healthcare from the Inside, Today', *Harvard Business Review*, September 2005, p78-91

See also, Cindy Jimmerson, *Review*, Lean Health Care West, 2004

4   Lean Transformation Frameworks

Lean Transformation is the core topic of this book, yet if you are hoping to find a shortcut for your Lean journey here, we will have to disappoint you. While one tends to look for the '3 steps to heaven', unfortunately all Lean transformations are different, and there is no one 'golden bullet' recipe to follow. The individual settings in terms of customer needs and operational capabilities will set very specific priorities, which also will change as the transformation progresses. However, there are indeed a range of very useful frameworks at hand that can help guide the implementation efforts, and avoid the most common mistakes.

The two Frameworks are presented here – the Flow Framework and the Hierarchical Transformation Framework - are intended to help with the appropriate use of the tools that follow. These are not the only frameworks, and we will review some other proven ones in section 4.8. In addition, there are thousands of 'house of Lean' versions, and other, often more fuzzy, frameworks. Choose what you are comfortable with.

4.1   The House of Lean

First, let us look at the conventional 'House of Lean'. The original was developed at Toyota. A typical is shown below. Note the two pillars: JIT and Jidoka (Flow and Quality or 'Go' and 'Stop'. This is a necessary regulating mechanism – you need both.

Here is the good news about such houses: they are familiar and easy to understand. They seem to make sense. They may have a proven record at organisations like Toyota.

Here is the not so good news: they suggest you need to build from the foundations up - irrespective of situation. It is strongly tools oriented. But, when do you begin to build the walls? Where does the customer come in? Via policy deployment? But what happens if you are failing your customers due to poor delivery performance? How do you deliver value? And the need for 5S and standard work efforts can be

misinterpreted – tidy up, and screw down those out of control employees. Sustainability issues often result because employees see the tools as 'nice to do' rather than essential, and management becomes disenchanted because there is no impact on the bottom line, and little on customer satisfaction – maybe for quite some time. Basically, it is often a problem of local initiatives that yield local gains, but not necessarily systemic change.

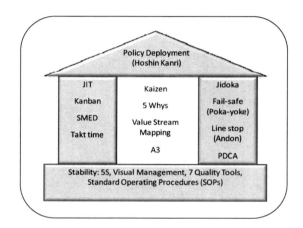

4.2   The Flow Framework

The Flow Framework, by contrast suggests broad-scale actions rather than tools. It presents a way of thinking about delivering value and getting to the heart of the issues. So whilst the House of Lean may have some good points, the Flow Framework, in our opinion, is far more widely applicable and useful. The Flow Framework is presented below – in outline and in more detail.

Kate Mackle of Thinkflow developed the Flow framework out of years of experience in Lean consulting, often involving companies that had lost their way on their Lean journey by cherry picking inappropriate tools or using an inappropriate model.

The flow framework begins with an analysis of your capacity, and how you 'present it in the market'. This means understanding the variation of demand, and whether you have the capacity to handle it. Bottlenecks, particularly if they are shared resources with other value streams, are obviously key.

The heart of the Flow Framework is **Create Flow, Maintain Flow, Organise for Flow**, and **Measure to support Flow.** Why 'Flow'? Because 'Flow' directly delivers value to the customer and money to the company. Because Flow is a central concept of Lean – reduction in lead time, just in time, one-piece at a time. If you focus on flow and lead time, you drive out waste.

It is important to start with 'create flow'. Many implementations start with 'Organise for Flow' This is a key reason for failure, because the need may not be recognised, and these activities have no short-tem bottom line impact.

So, the tools that are required come out of answering the questions, not because they seem nice things to do. Perhaps use some existing tools, but also you may have to develop your own appropriate toolbox. This is what Ohno did. When the reasons for using the tool are absolutely clear, implementation and sustainability issues become far less challenging.

The Flow Framework is also future oriented. The questions (see the diagram) remain and need to be regularly addressed.

Generally you give preference to the left hand side with Create Flow. This delivers value and cash. When you focus on creating flow, any problems with organising for flow become quickly apparent. Mapping will often be an early step to understand the barriers to flow. Value streams will need to be defined. Bottlenecks or constraints will be identified. Contribution per bottleneck-minute analysis may change your priorities – developing some products and processes and scrapping others. Note that you seldom begin with 5S, standard work and layout. If you leap in and do these first, you may be rearranging deckchairs on the Titanic. But when the need is clear, only then do you rearrange the layout and decide on the appropriate organisation.

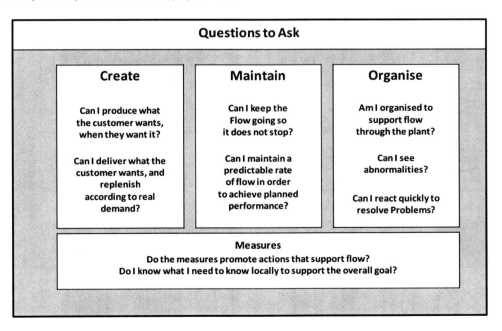

Questions to Ask

Create

Can I produce what the customer wants, when they want it?

Can I deliver what the customer wants, and replenish according to real demand?

Maintain

Can I keep the Flow going so it does not stop?

Can I maintain a predictable rate of flow in order to achieve planned performance?

Organise

Am I organised to support flow through the plant?

Can I see abnormalities?

Can I react quickly to resolve Problems?

Measures
Do the measures promote actions that support flow?
Do I know what I need to know locally to support the overall goal?

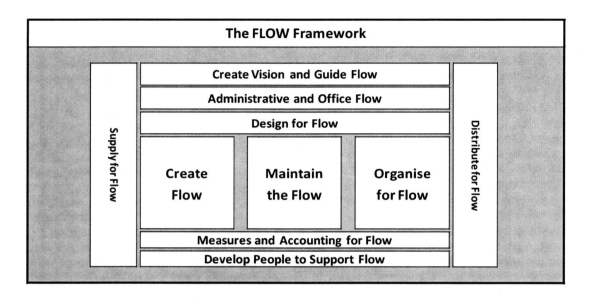

The FLOW Framework

- Supply for Flow
- Create Vision and Guide Flow
- Administrative and Office Flow
- Design for Flow
  - Create Flow
  - Maintain the Flow
  - Organise for Flow
- Measures and Accounting for Flow
- Develop People to Support Flow
- Distribute for Flow

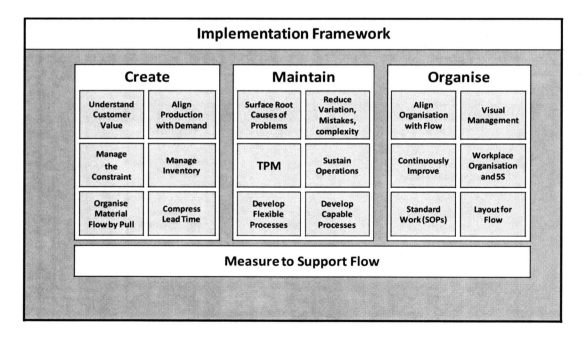

Implementation Framework

Create

Understand Customer Value	Align Production with Demand
Manage the Constraint	Manage Inventory
Organise Material Flow by Pull	Compress Lead Time

Maintain

Surface Root Causes of Problems	Reduce Variation, Mistakes, complexity
TPM	Sustain Operations
Develop Flexible Processes	Develop Capable Processes

Organise

Align Organisation with Flow	Visual Management
Continuously Improve	Workplace Organisation and 5S
Standard Work (SOPs)	Layout for Flow

Measure to Support Flow

Creating flow is one thing. Maintaining continuous flow, defect free is another. It may turn out that this is where the major problems lie. For some, perhaps capital intensive operations, maintenance of flow may be the priority. In other cases, quality may be the issue.

All the while you will need to examine the measures that are in force. Do they support Flow? For example there may be local measures or incentives that encourage the building of islands of inventory. Are measures compatible with the form of organisation – perhaps business units – that will be in force? And what about the measures relating to quality, to OEE, and to surfacing problems. The latter is a notorious area for measures that in reality conflict with flow.

Along the supply chain, supply for flow and distribute for flow means that you want a regular, smooth flow of materials through manufacture and out to customers. Avoid lumpiness if you can. If you can't, buffer your inventories such that flow is as smooth as possible between the buffers. Then work at reducing the buffer. Partnership along the whole chain is desirable.

Developing people to be capable of supporting flow, and administrative flow are also required. Once again, these tasks should not be done in isolation. Design for Flow refers to both the design process itself, and the products that come out of the process. The former may mean ideas such as the set-based approach, the latter incorporating concepts such as modularity and the possibility of postponement. Finally, last but by no means least, there must be an inspirational Vision of Flow. This in turn links with measures and accounting, probably through Policy Deployment.

4.3 The Hierarchical Transformation Framework

This framework is a more conventional, step-by-step approach, than the Flow Framework. It may suit a longer term implementation. However, there are overlaps with the Flow Framework so a manager may decide to adapt the overall Flow Framework but rely on sections of the Transformation Framework for the detail.

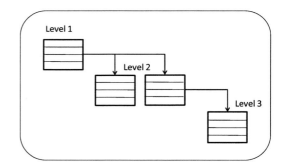

The Transformation Framework is intended to be hierarchical and iterative. The hierarchy is presented on three levels below. The steps in Level 1 are detailed in Level 2, and in some cases the Level 2 steps are further expanded on Level 3. The corresponding tools are given in Levels 2 and 3.

Level 1:

Gaining the Big Picture

Step	Activity
1	Understanding the principles
2	Understanding your customers
3	Strategy, planning and communication
4	Understanding the system, and mapping
5	Product rationalization and lean design
6	Implementing the foundation stones
7	The value stream implementation cycle
8	Building a lean culture
9	Lean supply
10	Lean distribution
11	Costing and measuring
12	Improving and sustaining

Level 2.1:

Understanding the Principles

No	Description and book Section
1	Lean Thinking
2	Fast, Flexible Flow
3	Five Lean Principles
4	Value and Waste
5	20 Characteristics of Lean
6	Time Based Competition
7	Gemba

Level 2.2:

Understand the Customers

Step	Activity	Book sections
1	Understanding customers	Lean Principles, Value, Kano
2	Segmenting customers	Supply chains
3	Understand demand	Demand management
4	Understand product characteristics	Operations Strategy (order winners), Quality Function Deployment

Level 2.3:

Strategy, Planning, Communication

Step	Activity	Book sections
1	Establishing the Strategy and Vision	Manufacturing Strategy, Value stream economics, Disruptive Technology, Value and Waste
2	Clarify tradeoffs	Time, Cost, Performance, Resources, Design(4 objectives)
3	Form focus factories	Product Family Analysis
4	Planning	Scenarios, Target Costing
5	Deployment	Hoshin Kanri/ Policy Deployment

Level 2.4:

'Check', Map and Develop Future State

Step	Activity	Book sections
1	Pareto Analysis	The Essential Paretos; Parts, Materials and Tools
2	Understand Demand	Understanding the Customer; Demand Management
3	Product Family Analysis	Product Family Analysis
4	Basic Mapping	*Basic mapping tools*
5	Check Capacity and Load (especially shared resources)	Theory of Constraints; Factory Physics; Shared Resources and Lean
6	Secondary Mapping	*Detailed mapping tools*
7	Shop floor Mapping	Work Combination, Cell Layout
8	Develop Future State	Mapping and Implementation

Level 2.5: Product Rationalisation, Lean Design

Step	Activity	Book sections
1	Simplify parts	The essential Paretos; Part, material and tool simplification
2	Organize for Lean design	Set Based Design
3	Link products to Customers	Quality Function Deployment Target costing
4	Consider design objectives and tradeoffs	Four objectives and six tradeoffs. Concept screening, Value engineering
5	Design the product	TRIZ, Design for manufacture, Modularity and Platforms

Level 2.6:

Implement the Foundation Stones

The Lean foundation stones are applicable in all situations. Whilst they do not have to be fully or even partly implemented at an early stage, a weak foundation leads to a weak and non-sustaining general implementation.

Stone	Description	Relevant Sections in the book
1	5S Housekeeping	5S
2	Standard work	Standard work
3	Improvement cycles	PDCA, DMAIC, Kaizen
4	7 Tools of Quality	(In 'Six Sigma and the Quality Toolbox')

Level 2.7:

The Value Stream Implementation Cycle

Value Stream implementation is central, ongoing activity within a Lean enterprise. The main steps are given below, and some steps are detailed further in Level 3.

Step	Description	Relevant Tools
1	Organizing	The Implementation Team Lean Promotion Office
2	Pre-mapping workshop	Pre-mapping workshop
3	Basic Mapping	Value Stream mapping. Spaghetti diagrams, Demand amplification, Quality Filter maps
4	Current state data collection	Lean Assessment, Activity Sampling, OEE, Basic data collection
5	Current state workshop	Current state workshop
6	Short Term Actions	Kaizen 'blitz', 5S
7	Detailed Mapping	Detailed mapping tools
8	Future state workshop	Future state workshop, Constraints, Creating Flow, Cell and Line Design, Creating the Future State(2). See below(7.8)
9	Simulation, Measures, Costing	Costing and Performance measurement, Implementation cycle
10	Internal implementation plan	Internal implementation, Hoshin, People
11	External implementation plan	Lean Supply
12	After Action Review	After action reviews, PDCA

Level 2.8:

Building a Lean Culture

Step	Activity	Book sections
1	Lay the foundation	People basics
2	Understand the pitfalls	People pitfalls
3	Build team spirit	What is Lean?, Gemba, Visions
4	Skill up	Team Skills, Problem solving skills, 5S, Standard work, Kaizen
5	Take positive actions	Gemba, Communication, Waste walks, Audits, Hoshin
6	Tackle anchor draggers	Adoption Curve
7	Change behaviour	Culture
8	Build and sustain	Sustainability

Level 2.9:

Implement Lean Supply

Step	Activity	Book sections
1	Define the supply chain	Supply Chain Thinking: S-C Basics
2	Understand demand	Demand management, Runners Repeaters and Strangers
3	Rationalizing parts and materials	Essential Paretos: Part rationalization
4	Select channels	S-C Channels, S-Chain mapping
5	Rationalize the supply base	Supplier Strategy rationalization
6	Consolidate, Simplify and Schedule Supply	The Building Blocks, Ten Scheduling Tools, Milkrounds
7	Manage amplification	S-C Amplification
8	Develop suppliers	S-C Basics, Suppliers partnerships, Supplier associations

Level 2.10:

Implement Lean Distribution

Step	Activity	Book sections
1	Defining distribution channels	The Right Supply Chain Partnership, Trust
2	Establish build to order principles	Creating the Future State (#1 and #2)

Level 2.11:

Costing and Performance Measures

Step	Activity	Book sections
1	Understanding the basics	Throughput, Inventory, Operating expense; Constraints, Pareto analysis
2	Measuring the right things	Balanced scorecard. Performance Prism
3	Measuring things right	Performance measurement
4	Costing it out	Lean accounting

Level 2.12:

Improve and Sustain

Step	Activity	Book Sections
1	Surface problems	Kaizen
2	Use Scientific Method	PDCA, Four Rules
3	Humility	Leadership, 25 Principles
4	Continue to adapt	Sustainability

Level 3: In this section two aspects are expanded upon from Level 2 – Detailed scheduling, and Lean cell and Line Design

Level 3.1. Designing The Scheduling System

Detailed scheduling system design is a late step in Lean implementation – recall, 'pull' is Womack and Jones' fourth principle. The main steps are given below, based partly on Rother and Shook's 8 steps to create a Future State from their book 'Learning to See'. Each step is briefly explained below. More detailed explanation of the tools is given in relevant sections. These sub-sections detail the necessary steps within section 2.7 above – the Future State Workshop. These are the detailed steps for Pull System design (this section), and Cell Design (next section).

Step	Activity	Book sections
1	Ensure demand is smoothed as far as possible	Demand Smoothing
2	Product family identification	Product Family Analysis
3	Value Stream Mapping	Value Stream Mapping Identification of constraints
4	Strategy and subcontract issues	Value stream economics, Manufacturing strategy
5	Segment the map into Value Stream loops	Value Stream Loops
6	Calculate takt time	Takt time
7	Identify constraints convergences and variation	Theory of Constraints, Factory Physics, Six Sigma.
8	Decide container size or move quantity	Move quantity, Kanban
9	Decide pitch increment	Takt and Pitch time
10	Build to finished goods or directly	10 Scheduling Concepts. Supermarkets
11	Investigate continuous flow possibilities	10 Scheduling Concepts
12	Locate supermarkets	Supermarkets
13	Decide on the pacemaker	Pacemaker, Material Handler
14	Level production at the pacemaker	Heijunka, Mixed Model Production, Material Handler
15	Calculate batch sizes at changeover stages	Batch sizing, Priority Kanbans, EPE (every product every)
16	Design kanban loops	Kanban, CONWIP, DBR
17	Design material handling routes	Material handling route
18	Form cells	Cell and Line Design
19	Improve	Kaizen

Level 3.2 Cell and Line Deign

Cell and line design is a hierarchical process, from factory layout to detailed workstation ergonomics. More detailed explanation of the tools is given in relevant sections. Note that these steps are done at the 'focus factory' level.

Step	Description	Relevant Tools
1	Product family or value stream identification	Product Family Analysis P-Q analysis
2	Value Stream Mapping, and shared resource analysis	Value Stream Mapping Spaghetti diagram Constraint Analysis
3	Strategy and subcontract issues	Value stream economics Manufacturing strategy
4	Plant layout and location of supermarkets	Lean plant layout, Supermarkets, Building Blocks
5	Activity timings and sampling	Activity timing Determine VA, NVA, NVAU Activity sampling
6	Calculate takt and cell cycle time	Takt time Cell cycle time
7	Identify any constraints	Theory of Constraints, Drum Buffer Rope, CONWIP
8	Paper kaizen	Paper kaizen and waste analysis
9	Theoretical minimum operators	Minimum operator calculation
10	Build to finished goods or directly	10 Scheduling Concepts. Supermarkets
11	Cell reference cost and Savings calculation	Cell Costing Lean Accounting
12	Cardboard simulation	Cardboard simulation
13	Cell layout design	Cell layout, 5S Pokayoke Material handling / Runner
14	Operator balancing (and plus one and minus one analysis)	Cell Balancing Yamazumi board Work Combination chart Cell layout chart Standard work
15	Workstation ergonomics and workstation design	Ergonomics Cell workstations
16	Pull system design	Kanban and Pull, Heijunka

4.4 General Approaches to Lean Implementation.

Almost every 'Lean Guru' and consultancy has their own approach to Lean Transformation. There is no Six-Sigma-like DMAIC agreed process. Inevitably, some are better than others, all claim to work, most of them can quote at least one successful implementation, sometimes many. The point is a 'horses for courses' message: there is no right or wrong. We cannot recommend one over the others in all circumstances. The approach should be chosen based on need: if strategy is weak, start there. An audit approach can be a good foundation, but it is essentially 'point kaizen'. There are also differences between manufacturing and services. For example, Peter Hines' strategy approach works well in small-or-medium-sized enterprises; in larger organisations changing the strategy will not be an option. Remember also, that whilst the Toyota system is undoubtedly effective for short-cycle repetitive manufacturing this does not mean that all tools will work in the pharmaceutical, or in aerospace or in low volume custom environments. The table attempts to summarise some of the better approaches known to the authors.

	Traditional Lean	Strategy	Systems	Flow	Audit	Consultant Blueprint
Approach	Toyota the exemplar	Strategy	Systems view	Flow	20 Keys or similar	Blueprint
Leading authority	Womack & Jones Liker Meier	Hines	Seddon	Mackle	Kobayashi	Big consultant
Method	Prescriptive	Prescriptive	Contingent	Contingent	Contingent	Prescriptive
Way in	5 Principles, 14 Principles	Strategy & Policy Deployment	'Check' Plan Do	Create, Maintain, Organise for flow	Audit	Start with top mgmt; use standard blueprint
Man/Serv	M	M & S?	S	M	M & S?	M & S?
Direction	Top down, Lean enterprise	Top Down	Listen to customer, involve people	Look at b/neck; DBR?	Audit by expert	Top down 'Exploring opportunity'
Early step (1)	Walk and i/d wastes; map	Management team meets	i/d purpose; understand Demand	Understand ability to meet customer needs	Spider diagram of strengths weakness	Map; Kaizen events
Early step (2)	Waste (maybe Muri, Mura)	Attain alignment; Deploy measures	i/d failure demand	Loading of b/necks; capacity issues	i/d priority keys	Evaluating change capability
Mapping	Early; Classic VSM	Quite early; 'Big Picture' & other mapping tools	Downplay; Outline only; dirty data	Quite early; info flows and financial aspects	Hardly	Early; classic VSM
5S and std work	5S, std work early; takt time	5S usually Early	No 5S; no / little std work	5S and std work drawn with need	5S usually early	5S part of 'demonstrating change'
Tools	Used	Used	Emerge	Emerge	Used	Used
Concerns	Expand to s/chain; extend to enterprise	Sustain -ability	Intervention	Lean accounting		Change management
Limitations/ Weaknesses	Automotive / Toyota applies everywhere	People? B / necks	Call centre / 'break – fix' Dominate	Complex schedule?	Little adaptation / point kaizen approach	Blueprint approach applied everywhere
Notes	1	2	3	4	5	6

Notes on table:

1. Womack and Jones, and Liker Meier are authors not consultants or active implementers. Both groups strongly champion Toyota. The 'House of Lean' may be one model that is used.

2. Hines is a leading Lean academic and part time consultant. Begins in the Board room.

3. This is an attempt to capture the Vanguard methodology. John Seddon is a leading figure and author in Lean Service, with emphasis on systems.

4. This is an attempt to capture the 'Thinkflow' model. Kate Mackle is a leading Lean thinker and consultant particularly in process industry and complex environments. Combines Lean and Theory of Constraints / DBR (drum buffer rope).

5. There are several Audit approaches used by various consultants and organisations like Ford and GSK. Kobayashi's 20 Keys is probably the original. Often attractive to top managers who like a simple score, but a danger is 20 or so separate projects or kaizens.

6. Several large consultancies use a fairly standardised Lean roll-out procedure, beginning with top level contact

4.5 The Failure Modes of Lean Implementations

Over the years, we have seen many implementations of Lean – initially in automotive, then in manufacturing in general, and more recently in service and healthcare operations. Many of these initiatives were successful initially, but all too often they were not sustained over a longer period of time. In the following section, we will outline, from our own personal experience, why these implementations have failed. A survey of the leading 1,000 Canadian manufacturing companies in 2007 revealed that 'backsliding to old ways of doing things' and 'lack of implementation know-how' were by far the greatest obstacles to Lean!

4.5.1 Senior Management Buy-in and Support

The first failure mode of Lean implementation relates to senior management support and buy-in. While we would like to believe that Lean can be implemented 'bottom-up', by doing good stuff on the shop-floor, then rolling it out to the entire organisation, this is a myth. The basic problem is that implementing Lean very soon requires changing not just the layout of the facility or lines, but also requires changes in the work organisation (by introducing teams, for example). This requires changing people's responsibilities, rewards and incentives, and very often also their pay. Changing front line manager tasks are key to Lean. To do that, senior management support is needed. If the measures, rewards and incentives are not changed, very often you will get some Lean enthusiasts who contribute, but many others will not see a need to contribute, or might even sabotage the Lean efforts if they feel threatened. Having specific and public support from a senior manager may resolve this issue. An open letter to all employees explaining the plans for Lean would be a good start. Even better would be active participation.

4.5.2 Dealing with Competing Initiatives and Initiative Overload

In many organisations we have seen multiple improvement programmes over time, or sometimes run in parallel. Such initiative overload leads to confusion ('Is all the good work we have done so far no longer valid?'), to apathy or sitting-it-out ('Why should I take part in Six Sigma: in two months time we will do something else anyway..'), to open resistance ('These new programmes don't change anything: within a few weeks everything is back to where it was anyway').

So how do you deal with 'competing initiatives'? The most important point to remember is that most improvement approaches have similar goals: value enhancement, lead time reduction, reduction in defects or variability, and ultimately, cost reduction. The problem is not **where they want to go**, but **how to get there**.

When introducing a new initiative, it is hence vital to make people understand how it fits in with the existing Lean programme, and that it does not replace the existing improvement programme, but will add to it. The name is important, as many perceive 'Lean' to be a competitor to 'Six Sigma', whereas in fact they are largely compatible. It is not by chance that, say, Unipart doesn't call its production system 'Lean', but the 'the Unipart Way'.

4.5.3   *Lean, Six Sigma or Lean Six Sigma?*

We now know that Lean and Six Sigma can work well together. However, they are not the same! They are two very different concepts, with very different strengths.

Consider this analogy: Lean is like the Public Health Engineer, Six Sigma is like the surgeon: both want to improve the health of the population, yet the way they go about it is very different! The Public Health Engineer will see to providing clean drinking water, a working sewage system and efficient garbage collection. He provides the environment or framework, and indirectly saves hundreds of lives every day. The surgeon on the other hand is called in to solve a particular issue, and does so in a single high-profile project (the operation) using his specific instruments and skills. Overall, society needs both. What does this imply?

Do not do Six Sigma before you understand the end-to-end process! Start with Lean, develop a complete understanding of value, and the value stream, then pull in Six Sigma as and when you need it.

But, should you even call it Six Sigma? Think carefully about the negative consequences of creating an apparent 'elite'.

Lean and Six Sigma are complementary in many aspects: Lean sets the philosophical background of value-focused thinking, Six Sigma provides a powerful toolkit to address specific issues and problems that have been identified. It is important to understand this distinction! We often come across situations where firms want to implement Lean, but either already have a Six Sigma or TOC programme, or at some point later in time want to implement these. This situation is dangerous, as the attention shifts to a new 'buzzword' and many improvement can be quickly lost. The question then arises: can Lean, Six Sigma and TOC be merged? Does a 'Lean Six Sigma' exist? Don't let these buzzwords confuse you. Remember here that many improvements approaches are complementary to implementing Lean, and that the focus on the reduction of variability in Six Sigma is also a key part of Lean:

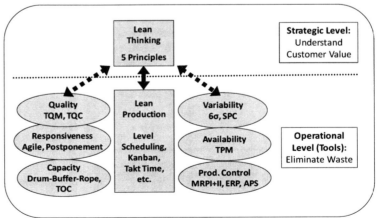

Lean creates stability through:

- Heijunka or level scheduling
- Small batches in production and shipment
- Kanban card that control the total inventory
- Time compression: SMED, minimal inventory
- No forecast-driven operations

Thus, Six Sigma – as a tool for variability reduction – fits very well into the wider umbrella of Lean. The one issue where they differ is where to start: Lean starts with analysing customer value, and a value stream map. Six Sigma will start with defining a process, and measuring its performance in terms of opportunities (defects). Be aware of the fact that Six Sigma has a strong methodology (DMAIC), which is powerful, but can also be very inflexible, and most importantly lacks the strategic perspective that Lean can provide.

Also, Six Sigma has a very rigid training regime with 'green belts' who are trained in basic statistics, and 'black belts' who receive several weeks of training and need to complete a certified improvement project at the end. This can be useful as it creates a common set of skills, but it can also be elitist! Lean only works if the ownership for improvements rests at worker level, and not with a select number of 'belts'.

Always consider Lean at the strategic level first by focusing on customer value, the value stream, and on demand, and then decide on the operational level what tools or concepts you need to reduce waste, and enhance customer value. Six Sigma is best used as pull, not push. Always question buzzwords – very often when you get to the philosophy behind these you will 're-discover' Lean ideas.

4.5.4 Abusing Lean as Short-term Fix

Lean is not a means for cost reduction (!), but unfortunately is often abused to achieve exactly that. More specifically, the two most common mistakes are to use Lean to reduce inventory, and to reduce headcount.

Inventory is a common focus as it is visible on the shop-floor, it is easy to measure, and it is a direct operating expense. So often it becomes the focus of improvement programmes, with disastrous consequences. Inventory MAY be there for good reasons:

- To buffer against internal uncertainty (defects, breakdowns, variation)
- To buffer against external uncertainty (demand fluctuations, supply, quality)
- To buffer against imbalances in the capability of the facility

This is shown in the 'water and rocks' analogy (see figure). So when inventory is drastically reduced without solving the underlying issues, many problems surface at the same time, a fire-fighting frenzy erupts, and within a short time the initial process and inventory levels are restored.

The Lean way is to reduce inventory level gradually, to expose one problem first, solve it, then expose the next, and so on. Bill Sandras refers to this as 'one less at a time' implementation. Inventory reduction becomes a means of identifying problems, not a goal in its own right.

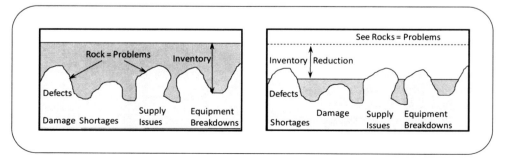

Headcount reduction has similar dangers. Labour has a direct bearing on productivity, and also generally is the most significant variable cost. So any reduction yields direct 'bottom-line' benefits. This brings a great temptation to lay off any labour that has been saved by Lean. However, doing this even once means that Lean will be perceived as a headcount reduction tool within your operation, and no one will 'improve themselves out of a job'. So the correct way is to improve the process, then use the 'spare' labour as a kaizen team initially, until you acquire more business for your firm to soak up the excess labour.

The cycle is: improve process, which results in better quality and lower cost, which increases market competitiveness, which increases business, which requires more labour, etc. It should be a virtuous cycle of growth, not one of reduction!

4.5.5   Misaligned Performance Measures

Basic game theory tells us 'You get what you measure'. In fact everyone, from CEO to shop-floor worker, will act so as to look good on his or her personal performance measures. Performance measures drive personal behaviour, and this is important to understand. If you want to make a change in a process, but the performance measures are not aligned to the new goal, there will be resistance. There might be intellectual acceptance, but misaligned measures will generally lead to counter-productive behaviour. Performance measures need to be aligned with what you want to achieve. Also remember that short-term measures will lead to short-term behaviour. See the sections on measures and policy deployment for more detail.

4.5.6   Lack of Ownership

Process improvement can be dictated, either by external consultants or by adopting a corporate template 'how things ought to be done'. The problem is that imposed improvements are generally not owned by the teams doing the actual work, and hence there is a strong resistance to being told what to do. The ownership of improvements needs to reside at process level, and workers need to feel that they have a stake in the future state process. Otherwise the new process is likely not to be sustainable. This is a particular problem when bringing in external consultants who make very good changes to the process, which are often not sustained because they are not 'owned' by the workforce. Team involvement in any process improvement is vital, and the process ownership issue must never be neglected.

Emiliani calls Lean-without-people-involvement 'Fake Lean'.

4.5.7   Keep the Momentum!

Lean does not support itself: the 'self-sustaining Lean culture' is a myth that is often propagated. The case of Wiremold (see below) should serve as a stark warning! Lean requires continuous support (and pressure) from the top, telling everyone in the firm that this is where the company is headed. Complacency is also a great danger, and one that Toyota is most afraid of – having a clear need to improve in order to ensure survival is a great motivator. Once you have reached the status of being an 'industry benchmark', such motivation is hard to come by. Then other goals need to be defined, such as becoming environmentally friendly for example. So keep the momentum alive! We'll come back to this in our section on Sustainability and Making Change Stick.

4.5.8   Should you call it 'Lean'?

In many ways Lean is an unfortunate word – sometimes carrying with it 'Lean is Mean' undertones, impressions of Japanese culture, sweatshops and failures. However false these may be, a bad brand name can be a killer.

Many have chosen to call their programmes 'The XYZ Production System' This too can be unfortunate because it is then seen as a 'production thing'. Best may be to either not give your program a title at all or to adopt an all-inclusive title like Spirax Sarco's LIFE (Little Improvements From Everyone), or 'The HP way'.

4.6   The Wiremold Case

Wiremold, one of the most-frequently cited cases of Lean implementation success, provides a stark warning how soon Lean can dissolve.

Wiremold was founded in 1900, a manufacturer of wire and cable solutions. Facing financial difficulties, the former CEO Art Byrne started an aggressive program of quality improvement and product introduction using Lean. After a decade of of hard work, the company was considered a model of manufacturing excellence – nationally and internationally.

When Wiremold was bought by Legrand, according to Emiliani, it soon transpired that the new owners had little interest in learning about Wiremold's process improvement capabilities. Art Byrne eventually retired in August 2002. Gradually so did many of the other original team members. So as the 'Lean leadership' left, Lean lost its support in the management team. Legrand, according to Emiliani, did not recognise the value of the Lean Management system, as it had a history of conventional 'batch-and-queue', which prevailed. Emiliani says, 'Lean requires constant attention and maintenance and improvement to keep it alive and healthy. It is [the] people that keep Lean alive through the daily practice of Lean principles, processes and tools.'

In the end, one of the most successful Lean implementations in North America faded.

So, the conclusion from this case is that there is no such thing as a long-term self-sustaining Lean system. Years of good work can be undone in a relatively short time if there is a lack of top management support or inappropriate KPI's.

Further Reading

The above case is a summary of Emiliani's Lean Blog of July 2007 on www.leanblog.com, and the epilogue to the second edition of his book: Bob Emiliani, *Better Thinking, Better Results,* The CLBM, LLC; 2nd edition, 2007.

4.7   A Warning on Lean Improvement

We should never be complacent about Lean or Lean improvement. Because some managers (and most professors!) learn the basis of their field from concrete models and case studies there is seldom disagreement over fundamentals. But….

'Individuals who break through by inventing a new paradigm are almost always... either very young or very new to the field whose paradigm they change. These are the men who, being little committed by prior practise to the traditional rules... are particularly likely to see that those rules no longer define a playable game and conceive another set that can replace them'.

(Thomas Kuhn, *The Structure of Scientific Revolutions)*

'It is easy to obtain evidence in favour of virtually any theory. By not pursuing instances where (the theory) does not work, or is not needed, we may be denying ourselves the opportunity of discovering or evolving a better theory'

..and..

'..of these new values that we have invented, two seem to me the most important for the evolution of knowledge: a self-critical attitude.......and objective truth...'

(Karl Popper, *All Life is Problem Solving*)

5 Strategy, Planning, Deployment

'The point is not to understand the world. The point is to change it'.

(Karl Marx)

5.1 Operations Strategy

The first reaction to the terms 'operations strategy' or 'manufacturing strategy' might well be to say: we have a strategy for the overall business, why do we need an operations strategy in addition to that?' This is a common perception, and not necessarily wrong. It is called the 'market view', where essentially customer needs and requirements filter down to operations, and determine exactly what operational capabilities are needed (cost, lead-time, quality, capacity to respond to demand swings etc.) to win orders in the market. Here, operations is *reactive* to corporate goals, where the 'strategic objective' is to achieve efficiency in production, that is to reduce unit cost, increase quality, etc.

However, there is a second view to this, called the 'resource view'. Here the argument is that the reactive view to operations strategy may omit some important contributions that manufacturing or operations can add to the overall business.

These are sometimes also referred to as 'dynamic capabilities'. Such dynamic capabilities take time to build up but are then hard to copy, unlike just buying a machine or even another company. Robert Hayes et al make the analogy with a golfer who buys the best set of clubs, but still cannot win a competition. This could be a distinctive technology or process that provides differentiation in the marketplace, as long as competitors can't match it, or how manufacturing supports the way in which products win orders in the marketplace. Think about Toyota, which is using its mastery of Lean (that was developed in manufacturing) in all parts of its business, or Wal-Mart and Dell, which both derive considerable competitive advantages from the way they run their operations. Toyota, Wal-Mart and Dell have all effectively developed strong operational capabilities that provide them with a competitive weapon, and thus 'feed back' into the strategy formulation process by contributing one or several critical factors. So, what an operations strategy does is to reconcile market requirements with operations resources and capabilities. Slack and Lewis use the following definition *'Operations strategy is the total pattern of decisions which shape the long-term capabilities of any type of operation and their contribution to the overall strategy, through the reconciliation of market requirements with operations resources.'* The alignment between the two is called 'strategic fit'.

Further Reading

Terry Hill, *Manufacturing Strategy: Text and Cases*, Macmillan, 2000

Nigel Slack and Michael Lewis, *Operations Strategy*, FT Prentice Hall, 2002

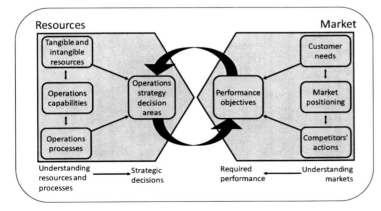

5.2 Tying in Operations Strategy with Lean

As Toyota and many others have illustrated, Lean can provide companies with a distinctive capability in the marketplace that contributes to the overall business strategy. In that sense Lean considerations can and should be part of any operations strategy discussion. In the next sections a number of general approaches to operations strategy will be discussed. These are generally longer term. For the short and medium term operations strategy can and should be linked with Lean approaches, discussed elsewhere in this book:

- **Value Stream Mapping** – particularly Value Stream Mapping, and Big Picture analysis will give a high level indication of the opportunities. Don't just think about the physical flow, also think about Marketing, Human Resources, and Financial Maps (see section below).

- **The Flow Framework**, and associated assessment, will give an excellent indication of the internal part of SWOT (strengths, weaknesses) analysis

- **A3 analysis,** which will assist with detailed identification of issues

- **Supply chain considerations** that link in strategically important supplier and distribution issues.

- **Gemba,** to get a real hands-on feel of customer issues

These can all be integrated into the topics discussed in the following sections 5.3 to 5.8. Use the models in section 5.8. to guide your choice of the sequence in which these should be used.

5.3 Understanding the Process: the Product-Process Matrix

A long established but still useful view is the Product Process Matrix. See the Figure. This aligns different types of Layout against volume. Laid out in this way, you get non-feasible regions (for example it would be folly to set up an assembly line for very low volume operations) and a feasible region that is in line with conventional thinking. Note, however, that the feasible region is quite broad.

The feasible region implies that the process or layout must evolve as the volume increases. Failure to adjust leads to a system that is out of alignment. Moreover there are appropriate, but overlapping, scheduling systems for each region. In the one-off project management area (e.g. new product introductions, large construction) critical path analysis is an appropriate tool. For job shops with shifting bottlenecks APS (Advanced Production Scheduling) is appropriate. A slow moving pulse line is often an appropriate form of layout for low volume, but regular production. Drum Buffer Rope (DBR) is often an effective scheduling tool for intermediate volume, particularly where there are bottlenecks (see section 10 on TOC for more detail).

Traditional Lean or JIT scheduling, involving kanban pull systems is appropriate in the higher volume repetitive area – this is particularly applicable where there are cells or assembly lines. In high volume, continuous flow operations such as process plants, optimisation scheduling tools such as Linear and mathematical programming is found.

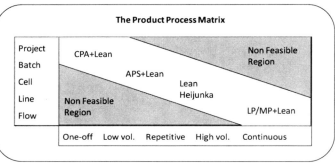

The Product Process Matrix

Note that Lean techniques overlay all these areas to a lesser or greater extent – for example 5S and SMED apply throughout.

In more recent times, many organisations have attempted to 'get the best of both worlds' (the advantages of higher volume low cost with increased customisation), by adopting 'mass customisation' approaches such as postponement, platforms, and modularity. This effectively widens the feasible region towards the bottom left corner of the product process matrix.

5.4 Understanding the Customer

5.4.1 The Kano Model

Dr. Noriaki Kano is a Japanese academic who is best known for his excellent 'Kano model'. The Kano Model has emerged as one the most useful and powerful aids to product and service design and improvement available. The Kano model relates three factors (which Kano argues are present in every product or service) to their degree of implementation or level of implementation, as shown in the figure. Kano's three factors are Basic (or 'must be') factors, Performance (or 'more is better') factors, and Delighter (or 'excitement') factors. The degree of customer satisfaction ranges from 'disgust', through neutrality, to 'delight'. A Basic factor is something that a customer simply expects to be there. If it is not present the customer will be dissatisfied or disgusted, but if it is fully implemented or present it will merely result in a feeling of neutrality. Examples are clean sheets in a hotel, a station tuner on a radio, or windscreen washers on a car. Notice that there may be degrees of implementation: sheets may be clean but blemished. Basic factors should not be taken for granted, or regarded as easy to satisfy; some may even be exceptionally difficult to identify. One example is handouts that a lecturer may regard as trivial but the audience may regard as a basic necessity. If you don't get the basics right, all else may fail - in this respect it is like Maslow's Hierarchy of Needs: it is no good thinking about

self esteem needs unless survival needs are catered for. In fact, the Kano model is based partly on the Herzberg Motivator-Hygiene theory of motivation. Market surveys are of limited value for basics (because they are simply expected). Therefore a designer needs to build up a list by experience, observation and organised feedback.

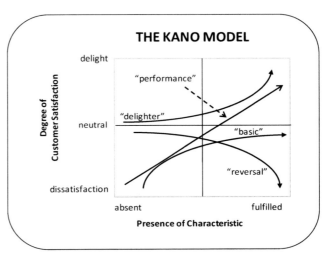

To test if a characteristic is basic, performance or delighter, ask two questions:

1. How do you feel if (the characteristic) is absent?
2. How do you feel if (the characteristic) is present?

   →If 1=bad, 2=neutral, it is a basic

   →If 1=neutral, 2=good, it is a delighter

   →If the answer is 'It depends', it is a performance factor.

Notice the non-linear shape of the curve. This is in line with economic theory which suggests that most people have such non-linear responses. If you had a 50/50 chance of winning or losing $1 million would you do it? Most people would not accept a bet with a 90% chance of winning $1 million if they had a 10% chance of losing $1 million, but for $10 there would be more takers.

A Performance factor can cause disgust at one extreme, but if fully implemented can result in delight. This factor is also termed 'more is better' but could also be 'faster is better' or 'easier is better'. Performance factors are usually in existence already, but are neutral, causing neither disgust nor delight. It is not so much the fact that the feature exists; it is how it can be improved. The challenge is to identify them, and to change their performance. Examples are speed of check in at a hotel, ease of tuning on a radio, or fuel consumption. Performance factors represent real opportunity to designers and to R&D staff. They may be identified through market surveys, but observation is also important, especially in identifying performance features that are causing dissatisfaction. Creativity or process redesign is often required to deliver the factor faster or more easily, and information support may play a role as in the 'one minute' check in at some top hotels.

Finally, a Delighter or Excitement factor is something that customers do not expect, but if present may cause increasing delight. Examples are getting a bottle of water from a hotel doorman when you return from a jog, or a radio tuner that retunes itself when moving out of range of a transmitter. By definition, market surveys are of little use here. Once again, it is creativity, based on an appreciation of (latent) customer needs that can provide the breakthrough. But we need to be careful about Delighters also: a true Delighter is provided at minimal extra cost - it would certainly cause customer delight to give them all a complimentary car, but would be disastrous for company finances. Therefore, perhaps a more appropriate hotel Delighter would be to give guests a choice of sheet colour, pillow type (English or Continental), and sheet type (linen, satin, or cotton). There are risks with providing Delighters – customers may come to expect them – as may have happened with Ford and GM car discounts.

Kano factors are not static. What may be a Delighter this year may migrate towards being a Basic in a few years time. Also, what may be a Delighter in one part of the world may be a Basic in another. Thus it is crucial to keep up to date with changing customer expectations. Benchmarking may be a way to go. From Kano we also learn that a reactive quality policy, reacting to complaints, or dissatisfiers, will at best lead to neutrality but proactive action is required to create delight.

The Kano Model works well with Quality Function Deployment and is increasingly being used with TRIZ. Basics should be satisfied, and delighters can be explicitly traded off in the 'roof' of the QFD matrix. For example, fuel consumption may suggest a lighter car, but safety suggests a stronger one - so the quest is to find a material that is light, strong, and inexpensive.

Further Reading

Special Issue on Kano's Methods: *Center for Quality of Management Journal*, Vol 2, No 4, Fall 1993 (several articles, including administering Kano questionnaires)

Joiner, B.L., *Fourth Generation Management*, McGraw Hill, New York, 1994

Lou Cohen, *Quality Function Deployment*, Addison Wesley, Reading MA, 1995

5.4.2 Scenarios and Predictions

Why be concerned about Scenarios in a book on Lean? Because we live in a world of discontinuity - think about the recent hikes in energy costs. Because conventional forecasting using extrapolation is more likely to lead to inappropriate size and location decisions - Mass instead of Lean. Scenarios are better at picking up possible discontinuities, thereby encouraging a more flexible, Lean approach. When making longer term plans or decisions, a forecast can be notoriously unreliable. Consider the effects and difficulty of predicting the long-term outcome of the Iraq War, the impact of China and India, the pensions crisis, and stock market turbulence directly affecting plant investment. So instead, use a few scenarios. For instance, most medium and larger companies in the UK should be considering various inflation scenarios, oil prices, environmental scenarios, and the impact of China on world resources.

At Shell, all major projects are judged against two or three scenarios. All projects are expected to demonstrate robustness against all the scenarios. This is a powerful supplement to traditional financial evaluation such as payback or ROI / EBITDA, where figures are sometimes (always?) massaged. The use of scenarios instead of forecasts is also consistent with 'pull' decision-making (starting in the future and working back) rather than 'push' (starting from the present and selecting from apparent alternatives).

Ilbury and Sunter have developed a most useful 2 x 2 matrix, referred to as the 'foxy' matrix – since foxes are supreme adapters. It has two axes, certainty – uncertainty, and absence of control – full control. This gives four cells that turn out to be a useful way of developing and using scenarios but also for general strategic thinking. We will use an example of a food packaging manufacturer considering Lean and Six Sigma.

1. The first cell and starting point is **certainty and absence of control**. Here the 'rules of the game' are established. These are the non-negotiables and givens of the situation. This includes the developing business environment and demographics, and the legal requirements for food packaging. The first cell may include benchmarking on cost, quality, lead-time, delivery, and inventory turns. Non-negotiables may be no compromise on hygiene, and not moving operations to Eastern Europe.

2. The second cell is **uncertainty and absence of control**. Here there are two activities – identifying the key uncertainties and developing the scenarios. In manufacturing key uncertainties may be alternative products and influence of China. Scenarios may include a narrative on growing public reaction against non-degradable packaging,

growing supermarket insistence on involvement in the supply chain and integration with non-food manufacturers, and the shift in demand for types of food as populations age, decline and become more health conscious. Developing the scenarios involves lateral thinking.

3. The third cell is **uncertainty-control** where options are identified. Options may include Leaning the plant and supply chain, implementing Six Sigma but not Lean, sourcing of some materials from China, and developing a complete new product line.

4. The fourth cell, **control – certainty** is where decisions are made. The food packager decides the appropriate response, weighing up the work in the earlier cells.

Ilbury and Sunter claim this matrix switches emphasis from SWOT analysis (strengths, weaknesses, opportunities, threats) to OTSW – thinking about opportunities and threats first because they are outside our control, and then matching them with strengths and weaknesses. In the third cell, options are expanded from TINA (there is only one alternative) to TEMBA (there exist many better alternatives).

Peter Schwartz in *Inevitable Surprises* says that the future is not as unknown as we think. He lists some 'future shocks' that should not surprise including where the proportion of over 50's (with money and time but also medical issues), prices of materials, availability of water, and patterns of human migration from less wealthy to wealthy nations. Demographics is, of course, the great underlying but predictable force.

According to van der Heijden there are three ways to build scenarios: Inductive, Deductive and Incremental. The inductive method builds stepwise from existing data, stringing known and chance events together. For instance, in the car example new models may be scheduled at particular points in the future. But the levels of demand are unknown. A framework of coloured cards (one for each level of demand) that maps new model introduction is assembled. Then an external trend or event is introduced, and the scenario team talks through the consequences. Tight questioning is desirable ('why does that happen?'), but destructive criticism ('that's stupid') is not allowed. Also, scenarios must be equally plausible. The test of this is whether the events are worth planning for; an earthquake that destroys the plant is not, unless you live in California or Kobe.

The Deductive method begins by identifying a series of events about which there is uncertainty, and then uses a 'decision tree' as a framework. For instance in the car example: Do we win the Ford business? If yes, does the market grow? And if yes again, will car manufacturers require first-tier suppliers to assemble on site? The 'no' branches are also explored. Probabilities can also be attached to each branch, thereby preventing the exploration of very unlikely scenarios. The Incremental method simply uses conventional forecasts and projections.

Projections and Reference Projections:

Russell Ackoff favours the use of 'Reference Projections' as a way of challenging business assumptions. This projects ahead assuming nothing else will change. If you have been growing at market share at 5% per year, within 20 years you will have the entire market. Is that likely? Or if costs and prices of your products are diverging, how long is that sustainable? The power of the method is that everyone realises that things have to change.

Clayton Christensen discusses what is called 'discovery driven planning'. Here, projections are made and the assumptions that will allow those projections to happen are monitored. This then allows a company to develop anticipatory plans ahead of time, rather than reactive plans.

Real Time Data

Leading companies are now beginning to use the power of real time data, not to forecast but to anticipate customer requirements. Tesco is an outstanding example, using Tesco cards to segment the market into, for example, frequency and amount of spends. This is combined with knowledge of purchases thereby giving incentives for lagging customers to increase their spending on known usage. Real time data also enables Tesco to specifically tailor individual stores to local requirements both by product type and by volume. When linked with point of sale inventory and frequent 'milk round' delivery, Tesco is able to stock a wide range of specifically targeted products but with low shelf inventories, at each specific store. Trends are quickly noticed.

Similar real time business practices are emerging in other sectors. Google, for example, does continuous large scale experiments with web page design.

Until recently we thought that 'build to order' (like Dell) was the ultimate Lean supply chain. It may well be for some. However, our recent work with car manufacturers shows that the maximum profitability requires a differentiated treatment for each product and market. Even Dell is reconsidering its supply chain strategy, with less flexibility offered to some customer segments. In order to maintain choice to customers, there is now the possibility of anticipating needs using real time data, continually tracking and adjusting.

Further Reading

Kees van der Heijden, *The Art of Strategic Conversation*, John Wiley, Chichester, 1997

Chantell Ilbury and Clem Sunter, *The Mind of a Fox*, Human and Rousseau, Cape Town, 2001

Russell Ackoff et al, *Idealised Design*, Wharton Publishing, 2006

Vivek Ranadive, *The Power To Predict*, McGraw Hill, 2006

Ian Ayres, *Supercrunchers*, John Murray, 2007

Clayton Christensen et al, *Seeing What's Next*, HBS Press, 2004

Thanks to Barry Evans of LERC for the Tesco examples.

5.4.3   Disruptive Technologies

Two of the primary rules in Lean are listening to your customers, and continuously improving. Also, benchmarking internally against a low waste future state and against tough competitors. But, are there situations in which this is not only misguided but also deadly? Perhaps so – where there are so called 'disruptive technologies' at work.

Clayton Christensen of Harvard has produced an inspiring analysis. Christensen distinguishes 'sustaining technologies' from 'disruptive technologies'. A disruptive technology is one that classically starts small, is simpler than the existing technology, and is ignored or even scorned at early stages by customers and managers alike. But the technology develops until seemingly overnight it becomes a serious proposition for customers to consider. Customers 'don't know that they want it until they want it'. Meanwhile the sustaining or established technology continues to improve, often outstripping the needs of many customers. There is a danger that companies become so focused on competing by continuous improvement, often putting their best people onto this, that they ignore outside challenges, especially those that are initially not seen as challenges. Witness vacuum cleaners and Dyson, mainframes and PCs, sailing ships (like the Cutty Sark) and early steamships. All these were initially seen as a non-threat.

Reading Christensen is a strong reminder that the current attention to sustainability may be misplaced. Adaptation and flexibility are required, not constraining sustainability.

Improvement can create a void that the typically low cost disruptive technology fills. Customers think they want it, but don't know about the alternatives. By then it is often too late for companies with the sustaining technology to catch up. Witness Amazon.com as against established booksellers offering comfortable seating and coffee. Or department stores displaced by discounters or on-line sellers.

Christensen states that with a disruptive technology many of the normal rules of business don't apply. Thus market research, allocating resources, killing off low return business, investment hurdles, and continuous improvement are all good policy for sustaining technologies, but may be the very policies that prove deadly in the presence of a disruptive technology. 'Markets that don't exist can't be analyzed'. This is not a failure of poor management; it is the very fact that they have done everything right that causes them to fail. The 'innovator's dilemma' is that continuing with existing lines of innovation, listening to customers and going after more lucrative developments is precisely wrong. Precisely because a disruptive technology displays minimal initial impact on corporate growth or existing markets it fails to attract the interest of executives who must look for far bigger gains. Christensen suggests that the way to deal with disruptive technologies is to establish a completely separate division, perhaps geographically separated but certainly organisationally separated from the parent, where small innovations are still viewed with excitement. This happened in the successful start-up of IBM's PC division or HP's inkjet printer division. Christensen suggests that the management of disruptive technologies requires different resources, different processes, and different values to those required for sustaining technologies. Strong visionary leadership is required, but also a different sort of leader to those skilled at managing sustaining business. Reading Christensen's brilliant analysis leaves the open question, 'Is that why so many kaizen or Lean initiatives fail to deliver?' Because the mindset is about sustainability, not radical change? Christensen points out that imitation is sometimes precisely the wrong thing to do. It may build only yesterday's competitive advantage. Successful strategies need to have a deep understanding of the processes of developing competition, not the transient 'solutions'.

Further Reading

Clayton Christensen, *The Innovator's Dilemma: When New Technologies Cause Great Firms to Fail*, Harvard Business School Press, 1997

Clayton Christensen and Michael Raynor, *The Innovators Solution*, Harvard, 2003.

5.5 Value Stream Economics: What to Make Where

Questions of what to make, where to make, and what to outsource have become increasingly pressing questions with the rise of low cost, high capability manufacturing countries such as China, India or Eastern Europe. Here we consider internationalisation and the issue of what to make where.

5.5.1 The Outsourcing and Offshoring Question

The first principle of value stream economics is to seek to adopt a Systems Approach. In other words try to maintain a holistic view for both now and the future. Consider the interactions. A system is like a child's mobile – touch one part and everything moves, some in unexpected directions. Build a scenario? – see the earlier section. Also, take a longer-term view. The following is a tentative checklist of factors that need to be added to the ex-works direct cost of a product made abroad.

Note: direct costs means that all overhead is excluded.

1. No doubt, the biggest factors are often lead time and flexibility. How much does the business compete on lead-time and responsiveness? What opportunities are to be gained, and what lost by a change in lead-time. Consider, at least, the counter argument to outsourcing – what opportunities will be created by insourcing?

2. Outsourcing and overseas have been fashionable in the 2000s. Beware of following the herd. A re-think due to material supply continuity, and a realisation

of 'failure demand', is occurring.

3. Loss of core competence.

4. Normal transport costs from the supply site to home markets – less any costs saved where the new location serves an overseas market.

5. Extra transport costs (e.g. airfreight) and an estimate of the frequency with which this will occur as a result of quality problems or imperfect forecasts (seldom zero). Note this may include shipping back of defective item for rework.

6. Loss of control, and thereby reputation, resulting from quality, scheduling, government regulation, etc.

7. Staff redundancy costs – possibly amortised.

8. Overhead costs – will the overhead be saved, or will it have to be allocated to another product? What overheads will be incurred at the new location, and how will they be allocated? What are the actual differences in flows of real money?

9. Will any space savings actually result in a saving – can the space saved be used for another purpose? If not, there is no saving.

10. Currency fluctuation insurance costs.

11. International taxes and duties. Note that these may vary in relation to international agreements, but also in accordance with 'tariff wars' that may be imposed at short notice. How long will the tax benefits last?

12. Customs clearance time and the resources need for the extra paperwork.

13. The extra inventory held in the pipeline and the one-off cost of that inventory – possibly amortised over the life of the product.

14. The extra inventory held as a result of demand uncertainties over the longer lead-time horizon.

15. The extra inventory held initially whilst ramp-up takes place. Is it reasonable to assume that there will be no hiccups in

early days? Will utilisation be as good?

16. The extra inventory held as a result of quality problems or damage. Remember the vessel containing a full cargo of new 7 series BMWs that sank in the English Channel in 2003, or the looting from the freighter that ran aground off Devon in 2007?

17. Hence insurance costs.

18. Increased obsolescence risks – e.g. computers with chips in transit.

19. General costs of quality: internal failure costs (scrap, rework) and external failure costs, including increased warranty costs.

20. Ramp-up costs in general – including training, sorting out problems, visits by home engineers.

21. Loss of customer goodwill, or loss of customers – period – as a result of quality service, support, customisation opportunity, etc. In 2007, the paint on some Chinese toys was found to have toxic levels of lead. In 2008 another toxic chemical was found. Apart from scrap costs, what will this do to the reputation and loss of future sales of the companies concerned?

22. Marketing costs – can you sell a longer lead-time?

23. Costs of loss of design expertise for future products. Difficult to assess.

24. Costs of loss of manufacturing expertise. Difficult to assess, but real – think about the next product that has to be developed from afar.

25. Political risk costs (naïve to assume zero?). And bribes?

26. Finally, allow a percentage for data inaccuracy – several of the above will be guesstimates.

27. Against this list, you can subtract any incentives gained and opportunities in the new market. A 'benefit' may be transfer pricing – (cooking the books?) – so that the profit is made in low tax locations. Lean accounting?

Another approach on how to assess the cost of offshoring is to break the cost down into three elements: static, dynamic and hidden cost. This logic can be applied to assess the viability and risk of global sourcing and offshoring strategies.

1. **Static costs** are the obvious costs that occur in manufacturing and transportation, which include materials, labour, energy (and the cost of the capital investment if you offshore), as well as the transportation and any customs clearing cost. Local sourcing and manufacturing generally features higher labour cost, offshoring and global sourcing generally incurs a penalty on transportation costs. The problem is that this static perspective does not consider any additional costs in the supply chain.

2. **Dynamic costs**: these include any additional pipeline inventory, as well as the higher risk of obsolescence and lost sales, as well as the cost for additional buffers needed to deal with uncertainty. Often these dynamic costs are not considered up front, but from experience always occur! Consider also how likely the event of expedited shipments is if problems occur. Airfreight is very expensive in comparison to sea- and land-based transport.

3. **Hidden cost**. These costs generally emanate from changes in exchange rates, political risks, or rises in labour and energy cost. These costs are not so easy to quantify, but can be equally important in the longer term: while many calculate the current savings if sourcing globally or offshoring on current labour costs, fewer consider the possibility of inflation and currency fluctuations. Also, think about rising oil prices, and the implications on transport cost. In many emerging markets, such as India or in Eastern Europe, double-digit percent rises in wages are being observed. Use a Net Present Value (NPV) calculation

to discount future savings. Also consider risks related to the loss of intellectual property (IPR), as well as political risks.

The table below is a summary of the main costs that should be considered when sourcing globally, or offshoring operations:

and accounting complexity. But at what level in the bill of material is a split appropriate? An interesting case is Dell, which chooses to insource assembly, as opposed to several less successful computer manufactures that choose to outsource – by using 'contract manufacturers'.

Static Cost	Dynamic Cost	Hidden Cost
• Purchase price ex factory gate	• Increased pipeline and safety stock due to demand volatility	• Rise in energy or transportation cost; carbon offset costs
• Transportation cost per unit, assuming no unexpected delays of quality problems	• Inventory obsolescence due to long logistics lead-times, e.g. in case of quality problems	• Currency fluctuations, in particular for artificially pegged currencies
• Customs and duty to clear one unit for export	• Engineering time needed to address quality and warranty issues	• Remaining overhead at the headquarters (Purchasing, technical assistance, R&D, product development)
• Insurance and agency fees	• Expedited shipments, e.g. air freight, to ensure uninterrupted supply	• The loss of intellectual property to contract manufacturers, as well as legal risks in terms of ownership of facilities and market access
	• Cost of lost sales and stock-outs, as the supply chain is unresponsive	• Labour cost inflation
		• The strategic risk of political instability and change

5.5.2   The Location Issue – Vertical Splits

Think about products and channels. Consider relocating that part of the plant that deals with stable, runner products whilst retaining locally products requiring a more flexible shorter lead-time. Or a mixed approach that involves retaining products locally during ramp-up or de-bugging (to sort out quality, standard work, kaizen, cell design), and only then relocating the stable or mature line.

5.5.3   The Location Issue – Horizontal Splits

Locating subassemblies away from base involves adding to scheduling and coordination complexity whilst possibly reducing engineering

5.6   The Essential Paretos

Pareto Analysis or the 80/20 rule has been called the single most important management concept of all time. Quite a claim! There are four essential Pareto analyses that every operations manager, and especially every aspiring Lean manager, should be aware of in relation to their own plant, and a fifth that every marketing manager should be aware of. Pareto's Law, about the vital few and less important many applies to inventory, quality, marketing, layout, warehousing – and much more.

The 'physical' Paretos will be discussed here. But there is another type – the type that applies to change. A small proportion of people have a very large influence on the success of a Lean transformation. This will be looked at in the People section of the book.

5.6.1   Inventory ABC Analysis

Inventory ABC analysis is a long established procedure, but vital for the Lean manager as an aid to inventory control and a guide to the selection of inventory systems (tight and loose kanban, 2 bin, VMI and the like – when combined with the runners, repeaters and strangers concept). The system selection question is discussed in the Scheduling section of the book. Here we discuss basic inventory control issues. You simply cannot run a Lean system with poor inventory control, but parts warrant different level of attention.

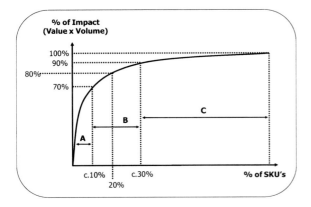

At the part or component level, A parts are high value items, B intermediate and C low cost. Typically A parts are a small percentage by number but a large percentage by value. C is the reverse. Of course you need all types of part to complete a product, but you can afford to be less tight (more safety stock) with C items. To construct an ABC analysis, simply rank the parts from highest to lowest unit value in a long list. The top perhaps 15% of parts by value warrant special attention (tight kanban, low safety stock, careful monitoring and demand forecasts), the middle 50% less so, and the remainder even less (2 bin system, more safety stock, automated monitoring). This should be modified in two ways:

- by the Runner Repeater Stranger concept (see below)
- by the risk factor – the probability of supply disruptions, quality issues, etc. If an item is prone to this, it should be upgraded – from C to B, or B to A.

This ranking should also be the basis for cycle counting – count a few parts every day, such that

A's are counted monthly, B's quarterly, C's once or twice per year. Ignore this at your peril – whilst not as glamorous as a Six Sigma project it is at least as important.

5.6.2 P-Q (or Product Quantity) Analysis

P-Q analysis is a Pareto procedure that simply ranks parts or products by volume, or quantity. It is an essential early step in deciding on layout. P-Q is also linked with the Runners, Repeaters and Strangers concept. Top end items (Runners) can justify dedicated equipment. Mid range parts (Repeaters) may have to share cell facilities with other parts. Often some routing changes will be required. Parts in the tail of the Pareto (Strangers) are problematical for cells. The best case is that by clustering routings they can be brought into a cell. The worst case is that these parts will have to go into a residual cell or job shop – or they are candidates for outsourcing.

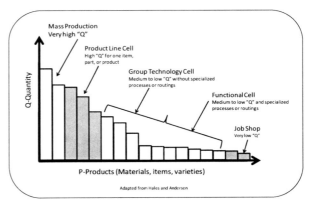

An important issue is at what level should you do a P-Q Analysis? A general answer is – the same level at which the Master Schedule is developed. There are two types of Pareto P-Q analysis – value based and volume based – so the vertical axis above may be value or volume. The value based analysis may identify a small number of products where dedicated lines may be worthwhile irrespective of volume – for example a cosmetics line that is run on average one day per week, but thereby allows low inventories and extreme flexibility. A volume-based analysis may justify dedicated lines based on volume.

But not so fast! By rationalisation methods (design, modularity, etc.) stranger parts can be made into repeaters. Recall also that while end items or products may be unique, their subassemblies or components may not be. Therefore an upstream cell may be justified, even though a downstream cell may not be. Or, if you have a decoupling point in your supply chain, you can have part family cells upstream, but downstream the cells may be customer specific. The parts Pareto section below is of relevance.

Not so fast, part 2 – the direction of movement should also be noted. In other words a low volume product that is growing is more important than a slightly higher volume product that is declining.

Another derivative is P-Q-R, where in addition to product quantity you also consider routing. Here, the complexity of the routing for a product or item is used as a third decision variable in what layout to use: the more standard the routing, the more likely a line or cell layout is justified. The more individual the routing, the more likely a job shop is appropriate.

References

H Lee Hales, Bruce Anderson, 2002 *Planning Manufacturing Cells,* Society of Manufacturing Engineers, Dearborn MI

5.6.3   *Contribution Analysis*

Contribution is selling price minus direct costs (i.e. contribution to overheads). It is essential to know which products are making your money. A cumulative contribution analysis would look similar to the Pareto / ABC analysis for inventory shown above, except that the vertical axis would show the accumulated contribution. The analysis needs to be done for the current state and for the future state. Again, there are two sub-categories. The first is total contribution. Which products are making the greatest contribution,

and which if any are making a loss? There may be strategic issues here – some products may have to be retained as loss leaders.

The second, and possibly more important, is 'contribution per bottleneck minute'. If you have a clear bottleneck process (where you could make more money if you had more capacity) or a stage that is a near bottleneck (or constraint), then you need to divide the unit contribution by the time spent on the bottleneck. Clearly, you don't want to have products that make small contribution *and* that tie up your precious bottleneck capacity. But do take care with this type of analysis – if you cut products, be sure that all the 'direct' costs you have assumed will actually come down.

5.6.4   *Parts, Materials, and Tools Paretos*

Parts, materials and to a lesser extent tool rationalisation is a huge, and often untapped, field of opportunity in many companies. In fact, the potential often far exceeds taking out inventory as part of a scheduling, kanban, or value stream implementation. A series of Paretos is the way in. Where rationalisation has not been done or specifically controlled for several years, part proliferation may be rampant. Getting after proliferation is big task, usually requiring a specific team, but with big payoff in inventory savings. But, more importantly, this activity allows the business to become much more flexible and in some instances literally to change the business model from long lead-time, high inventory build to stock, to short lead-time, build to order.

The conversion of 'strangers' into 'repeaters' and 'repeaters' into runners' should a fundamental part of every Lean transformation. For example, in one organisation that one author was involved with, over 50 types of bar stock were reduced to just 5; in another 28 types of fastener were reduced to just 6. In a third case, implementation of NC laser fabric cutting in-house (on 4 material types) enabled a short-lead-time pull system to be introduced with one pacemaker, replacing vast quantities of different material inventories (well over 100 part numbers), forecasting, purchasing, MRP transactions, long lead-times, and part shortages - as Zorba would have said, 'the full catastrophe!'.

Rationalisation may involve upgrading existing specifications, and so appear 'uneconomic' but the payoff is in simplicity of control, in flexibility, in inventory – but more importantly, in business opportunity.

Rationalisation requires high-level support – and will typically meet resistance from both accountants concerned with unit costs (but not system costs), and from designers concerned with optimal product design and product 'waste'. Of course, while individual product waste may increase, overall system cost is slashed.

The first step is to list the categories of major parts, components, materials, and tools. For example part categories may include fasteners, bushes, housings, and gears. Components may be motors, printed circuit boards, transformers, and containers. Materials may include coil steel, bar stock, plastic for injection moulding, fabrics, gloves, and packaging material.

The second step is to decide which categories to tackle first. Set up a multi-function group from the Lean promotion office, manufacturing, design, purchasing, accounting, and possibly marketing. Anticipate a stormy session, and get it chaired by a sympathetic system thinking, top manager.

The third step is to draw up usage Paretos for the chosen categories. Rank by annual usage and also by current inventory holding divided by annual usage last year. This second ranking will sometimes result in a few 'infinite' categories where usage has ceased. Examine the tail of the Pareto in the former case, and the head in the latter case.

Get the team to examine systematically each item from these two ends, always looking at the possibility of eliminating or combining items. This standardisation and rationalisation activity must be ongoing. Part proliferation slowly creeps in, in

the Design office and via Marketing. These functions do need to appreciate wider system economics.

A final word: Richard Schonberger has provocatively said that an organisation can go a good way along the Lean path by not going near the factory floor, but by instead focusing effort on Design!

5.7 Formulating an Operations Strategy

To some, strategy and deployment is all the same thing. This view is probably correct because these topics form an ongoing process. Manufacturing strategy is 'emergent' with strengths in Lean enabling a manufacturer to adopt strategies that are simply not open to the non-Lean. But no manufacturing strategy can compensate for an incompetent operation.

Today one still encounters the attitude that there is no such thing as operations or manufacturing strategy – there is only corporate strategy and marketing strategy. Operations are strictly subservient to these and must simply produce the goods at the lowest price and acceptable quality. Then there is the view that manufacturing strategy is simply concerned with trade-offs: low cost vs. long lead times, or high cost and greater flexibility against low cost and less flexible.

Both these views are outdated. Lean has shown that some traditional thinking on trade-offs is simply wrong. For example, you can have high quality and low cost – in fact those two go hand in hand. Today manufacturing strategy, when done correctly, is a 'formidable competitive weapon' (to quote Wickham Skinner) and an equal partner to marketing and finance.

One of the classic models how to formulate a manufacturing strategy was proposed by *Terry Hill*. His approach begins with corporate objectives – manufacturing strategy is seen as contributing towards these objectives. Corporate objectives lead to marketing strategy. Marketing identifies appropriate markets, product mix, services, and the degree to which the company needs to customise and innovate. Marketing strategy leads into the identification of required 'order winners' and 'order qualifiers'. This is similar to the Kano model dimensions – see later. Order winners and qualifiers lead to prioritising the necessary capabilities or 'process choice' as Hill refers to it. One problem that several users have found is that it is difficult to identify clearly the 'order winners' amongst different customers. Use the Kano model for this. Another problem with the approach is that manufacturing is seen to be subservient to marketing, which neglects the dynamic capabilities that manufacturing has to offer (see above).

The *Platts and Gregory* procedure begins with a SWOT analysis, a market evaluation, and a manufacturing audit. From this a profile is developed for the critical capabilities. Gaps are identified. See the diagram below – this uses the Hayes et al capability dimensions.

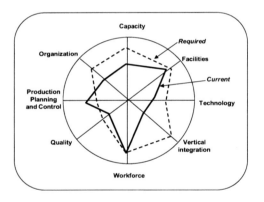

Slack and Lewis arrange the capabilities differently. According to them, there are five performance objectives – quality, speed, dependability, flexibility, and cost.

These five are the enablers for market competitiveness. These five objectives are attained through four decision areas – capacity, the supply network, process technology, and organisation and development. This gives a matrix containing 5 x 4 = 20 cells, showing the impact of decision areas on the performance objectives. Some of the cells will be very critical to success, others critical, but several secondary.

In fact one can arrange these five objectives (the 'whats') and four decision areas (the 'hows') into a Quality Function Deployment (QFD) matrix. See separate section. This gives added information and considerations about, for example, possible contradictions between decision areas, relative ranking of the 'hows', and weighting and competitive scores of the performance objectives.

Christensen et al say that the Resources, Processes, and Values (RPV) evaluation is fine for existing or 'sustaining technologies', but that Jobs-to-be-Done Theory or Value Chain Evolution (VCE) Theory can be more suitable in dynamic situations. The former looks at what jobs customers need to solve, and then tries to find solutions – that may come out of a completely different industry. This is like TRIZ – are you in the lawnmower business, or in the easy-to-keep lawns business? VCE says that there are different architectures (interdependent or modular) depending on the stage of product development. Is your product part of a system, or can it stand alone? Should you make whole cars (like Ford) or exclude the engine (like Lotus)?

Further Reading

Robert Hayes, Gary Pisano, David Upton, Steven Wheelwright, *Pursuing the Competitive Edge*, Wiley, 2005

Terry Hill, *Manufacturing Strategy: Text and Cases*, Macmillan, 2000

DTI, *Competitive Manufacturing*, 1990, (for Platts-Gregory procedure)

Nigel Slack and Michael Lewis, *Operations Strategy*, FT Prentice Hall, 2002

Ken Platts and Mike Gregory, 'Manufacturing Audit in the Process of Strategy Formulation', *International Journal of Operations & Production Management, Vol. 10(9), 1990, p.5-26*

Clayton Christensen et al, *Seeing What's Next*, HBS Press, 2004

5.8   Policy Deployment / Hoshin Kanri

A typical planning process looks like this:

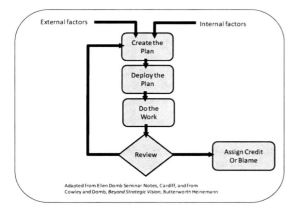

Adapted from Ellen Domb Seminar Notes, Cardiff, and from
Cowley and Domb, *Beyond Strategic Vision*, Butterworth Heinemann

There are several problems with this traditional approach to planning. Firstly, it seems common practice to take management to a hotel for a weekend, stick everyone into a meeting room with a flip chart, then pull ideas out of thin air. Most use opinions only, but no analysis. A SWOT analysis done on brown paper in a hotel room, based on people's perceptions, is the most common 'outcome' of such an approach. Then everything continues as before. Does it make a difference to the daily work? Next year, are last years' plans looked back at? And is credit or blame assigned without differentiating between common and special causes? Year +3 always looks brilliant, but seems to keep rolling forward! Little learning occurs.

The Hoshin process is different. First, plans are created by deliberate research and learning from last year. Only a few areas are selected. There are three feedback loops. The work is reviewed in line with the plans, the plans themselves are reviewed at the end of the period, and the planning process itself is modified if necessary.

The plans are deployed, level by level, in a participative way.

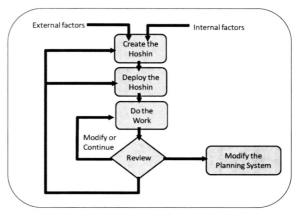

That is Policy Deployment. Now, more detail:

5.8.1   Policy Deployment: How to implement

Policy Deployment is just that – it is an effective way of deploying policy – making sure that all are not only moving in the same direction, but fully committed and bought into the steps that need to be taken en route. In essence Policy Deployment (or 'Hoshin Kanri') is about 'nemawashi' (consensus building) and 'ringi' (shared decision making). Literally translated, 'hoshin kanri' means 'the captain steers the ship', meaning that the captain gives orders on course and speed, which then 'cascade' down to the various functions in the ship: the engine room is told how fast the engine should go, the person steering the ship is told whether to turn the rudder to port or starboard, when and how far. Each function receives exactly the part of the order that is relevant to them, and is measured on executing 'their bit'. That way everyone works towards the same overall goal.

Policy Deployment has become a well-accepted way of planning and communicating quality and productivity goals throughout a Lean organisation. Some Hoshin Kanri work (e.g, by Jackson, and by Hines) is very strongly strategy related. Others accept that strategies are made by some other process and policy deployment is about deployment. In this section the concentration will be on deployment. It may be preceded by strategy as described in the last section.

Here we will use 'Hoshin' and Policy Deployment interchangeably, although strict Japanese style Hoshin is hugely bureaucratic and form-intensive so Policy Deployment is a preferable phrase to distinguish it from the rather non-Lean Hoshin.

It is the deployment process that should be central. To quote Pfeffer, 'What is extremely difficult to copy – and what therefore does provide competitive advantage – is the way a company implements and executes its strategy. Anyone can talk about being the technology leader or providing outstanding customer service. But few organisations can actually make good on that promise. That's why Wells Fargo CEO Richard Kovacevich once said he could leave the company's strategic plan on a plane and it wouldn't make any difference: 'Our success has nothing to do with planning. It has to do with execution.''

Hope and Fraser attack the wastefulness and dysfunctional behaviour that the traditional budget process engenders. The internal competitive behaviour that is frequently generated is the very antithesis of the cooperative systems approach advocated throughout this book. 'So long as the budget dominates business planning, a self-motivated workforce is a fantasy, however many cutting-edge techniques a company embraces'. So don't begin with the financial and issue top down command and control edicts. Instead, begin with the processes that can deliver the aims and therefore bring about financial results. This is what Policy Deployment enables one to do.

At the outset we should say that Policy Deployment can be used in two ways. First (undesirable) in a 'command and control' way – top down objective setting with little discussion – starting with the financial or other results in mind. This can and will encourage all the 'game playing' and pseudo measurement practices that are so common in many command and control

organisations. It is much better to use it in a systems way – understanding the needs of the system and then developing more detailed participative plans to meet the system requirements. Results come out of the process, not the other way around.

Policy Deployment is in fact the PDCA cycle applied on an organisational level. Witcher and Butterworth talk about the FAIR model beginning with the Act stage. Focus (act), Alignment (plan), Integration (do), and Responsiveness (check).

Cowley and Domb use a useful metaphor. This shows a road leading from the present position (the current state) to the destination (the vision or future state). The road is the plan or action plan. Along the road are scattered small rocks and large boulders – which are obstacles or problems. PD is used to remove the boulders one at a time. Kaizen is used to remove the smaller rocks, and importantly this is not what we are concerned with here. Also, there are boulders which are off the road. These should not be tackled – PD is about focusing on the 'vital few'. The vision must be shared, level by level. What *not to do* must be agreed. And alternatives must be developed. The road metaphor is even more useful when used with Maps – current and future state.

The traditional approach, as shown by the first flowchart above, are all symptomatic of a non-Lean planning process – what Ellen Domb calls 'Phoney Deployment'. By contrast, real Policy Deployment involves three stages of feedback as well as the following important differences. Refer to the second figure.

- The plans (or Hoshins) are created following real research into customer needs, good assessment from internal performance based on value stream appreciation, and by looking back on last year's performance and explaining the difference between what was planned and what was done.

- The limited number of Hoshins are deployed, level by level, in a participative way. At each level the higher level puts forward the 'whats' and 'whys', and the lower level proposes the 'hows'. Consensus is reached.

- Work is done that actually reflects the Hoshins, and is regularly reviewed and modified.

- When plans are not achieved – and there is the expectation that no plan is perfect so will not hit the target 100% of the time – the plan and the planning process itself is examined, rather than blaming the people. This, of course, is Deming's '94/6' rule.

- The whole process is one of learning – to understand the system and its environment better – rather than a 'command and control' procedure.

A 'Hoshin' is a word that is increasingly being heard in Western companies, to mean those few breakthroughs or aims that are required to be achieved to meet the overall plan. At the top level there may be only 3 to 5 Aims. But at lower levels, the Hoshins form a network or hierarchy of activities that lead to the top level Hoshins. They are developed by consultation. Hoshin objectives are customer focused, based on company wide information, and measurable.

Imagine a tree deployment process. For example, if paint quality is seen to be a problem, then there is a role for manufacturing (to improve the application process), for quality (to develop better standards), for R&D (to develop better formulations), for engineering (to look at paint application), for logistics (to ensure scratch-free delivery), for marketing (to understand customer requirements and the environments that products will be exposed to).

A good concept, with applications far beyond PD, is to use the Toyota concept of requiring all plans or projects to be written up as an A3 – see figure, and section 12.7 on A3.

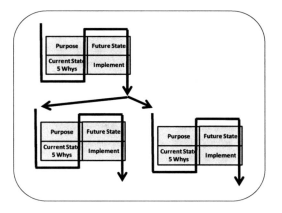

PD starts with the concept of homing in on the 'vital few'. Where there is little change in operating conditions, a company still needs to rely upon departmental management, and top management planning is not required. However, where there is significant change, top management must first understand the purpose of the system. Debate what the system requires. This requires strategic planning (for future alignment to identify the vital few strategic gaps), strategy management (for change), and cross functional management (to manage horizontal business processes).

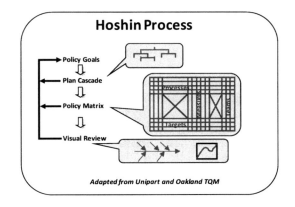

Hoshin Process

Policy Goals
Plan Cascade
Policy Matrix
Visual Review

Adapted from Unipart and Oakland TQM

Departmental management should be relied upon for 'kaizen' (i.e. incremental) improvements, but breakthrough improvements which often involve cross functional activities and top level support, should be the focus for PD planning.

An important aspect is learning. The future is uncertain, although the destination is clear. Treat every unexpected event as an opportunity to learn and to adapt.

5.8.2   'Nemawashi' and 'Catchball'

Once the vital few strategic gaps have been identified by top management, employees and teams at each level are required to develop plans in order to close the gaps. The premise is that insights and ideas are not the preserve of management. Moreover, commitment will be built by participation. This requires that employees have access to adequate up-to-date information - breaking down 'confidentiality' barriers found in many Western organisations. There must be a clear link, or cause and effect relationship, between the organisational aims, projects, delivery, and results. The employees themselves propose measures, including checkpoints. At each level, Deming's Plan, Do, Check, Act cycle operates. And, there is strong use of A3 methodology.

The main stages, explained by Cowley and Domb, are 'What do we need to do?', 'How should we do it?', and 'How are we doing?' The first stage is strongly linked with the strategy process, but benefits from feedback from later stages. What did we learn from the last time? This stage also involves the identification of Hoshins and other actions that are delegated to 'Daily Management'. In the second stage the nemawashi and catchball process then deploys the Hoshins. Catchball is a phrase from netball to indicate throwing the ball between team members before throwing it over the net. There is both vertical and horizontal catchball. Another analogy is the knock up in tennis where players hit the ball to each other, but without winners or losers. As deployment proceeds a group meeting takes place at each level. Ideas flow from all directions, and agreement is arrived at by consensus and negotiation, not authority. If a goal is really infeasible the upper tier is informed. A Japanese word for this is the 'Ringi' system. Much use is made of affinity diagrams and post-it notes. The process can and should be tied in with the Future State and Action Plans developed under Value Stream Mapping.

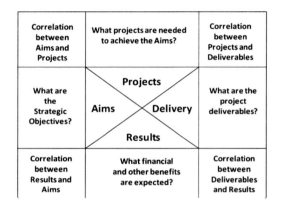

PD uses the 'outcome, what, how, how much, and who' framework. A Policy Matrix is useful here. See the figure. At Board level, a visioning process covers the key questions of what is required by the system, (the purpose and the design), what is to be achieved (e.g. reductions in lead time), how is it to be done (e.g. extend Lean manufacturing principles), and by how much (e.g. all areas to achieve 5S by year end). Specific quality and productivity goals are established. Then, the 'who' are discussed. Normally there will be several managers responsible for achieving these objectives. Appropriate measures are also developed.

The PD plans are cascaded in a Tree Diagram form. This cascading process is also different to most traditional models. In traditional models, cascading plans come down from the top without consultation, and there is little vertical and especially horizontal alignment. In PD, people who must implement the plan design the plan. The means, not just the outcomes, must be specified. And there are specific and ongoing checks to see that local plans add up to overall plans. The matrix is used to assure horizontal alignment.

As the plans are cascaded, projects at one level become the aims at the lower level.

A final stage in the cycle is the Hoshin Review where achievements against plan are formally rolled up the organisation. This uses visual results where possible. Exceptions are noted and carried forward. Hewlett Packard does this very formally once per quarter, 'flagging up' (by yellow or red 'flags') problem areas. Intel uses, against each Hoshin, a classification showing highlights, lowlights, issues, and plans. Again, root causes are identified. At Unipart in the UK, a policy deployment matrix like the one shown below can be seen in each work area, Staff are able to reconcile what they are doing with the wider organisations aims.

So PD is in essence an expanded form of 'team briefing' but requires written commitment, identification of goals, the setting of measures, and discussion at each level. In Western companies, top management sometimes spends much time on corporate vision but then fails to put in place a mechanism to translate the vision into deliverables and measures, at each level in the organisation. Hoshin may go some way to explaining why in better Japanese companies the decision making process is slower, but implementation is much faster and smoother.

Our sincere thanks to Ellen Domb for her help and inspiration.

Further Reading and References

Y. Akao, *Hoshin Kanri: Policy Deployment for Successful TQM*, Productivity Press, Portland, 1991 (apparently the standard reference for Hoshin, but virtually unreadable)

Michael Cowley and Ellen Domb, *Beyond Strategic Vision: Effective Corporate Action with Hoshin Planning*, Butterworth Heinemann, 1997

Ellen Domb, *Hoshin Planning*, Material presented at Cardiff Business School seminar, 2005

Michele L Bechtell, *The Management Compass: Steering the Corporation Using Hoshin Planning*, AMA Management Briefing, New York, 1995

Unipart, *Policy Deployment*, Material presented by Unipart staff during MSc Lean Ops modules, Cardiff Business School, 2007, 2008

Pascal Dennis, *Getting the Right Things Done*, LEI, 2006

Thomas Jackson, *Hoshin Kanri for the Lean Enterprise*, Productivity Press, 2006

Peter Hines, *Policy Deployment*, Material presented during MSc Lean Ops, Cardiff Business School, 2006

Jeremy Hope and Robin Fraser, 'Who Needs Budgets', *Harvard Business Review*, February 2003

Jeffrey Pfeffer, *What Were They Thinking?*, Harvard Business School Press, 2007, Chapter 25.

6   Preparing for Flow

The topics in this section are an integrated set. Takt time and activity times are basic building blocks for Lean flow. 5S provides the housekeeping basis. One of the 5Ss - perhaps the most important - is Standards. The 5Ss and particularly standards are closely related to the TPM methodology. Visual management is enabled by 5S, and makes TPM and standard work more effective. 5S, standard work, and TPM are also the basis for fast, consistent changeover operations. Finally, demand smoothing and small machines allow all the others to be more effective. Together, the topics are the foundations for fast, flexible flow. In combination these topics are an effective attack on Muda (waste), Muri ('overburden' or difficult work), and Mura (unevenness). But wary of making 5S or TPM an end in itself is huge waste! Convert to flow as soon as possible. Then come back and do more preparation.

These are really tools, so should not be used in isolation. They should only be used in conjunction with the wider Lean Transformation framework. They are not stand-alone ideas, although they are frequently (mis-)used in this way. 5S is a typical case in point.

6.1   Demand Management

'In the beginning, there was need', said Ohno.

Understanding demand should be the first, or at least a very early, tool to use with Lean implementation.

Study the demand patterns. For instance:

- Listen to what customers are actually saying or requesting. Write down a good sample of actual words. Then use an affinity diagram to arrange by request types. Then see if this matches what you thought customers were wanting.

- List the sources of demand. Certainly by customer type. Perhaps geographically.

- Analyse the demands into 'Value Demand' and 'Failure Demand'. Failure demand results from not doing something or not doing something right. This powerful idea from John Seddon is a variant of Cost of Poor Quality – but here you are tracking volume not cost. Failure demand should be eliminated, because it ties up capacity for no return on investment.

- What is the variation of demand over time? Perhaps by hour, day, week, month, year. What is an appropriate horizon for capturing demand? This will be linked in with what customers actually expect in terms of delivery.

Draw control charts of demand volume over appropriate time horizons. Draw the control limits. See if demand is 'in control'. If there are 'out of control' peaks try to find out why.

Then, early on, measure end-to-end response time. This is the time taken actually to meet customer demands, end-to-end. Not promised, or shipped, or first time delivery. But getting what they actually want.

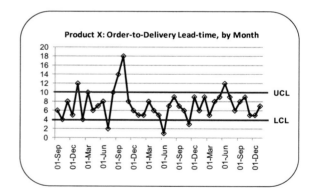

Levelling Demand

Generally, the smoother the demand, the better the flow. Demand can never be entirely smooth – but at least do not make instability worse by your own actions. What is attempted here are a few pointers, for internal demand and external (supply chain) demand. External demand can be converted to internal demand if what you regard as your supply chain changes.

External Demand Management

- Try to avoid policies such as quantity discounts (rather give discounts for regular orders), or monthly sales incentives (rather give incentives for obtaining regular orders). Manage order inflow with incentives, not stock outflow with rebates.

- Use the 'variety as late as possible' concept. Do not add variety until the last possible moment. Design has an important role.

- Develop a build to order (BTO) policy with appropriate trigger points, where push meets pull. In cars this stage is body in white – at that stage orders are firmed

- Use 'Yield Management' or 'Revenue Management' concepts – like hotels and airlines. Book ahead and get a discount; book late and pay more.

- Related to yield, segment demand into bands – fill in the troughs with longer lead time items.

- Offer customers upgrades. For instance Dell offers customers free or bargain upgrades thereby helping to smooth variety and to shift inventory. Both customer and manufacturer benefit.

- Manage demand variation, report it, discuss it, and make people responsible for it - especially Sales and Marketing. Know the tradeoffs between promotions and 'everyday low prices' (as Proctor and Gamble were surprised to discover).

- Avoid supply chain 'gaming' – for example running up inventories to appear more favourable at the next end of quarter statement. Have measures that work against such behaviour.

- Communicate along the supply chain. Try to make at the ultimate customer's rate of demand. Try to persuade supply chain partners to share information, and be willing to share it yourself.

Internal Demand Management

- Work further down the bill of materials. Perhaps the demands for various end items are erratic, but do these products share subassemblies, the demand for which may be much more regular? A related point is to aggregate demand. Demand for 1.6 litre, green, leather trim, sunroof Ford Focus may be erratic, but demand for Focus is much more stable.

- Have a policy to convert 'strangers' into 'repeaters', and 'repeaters' into 'runners'. See Design and Essential Paretos.

- Stabilise manufacturing operations by appropriate supermarkets.

- Use a single pacemaker, preferably for a whole supply chain.

- Reduce changeover times to make customer pull more possible. Much underlying demand is fairly stable, but becomes unstable when supply chain members distrust response times and available inventory.

- Use control limits, much like an SPC chart. As long as demand stays within these limits, don't change the plan. Or, use a CUSUM chart to detect changes to underlying demand. A CUSUM is one of the most effective ways of detecting shifts in demand patterns. See *Six Sigma and The Quality Toolbox* for a section on CUSUMS.

- Use 'below capacity scheduling' to make sure that you hit the production target. This means not scheduling to full capacity, but allowing a buffer period to catch up on problems. If there aren't any problems, do continuous improvement.

- Remember takt time is derived from customer demand and available production time – so it is partly under your own control. Don't over-react to changes in customer demand.

- Stabilise production at the right level in the bill of materials. Perhaps stabilise at the MPS level, and call off via the final assembly schedule.

- Give priority to regular orders. Don't let bad drive out good. Filter the erratic orders out, and make them in their own slot with lower frequency. Beware of large orders disrupting the regular schedule – split up the large order into smaller batches – and first ask the customer if the delivery requirement is really what is required.

- Use the 'available to promise' logic found in most Master Scheduling packages.

- Work according to medium term forecasts rather than short-term call offs. The medium term will be more stable and probably more reliable. Test the reliability of different forecast horizons, and don't be afraid to ignore short-term forecasts.

- Gear the incentives of distributors to work towards smooth demand.

- Move to 'milk round' deliveries – whereby several small batches are delivered on a single vehicle more frequently, rather than a big batch less frequently, meaning that total number of loads remains unchanged.

- Overproduction is the greatest enemy. Do not fall into the trap of just making a few more while things seem to be going well. This causes disruption in all later stages.

- Most of all have a vision of regular, smoothed demand. Identify the barriers that are preventing this from happening, and make appropriate plans.

Further Reading

Matthias Holweg and Frits K. Pil, *The Second Century: Reconnecting Customer and Value Chains through Build-to-Order*, MIT Press, Cambridge MA, 2004

James Fitzsimmons, *Service Management*, Third edition, McGraw Hill, New York, 2001

John Bicheno, *Fishbone Flow*, PICSIE Books, Buckingham, 2006

Yasuhiro Monden, *Toyota Production System*, (Second edition), Chapman and Hall, London, 1998

6.2   Total Productive Maintenance (TPM)

TPM can be regarded as integral to Lean. Certainly no Lean implementation can be a success with a high level of breakdowns. TPM goes well beyond breakdown issues to cover availability, performance, quality, as well as safety and capital investment through making best use, and extending the life, of equipment. TPM can be viewed in relation to the 'bathtub' curve below.

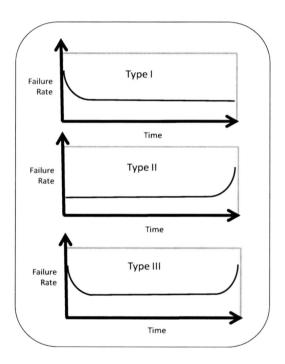

Not all machines have full bathtub curve failure characteristics. The types are:

Type I: Aero engines, where a lot of redundancy is built in ('fly to failure'). If one little piece fails, that is ok, because it keeps flying. Classic with electronics also, where a 'burn in' period means a high failure rate at the start, and stable operation thereafter.

Type II: Cars are a typical case here – they work well when new, but as wear and tear occurs, the failure rate rises. Also common for most engines and motors.

Type III: The traditional 'bath-tub' curve. Many products have a failure profile like this, from light bulbs to complex machines. Take a light bulb: when inserted in a socket it might fail straight away. Most bulbs will then begin to fail after a particular period.

With reliability-centred maintenance, the type of maintenance is adjusted according to the failure profile.

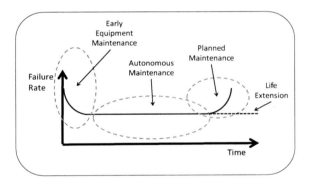

TPM has much in common with Total Quality. Everyone has a role to play, not just experts. In total quality there is the concept of 'the chain of quality'; the TPM equivalent is the equipment life cycle. Both aim at prevention. Both widen the scope to include the operator, the product, the process, and the environment. Both aim to 'spread the load' by getting front line staff to take over as much responsibility as possible thereby directing authority to where it is most effective and freeing up specialists to do more complex tasks (thereby creating a positive feedback loop).

Both use 'management by fact'. TPM can be viewed in relation to the 'Bathtub Curve', above.

TPM addresses all parts of the curve. It reduces breakdowns in the burn-in period by early equipment maintenance and by improved understanding of equipment usage. It reduces breakdowns during the plateau period by autonomous maintenance and the 9 step programme. The 9 step programme also extends the life of equipment. The upturn point is addressed through predictive and planned maintenance. Failures during the wear-out period are also reduced.

Generally, TPM activities are either EVENT BASED or TIME BASED. Think about driving your car. Do you check every time, when you clean, or every Sunday? Often, both are required.

6.2.1   The Six Big Losses and OEE

The six big losses and OEE concept is widely used in TPM. The six are divided into three categories (availability, performance, and quality), which are the basis for OEE. They are:

Availability: Breakdown losses are unplanned stoppages requiring repair. Any unplanned stoppage greater than 10 minutes is usually considered breakdown. These may be electrical, mechanical, hydraulic, or pneumatic. Changeover and adjustment losses occur when changing over between products. Often changeover is defined as the time from the last piece of the first batch to the first good piece of the next batch. But note that getting up to full production rate may take much longer, during which time adjustments may be required. The time lost is recorded.

Performance: Minor stops and Idling are often defined as stops taking less than 10 minutes. They result from a host of causes such as tip breakage, coolant top-up, jams, swarf removal, and small adjustments. When data is collected, minor stoppages are often revealed as (by far) the most significant loss. Since minor stops are so frequent and last such a short time they are often ignored because they are difficult to measure. Activity sampling may be an answer. Alternatively get the operator to just mark off each stoppage on a fence and gate IIII chart, and sample the time taken for 20 actual minor stoppages to give an average.

Reduced Speed losses result from the machine running at less than the design speed. Typical causes are flow restriction, program errors on a CNC machine, and worn tools, feeds or belts.

Quality: Defects result in scrap or rework and are the result of any problem that causes the machine to work outside of the specification limits. Start up losses result in scrap or rework during changeover. It is measured in pieces. Note that in some systems the time lost in attaining full speed is recorded here rather than as part of changeover losses.

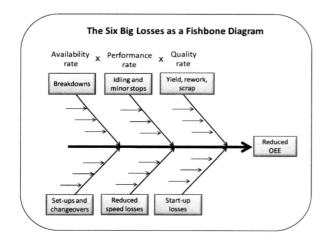

The Six Big Losses as a Fishbone Diagram

6.2.2 Overall Equipment Effectiveness (OEE)

OEE is Availability x Performance x Quality, expressed in percentage terms.

An example: A shift takes 9 hours. Working time is 8 hours – planned maintenance and meeting time is one hour. Breakdowns take 20 minutes. Changeovers take 40 minutes. The standard machine cycle time is 1 minute. At the end of the day 350 parts have been produced of which 50 are scrapped, 30 of them during adjustment.

- Availability: (8 x 60 – 20 – 40)/480 = 420/480 = 88%

- Performance: actual working for 420 minutes, so 350/420 = 83%
- Quality: 300/350 = 86%
- **OEE = 88% x 83% x 86% = 63%**

For non-process industry world class OEE is in the range 85% to 92%. **However,** there is no such thing is a universal world-class OEE figure. Note that figures in the 90's may indicate that insufficient time is being given to changeover, so batches may be too large. Each of the elements of OEE should be graphed as well as the overall figure. Keep these at Gemba. Also, below each chart, it is good to have a fishbone diagram showing possible causes. Even better, keep a CEDAC (cause and effect with addition of cards) where progress is recorded on coloured cards (red still to do, yellow done, notes on the back).

Cautions on using OEE

- OEE says nothing about schedule attainment. It is useless having a high OEE if you are making the wrong products!
- You can improve OEE in good and bad ways. Good ways are to reduce minor stoppages and decrease the length of a changeover. A bad way is simply to do fewer changeovers. OEE should not be used alone, but alongside schedule attainment which tracks how many batches were supposed to be made. So do not make OEE an end in itself.
- There is a cost factor – reducing changeover at great cost may be counterproductive.
- Do not measure OEE plant-wide. Combining machine performance is meaningless. Target only critical machines. Pareto!
- There are other ways to cook the books – simply don't make products with higher defect rates or more difficult adjustments.
- A boast like 'We have improved OEE by 20%' should be greeting with extreme caution: Is it overproducing? Is it a bottleneck? Is it appropriate? Is it because bigger batches are being made?

Should OEE be calculated shift-by-shift? Yes, because the principle is that you should only measure what you can do something about.

A serious drawback with OEE is that no measure of variation is included in the standard OEE calculation. In other words, two machines may have similar OEEs over (say) a week, but very different variation. Consider that two machines with the same overall OEE of 80% can have a very different impact on your manufacturing system, if one oscillates between 20-100%, the other between 78-82% OEE! The results, in terms of impact on the system, will in turn be very different. Of course, this is predicted by Kingman's equation. (See Factory Physics.)

We know from Kingman's equation how critical capacity utilisation is, how important it is not to overload a resource, the necessity to have a little reserve capacity, and the effects of variation on throughput. But beware of too narrow a view of capacity. Team leaders and supervisors at Toyota are known to have small spans of control. Surely this not Lean! Well, span of control and the resulting responsiveness, certainly affect capacity. So a machine with good performance characteristics may not be good enough unless supported by adequate maintenance and backup. This will show up in end-to-end OEE (what Peter Willmott refers to as door-to-door OEE) but may not show up in a machine's OEE.

6.2.3 Focusing TPM Activities

An excellent way to focus TPM activities is to graph performance as follows:

To construct: Start with the working day. Determine planned maintenance, break time, and planned idle time. Split the remaining time into the six big losses and actual effective working time. Divide by typical planned production in units. The result is a stacked bar chart as shown.

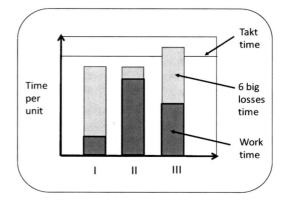

Consider the three machines shown: Machine 1 has the worst OEE. Machine 2 is closest to takt time. Machine 3 has a comfortable machine cycle but losses mean that total work cycle exceeds takt. Which to target? To target by worst OEE may be to miss the point. The question is, is this serious? To focus on machine cycle (as has been suggested in various value stream mapping publications) is not relevant as long as total time is below takt time. Machine 3 is the machine to focus on, even though it has moderate OEE and comfortable machine cycle time. And what to focus on, on Machine 3? Look at the 6 big losses. To take this analysis further, it would be good to repeat the analysis by days of the week, like a run diagram. And monitor OEE using SPC principles.

6.2.4 Willmott's 9- Step Model

Peter Willmott, UK TPM guru, has proposed a widely used 9 step model. The 9 step process is a long and thorough one. Absolutely not a quick fix. Short cuts generally don't pay!

Measurement Cycle

1. Collect Equipment History and Performance Analysis This step focuses the project (as above) and sets measurement objectives (e.g. cost, OEE, manning, material savings). Often this is for a machine or a cell, never for a plant – it would simply be too big a project. A small team is assembled. A TPM display board is set up at Gemba to show progress along the 9 steps. Collect manuals and drawings covering mechanical, hydraulic, electrical, and pneumatic. Assemble any available data on history – installation, working patterns, performance rates, replacements, and planned maintenance records. Also identify any safety black or brown points on the machine.

2. Define and Calculate OEE Clarify the meaning and interpretation amongst team members. Set up an OEE display board at Gemba. Brainstorm out possible causes and display on a chart. This step may involve several days, even weeks of data collection, to identify clearly the six big losses for different situations. Agree progress, methods and figures with plant management. A fishbone diagram is most useful here. See the figure above.

3. Assess Six Big Losses and Set Priorities This will involve an analysis such as shown in the last section. Agree and sign off the priorities with management.

Condition Cycle

4. Critical Assessment. This may be overlapped with step 3. Produce a list of all components of the relevant machines. Discuss and understand the role of each component and their interdependencies – not superficially, but in detail. Why exactly was it designed that way? Identify the critical components – and why they are critical. Discuss the optimal conditions for the operation of each critical component (e.g. temperature, lubrication, cleanliness, sharpness). Publish the optimal conditions on a drawing of each component. Then define the normal operating conditions. Finally discuss the sources of accelerated deterioration for each component – equipment based, operator based, environment based. This is a long stage of no immediate impact – but huge long term potential. Hang on in there. Note that only now, perhaps after several days or weeks, is the team ready to undertake specific improvement actions.

5. Initial Cleanup and Condition Appraisal. Agree the cleaning areas. Source all necessary and specified cleaning equipment. Photograph the current state. Systematically inspect every part of the machine in detail. Clean and inspect, capturing all problems found. At the end of cleaning, fill in an appraisal form for each component covering mechanical, hydraulic, electrical, and pneumatic. Develop a cleaning and inspection programme ('Cleaning is checking'). Identify sources of contamination (internal and external) and develop a plan to eliminate, isolate, prevent, or if necessary clean. Discuss with plant management.

The categories are S for safety, A for availability, P for performance, Q for Quality, R for reliability, M for Maintainability, E for Environment, and C for Cost. 1 indicates no impact, 2 some impact, 3 major impact

6. Plan Refurbishment Develop a phased refurbishment schedule covering item, labour hours, planned completion, and PDCA cycle stage. Discuss with plant management. Schedule the release of equipment. Implement elimination of contamination. Look into pokayoke and implement as far as possible. Examine the need for quick changeover and undertake this where necessary.

7. Develop Asset Care. Clearly define the role and the tasks of the operator. Produce a clean and check list, with appropriate frequencies. Develop a Kamishabi Board (see separate section), covering the phased and daily activities of maintenance, safety, quality, and operator checks. Identify, mark and colour code, all gauges, pipework, lubrication points, levels and sight glasses, nut positions. Etc.

Label all components and tools – with cross references to manuals. Indicate flow directions and motor rotations. Install inspection windows.

Problem Prevention Cycle

8. Develop Best Practice Routines and Standards. Taking all that has been learned in previous steps, assemble a best practice manual. Develop Single Point Lessons where necessary. See separate section under Visual Management. Review Standard Operating Procedures – amend where necessary, including pictures, in consultation with operators, and place at Gemba. Review the maintenance instructions.

Review equipment spares – what needs to be kept, where, how much. Index the spares and cross reference with manuals and SOPs. Develop a spares catalogue associated with each machine. Locate the manuals appropriately – not in the office!

Many good routines can be developed by using 'your God-given senses' to quote Willmott. Again, think of your car. At least 30 checks are possible without any mechanical knowledge. They will make a difference to the car, and maybe to your life!

9. Problem Prevention. This is an improvement cycle. OEE leads to the particular loss leading to the problems that are tackled by the 5 whys leading to solutions. A method is P-M analysis - similar to the quality matrix of variation, mistakes and complexity vs. the 6 M's (see Quality section). The preference is for low cost / no cost solutions, but also takes in technical solutions, and support service solutions. All of this feeds back to earlier stages to complete the cycle.

6.2.5 Some Special Features of TPM

'At its worst when new'. This provocative statement goes to the heart of TPM. Why should an item of equipment be at its worst when new? Because, it may not yet be quality capable, standard procedures not yet worked out, mistake proofing (pokayoke) devices not yet added, operating and failure modes not yet known, 6 big losses not yet measured or understood, and vital internal elements not yet made visible (through transparent covers) or monitored by condition monitoring.

Visibility. Like Lean, TPM aims to make what is happening clear for all to see. This means maintenance records need to be kept next to the machine, problems noted on charts kept next to the machine, and following a 5 S exercise, vital components made visible by replacing (where possible) steel covers with transparent plastic or glass. Also, following 5 S, any leaks or drips are more easily seen.

Red Tags. Red tags are a common form of visible TPM. Maintenance 'concerns' are written on red tags and hung on a prominent board on the shop floor. They remain there until action is taken. Red tags usually cover concerns that cannot be dealt with by operators.

Failure Modes and Scheduled Maintenance. In classic Preventive Maintenance, a 'bathtub' curve was often assumed, i.e. high failure rate early on, dropping to a low and continuing failure rate, then increasing at wear out. Routine maintenance is then scheduled just before the risk rate starts to increase. Today, we know that not all equipment has this pattern. Some may not have early high failure, some may not have sudden wear out, some may exhibit a continuous incline, etc. The point is you need to know the failure mode in order to undertake good scheduled maintenance practice. So, data needs to be recorded, the best way being automatically, for instance the number of strokes on a press for each die. And operators, who often have an inherent knowledge of failure modes, should be consulted. A maintenance cycle should be developed much like the cycle-counting concept, which visits important machines more frequently, allocates responsibility, and aims to improve not just maintain.

Condition Monitoring. Condition monitoring is a specialist function in TPM, but in some environments (for example heavy and rotating machinery) an important means to reduce cost. Methods include vibration detection, temperature monitoring, bearing monitoring, emission monitoring, and oil analysis. Today, there are hand-held, and computer linked, devices to assist.

Information Systems. Information systems were always an important part of PM, and remain so with TPM. However, their scope is extended from machines to include operator, safety and energy issues and also to allow for workplace data recording.

Design and Administration, and Benchmarking Today, TPM is beginning to be seen in administrative and white-collar areas. Of course, there are computers, photocopiers, and fax machines but there are also tidy (?) desks, filing cabinets, and refreshment rooms.

Progressive companies are beginning to cater for TPM in product design.

6.2.6   Dealing with Ramp-up losses

Ramp up losses, when a new line is started, can be significant. To get the line up and running reliably as quickly as possible means money straight to the bottom line. As such it should enjoy priority. The 9-step principles to be used here are basically the same as in Willmott's model.

Further Reading

Nick Rich, *Total Productive Maintenance*, Liverpool Academic, 1999

Seiichi Nakajima (Ed), *TPM Development Program*, Productivity Press, Cambridge MA, 1989

Seiichi Nakajima, *Introduction to TPM*, Productivity Press, Cambridge MA, 1988

Masaji Tajiri and Fumio Gotoh, *Autonomous Maintenance in Seven Steps*, Productivity, 1999

Peter Willmott, *Total Productive Maintenance: The Western Way*, Butterworth Heinemann, Oxford, 1994

John Moubray, *Reliability Centred Maintenance*, (Second edition), Butterworth Heinemann, Oxford, 2001

Our sincere thanks to Peter Willmott for his help and inspiration.

6.3   Takt Time and Pitch Time

Takt time is the fundamental concept to do with the regular, uniform rate of progression of products through all stages from raw material to customer. Takt time is the drumbeat cycle of the rate of flow of products. It is the 'metronome' (from the German origins of the word). Understanding takt time is fundamental to flow and mapping of repetitive Lean Operations.

Takt time is the available work time (say per day) divided by the average demand per day. Note that there are two variables – one to do with the customer, the other to do with the plant manager. Therefore, if demand changes a manager could maintain the same takt time by adjusting the available work time. The available time is the actual time after allowances for planned stoppages (for planned maintenance, team briefings, breaks). Demand is the average sales rate (including spare parts) plus any extras such as test parts and (we hope not) anticipated scrap. It is expressed in time units: e.g. 30 seconds or 30 seconds between completions. Where there are multiple parts going down a line, the overall takt time is calculated by dividing the available time by the total number of parts.

Example: Say you have A: B: C parts in the ratio 3: 2: 1 with daily demand 120A, 80B, 40C. Available time is 200 minutes per day. Takt time is 200 / 240 = 0.83 min. = 50 secs.

There would be a repeating sequence ABABAC every 6x50 = 300 seconds. In this regard, the ski chairlift analogy is useful – a constant flow of moving chairs but containing a mixed set of people.

Some companies allow for changeover times in calculating the available weekly work time. This is not really good practice because the takt drumbeat must be maintained throughout the plant, including changeovers. Nor is it good practice to allow for an overall equipment effectives (OEE) percentage. Building in these wastes leads to a lower takt time that translates to more people.

Where demand is seasonal or variable, selecting the period over which demand is estimated is important. Selecting a longer period will stabilise build rate but at the expense of more supermarket inventory to smooth out the bumps. Moving towards build to order means reducing the time horizon and having more frequent takt time. This means that lines may have to be rebalanced, which in turn means that operators should be involved in the task of rebalancing and know the concept of takt time. As you become more Lean and flexible you can reduce the period over which takt is calculated.

Note that 'takt time' is NOT 'cycle time', which is the time required to complete an operation or a machine cycle. Where such cycle times exceed takt, you have a constraint – requiring parallel processes or an additional shift. Also note that there may be several takt times within the same plant – for example one takt for cars, but one quarter of the time for the wheels. In some environments, such as chemicals made in tanks, takt may not be an immediately useful concept, but may still be useful for planning and control.

For machines and processes, working at the takt time may mean slowing down. Strangely and counter-intuitively, slowing down to achieve synchronisation may lead to a reduction in lead-time. This is because queues build up after machines that run faster than the takt time. This simple realisation, to try to get all machines in a plant running at the constant takt time, can have dramatic results. It changes the job shop into a pseudo assembly line. Takt time should drive the whole thinking of the plant and the supply chain. In a plant it is the drumbeat. Consideration needs to be given to the number of parts per product sold. Thus if there are four wheels per trolley sold, and wheels are all made on the same machine, the wheel machine needs to have a takt time of approximately one quarter that of the main assembly line. (Approximately because there may be special demands for additional spare wheels.) So takt times in a plant leads to overall synchronisation.

What about the takt time across a value stream that works different shift patterns – say a middle process always has to work weekends? Calculate takt at the downstream end, and synchronise using FIFO lanes.

Several takt times can be set for a line – and the line balanced for each takt time. Then, depending on demand, the appropriate takt and manning can be used.

Likewise, takt can be calculated daily in warehouse order-picking operations, to calculate the number of pickers. Here, there may be a morning takt and an afternoon takt.

What about takt time where there are shared resources between value streams? Generally, takt is not applicable here. Use a priority pull system or Drum Buffer Rope, with a supermarket. Likewise, in process industry where there are fixed-size vessels, or in offices where there is large variation in work content, takt may simply not be useful. In these cases, don't push it.

Pitch Time is the takt time multiplied by the container quantity or a convenient multiple of parts, typically 15 to 30 minutes. Instead of thinking the time to make one part, think of the time required to fill the standard container. The analogy of the ski lift is particularly appropriate – a constant rate of movement delivering 4 people at a regular spacing.

The pitch increment is the basic time slot used in Heijunka. The material handler in a Heijunka system should fit in with the pitch time. In that sense, pitch time is the vital drumbeat of the whole system, forcing regularity, visibility and flow.

6.4 Activity Timing and Work Elements

Activity timing is the long-established industrial engineering (or time and motion study) task of determining the duration of work elements. This is an essential input into cell balance ('Yamazumi') boards, value stream maps, scheduling, and costing. In Lean, timing is best done by operators rather than by I.E.s – thereby encouraging ownership and avoiding the pitfalls of suspicion and 'slow-motion' work. In time and motion studies, the rule is to sort out and standardise the motions before timing them.

Preferably, make a video of the tasks. This is better than live recording, because it allows backtracking, and slow motion. It also avoids the stress of several people with stopwatches standing over an operator. Be sure to video at least 10 cycles on each shift. If several operators are used, film each of them. A very useful learning experience is to get operators from different shifts together to see if there is variation between operators, and to agree on the best method. This should be an essential step in determining standard work. If you are an outside observer, it will take time to familiarise yourself with the exact tasks – best of all is if you can do the work yourself. This is another good reason for using operators themselves to do the timing.

With the correct methods established, begin the timing. If necessary re-video. It is good to video several operators, from different shifts. First, break down the work sequence into work elements each with a clear start and end point. Agree on these points. Keep manual (work) times, walk times, and wait times separate and record each of these under separate columns. Machine cycles should be separately recorded. Make a list of the sequence of activities that an operator goes through in a complete sequence. Some of the manual times will turn out to be non-value adding, or non-value adding unavoidable. At this stage, just record each separately. When balancing the cell, try to reduce or eliminate wait and walk times. Critically examine the possibility of reducing NVA or NNVA steps.

Time at least 10 good cycles of each – by a 'good' cycle is meant a cycle where nothing goes wrong. Discard 'outliers'. Time to the nearest whole second. For each work element time, take the lowest more frequently occurring time. In other words, in 10 observed cycles of 4, 5, 6, 6, 7, 7,8, 8, 8, 8 take 6.

Note: In traditional time study, the time used is the observed time plus an allowance for 'PR&D' – personal, rest and delay. Do not add this allowance. Rather, give operators more frequent breaks.

6.5   5S

5S is probably the most popular tool in Lean. But, should you start your Lean programme with 5S? Probably not!

There is the good news and the bad news. The good: it is easy to do, usually has a positive impact on quality and productivity and sends out a powerful message that Lean 'has arrived' and is for everyone. The not so good: 5S can be a diversion from real priorities, can be seen as merely tidying up, and can give Lean a bad name though over-zealousness.

First, be clear of the motivation for 5S. If the place is a mess, and needs a clean-up, please don't say you need a 5S program. Say that you need to tidy up! Why? Because if you establish in the minds of your people that '5S equals cleanup' (and that is all), then you risk misunderstanding of a very powerful concept, you risk sustainability issues, and you may even risk Lean being seen as something rather trivial – or worse – just a silly set of activities. Indeed, this has happened in several companies, spectacularly in various government offices. Later, when the real needs for 5S (reduction in variation, meeting the schedule, exposing problems, improving machine availability and performance, etc.) are recognised, THEN do 5S as a 'pull' activity. The real objectives of a 5S program should be:

- To reduce waste
- To reduce variation
- To improve productivity.

5S programs that work well are in situations where the need to achieve these three is well known and 5S is seen as the way to do it.

But 5S is also a mindset thing – changing attitudes from 'I work in an unorganised, messy office' to 'I work in a really well organised office where everyone knows where everything is and any out of place or missing item is seen immediately'.

The classic 5Ss are: Seiri, Seiton, Seiso, Seiketsu, Shitsuke. Most commonly these are translated into Sort, Simplify, Scan, Standardise, Sustain, but many other synonymous terms have been used, see below. A common alternative for 5S is the CANDO pneumonic – Cleanup, Arrange, Neatness, Discipline, Ongoing improvement.

6.5.1   SORT

Throw out what is not used or needed. The first step is to decide, with the team from the area, the sorting criteria – for example the team may decide that items that can kept at the workplace are

- items that are used every week

- items that are needed for important quick customer response
- items for health and safety

whereas less frequently used are kept firstly in cupboards, and secondly in the storeroom.

Then the team needs to classify literally according to the sort criteria. Touch every item systematically. If it can be kept at the workplace, is the quantity correct? If never used, or in doubt, then red tag or throw out. A Red Tag is a label with the date; if no one accesses it within a specified period it should be thrown out, recycled or auctioned. The sort stage should be done regularly – say once every six months, but as a regular activity not a re-launch of 5S. You know a 5S program is not working when it is really a frequently repeating sequence of 2S or 3S activities.

Be careful of over-zealousness and over-the-top. Within reason, permit some personal items to be kept at the workplace. Also permit personal discretion and location choice. Readers may recall the laughable situation rightly reported with some ridicule in *The Times* of December 2006 showing a taped-up location for a banana on a desktop. If that is your manager's idea of what Lean is about, resign immediately!

Some offices have been known to use Feng Shui to get the 'vibes' right. Is this OTT?

6.5.2   SIMPLIFY (or Set-in-Order or Straighten)

Locate what remains in the best place. A place for everything, using shadow boards, inventory footprints, trolleys or placing items at the right height, and colour matching equipment to areas – or simply good sensible locations. Like a kitchen where the family knows the location of cutlery and plates and do not have to be told. Are drawers and doors really needed? The best location is the place where it is silly to put it anywhere else. And everything in its place. If not in place and not in use, a problem is indicated. The standard is The 'Dental Surgery'. Why? Because everyone can relate to that standard of excellence, and know the consequences of failure.

Locate items by frequency to minimise stretching and bending. Repeat this stage whenever products or parts change. Use a spaghetti diagram for analysis. Ergonomic principles should play a role here, and an ergonomic audit may help. (There are many excellent, inexpensive books on office ergonomics – there is no excuse not to use them.)

6.5.3   SCAN (or Sweep or Shine or Scrub)

Keep up the good work. This includes physical tidy up, on an ongoing basis, and 'visual scanning' whereby team members are always on the lookout for anything out of place, and try to correct it immediately. Some companies adopt a 5 minute routine whereby operators work out a 5 minute cleanup routine for each day of the week so that by week end everything has been covered the required number of times. Designate exactly who is responsible for what and what the standard is. The stage will require suitable cleaning equipment to be suitably located and renewed. There may be a sign-off chart for routine cleaning. By the way, have a standard procedure for the 5 minute cleanup routine.

'Cleaning is checking,' means that these are integrated. You don't just clean up, you check for any abnormality and its root causes. The garage analogy is first clean up – this enables oil leakage to be identified. Continue to clean up any leakage that occurs. But then ask 'why is leakage occurring' and decide what should be done for prevention. On your car you don't check oil, water, tyres, and tyre pressure every time you drive, but you do check then when you clean your car. Same principle.

Scan may also include calibrating, keeping track, observing, monitoring, looking out for wastes, lubricating, dusting, computer monitor cleaning, and routine servicing.

It would be good to engender the -

- 'Mary Poppins' effect – making clean-up fun, or

- 'Tom Sawyer' effect – demonstrating to others that they are really missing out by not tidying up (!)

6.5.4   STANDARDISE (or Stabilise or Secure)

Only now is it possible to adopt standard procedures. This is the real bottom line for 5S. See the sections on Failsafing and Standard Work. But 5S standards also need to be maintained. So develop standards for the first 3 Ss. Standardising also includes measuring, recording, training, and work balancing.

6.5.5   SUSTAIN (or Self Discipline)

Everyone participates in 5S on an ongoing basis. Sustaining is about participation and improvement about making the other 5S activities a habit. Carry out audits on housekeeping regularly. Some award a floating trophy for achievement. Others erect a board in the entrance hall with current 5S scores. Yet others have a weekly draw out of a hat and then all first line managers descend on the chosen area to have a close look.

Some companies add a sixth **S – SAFETY**. Although good to emphasise, a good 5S program should stress safety as an aspect of each of the five stages. It may confuse to list safety separately. Safety procedures and standards should also be developed, maintained and audited as part of the programme. The removal of unsafe conditions should certainly be integral to 5S.

Some companies adopt regular walk about audits and competitions. The first line supervisor audits daily, the area manager weekly at random times, the section manager monthly and so on.

6.5.6   5S 'for Dummies'

Two useful, and easy to remember, mnemonics are used in the book *Organising for Dummies* by Roth and Miles. This is not a manufacturing or office book, but rather one about tidying up and clearing clutter in your house. Nevertheless, it has some highly practical advice. The mnemonics are WASTE and PLACE.

WASTE concerns questions to ask:

W: is the item Worthwhile to keep? Think possible appreciation in value. Cost of storage vs. Cost of replacement.

A: will it ever be used Again? Think probability.

S: is it also kept Somewhere else? And should it be?

T: should it be Tossed or Thrown out. Think consequences.

E: should the Entire quantity be kept? Is the likely usage coverage period excessive? Should less be kept?

PLACE concerns storage and control considerations:

P: Purge – look at the WASTE questions above.

L: Like with like – can variety be reduced? Part proliferation – see The Essential Paretos.

A: Access – how fast is the item required?

C: Container – how to store?

E: Evaluate. Does it need to be kept in a perfect state of readiness? Stock counting frequency, accuracy required and ABC category (e.g. cycle counting?)

Reprinted with permission of John Wiley & Sons, Inc. Copyright: Eileen Roth and Elizabeth Miles, *Organising for Dummies*, Wiley, 2001

6.5.7   5S as Root Cause

5S lies at the root of many issues in service and manufacturing. All staff need to be sensitised to this fact, and be encouraged to make improvements as soon as possible – not to have to wait for some kaizen event. For example, in service:

- when a nurse arrives at a patient, or a repairman arrives at a site, and discovers that there is something missing – it is a 5S issue

- when a document cannot be found – it is a 5S issue
- when a surgeon encounters an unfamiliar layout in an operating theatre – it is a 5S issue
- when customers ask for the location of an item in a supermarket or shop – it is a 5S (or visual management) issue
- when lecturers don't find working flip chart pens - it is a 5S issue
- when a stationery store runs out of paper – it is a 5S issue
- and so on….

In all these cases it is no good just to correct the error. The situation must be highlighted and a simple non-bureaucratic procedure put in place to solve the problem permanently.

6.5.8   *5S Sustainability*

Many companies now claim that they are doing 5S but are in fact doing 2S sporadically. Having no 5S sustainability is a waste, and the programme requires increasing effort to re-energise. The real productivity and quality benefit of 5S are in the later Ss, particularly standardisation, not the relatively easy-to-do first two.

Hirano suggests a host of 5S activities, carried out at various frequencies: Amongst others:

- A 5S month, once a year (?), to re-energise efforts
- 5S days, one to four per month including evaluations
- 5S seminars by outside experts – with lots of photos.
- 5S visits to leading outside companies
- 5S patrols, following a set route

- 5S model workplaces (this has become popular in NHS hospitals)
- 5S competitions
- 5S award ceremonies
- 5S exhibits
- 5 minute 5S each day

Doing all this sounds like overkill, but putting a few in place has proved helpful in sustaining 5S.

6.5.9   *Extending the 5S Concept*

Perhaps less recognised, the 5S concept can be powerfully applied to information and information flow. Sort and simplify the information transactions that flow around the organisation. Of course, it is appropriate to ask about e mail flow and to establish standardised rules about cc's and when to expect answers. But even more potent is to examine the decision processes. What is the minimum necessary information that is required for planning, scheduling, orders, invoicing, staffing, recruiting, and more? This overlaps with Lean Accounting (as opposed to Accounting for Lean), and with A3, but goes far beyond them. The saving and streamlining potential will often make physical 5S seem trivial. Sort should include not only wasted communication, but the time-accuracy tradeoff. Is 90% accurate in half the time not far better? Simplify should include the best means of communication. One Danish company continuously projects current data into the factory and office data projectors.

Scan includes the best way of updating. Standardise should cover all information flows. Sustain: %S physical audits are OK, but what about auditing all transactions?

	Inventory	Suppliers	Computer Systems	Costing Systems
Sort	Throw out all dead and excessive stock	Select the best two suppliers in each category. Scrap the rest.	Delete all dead files and applications	Do you need all those costs and variances? Prune them!
Simplify	Arrange in the best positions	Cut all wasteful, duplicate transactions	Arrange files in logical folders, hierarchies	Cut transactions. Review report frequency. Incorporate o/head directly
Scan	Regularly review dated stock and ABC category changes	Improve supplier performance by supplier assoc & kaizen	Clear out inactive files regularly	Audit the use made of costing reports and transaction size & frequency
Stabilize	Footprint, standard locations	A runner system, payments	Systems, formats	Adopt reporting standards
Sustain	Audit ABC, frequency of use.	Audit performance	Audit perform & response	Review and reduce.

Further Reading

Productivity Press Development Team, *5S for Operators*, Productivity, 2002

Hiroyuki Hirano, *5 Pillars of the Visual Workplace*, Productivity, 1995

Karen Kingston, *Clear Your Clutter with Feng Shui*, Piatkus, 2002

6.6   Visual Management

Visibility, visual management, or 'control by sight', is a key theme in Lean operations. Visual management should be integrated into 5S and standard work. In fact, visual management is the 'litmus test' for Lean – if you go into any operation and find that schedules, standard work, the problem solving process, quality and maintenance are not immediately apparent, and up to date, there is an excellent chance that the operation is far off Lean. Visibility has been joined by audio.

Visibility fits in well with several other Lean themes:

- Speed (no waste of time having to look for information), Improvement (progress should be for all to see, and celebrated)

- Up to date and clear schedules (via kanban, progress boards, automatic recording)

- Making problems apparent (via overhead Andon boards or lights)

- Involvement (clarity on who is doing what and who can do what)

- Teamworking (making visible the good work of teams, and skill matrix)
- Standardisation (keeping standards up to date by locating them at the workplace)
- Responsiveness (requiring quick response to maintenance and quality problems via for example a line stop chord or a red tag maintenance board).

Ford Motor Company makes a distinction between visual display, concerned with provision of information, and visual control, concerned with action. The idea is to be able to gain the maximum amount of operating information and control without having to go off the shop floor, or go into a computer system – a variation of Gemba management. This applies to operators, supervisors, and managers. A few examples of visual management follow:

- Machines: transparent plastic guards and covers used wherever practical to enable operators and maintenance people to see the innards of machines.
- OEE charts placed next to machines or at team meeting areas; there should be four graphs – one for overall OEE and one each for the three elements of OEE. Below each graph should be kept a fishbone diagram with the contributing factors.
- Changeover times should be graphed routinely, to prevent slippage.
- The Heijunka box is a visual display of the status of the day's schedule.
- Kanban priority boards used where there is changeover (triangle kanban) give a continually updated display of the urgency of the next products to be made.
- Lights to indicate status, with the overhead Andon Board linked to computer to record stoppages for later analysis (not blame!)
- Management: Is it possible to have the production control and scheduling office on the shop floor?

- Cost, Quality, Delivery performance should be a central trio on display, possibly joined by Safety, Lead-time, and Days of Inventory.
- Line rebalancing charts showing takt time, and with magnetic strips for each work element, kept in the team area.
- Keep teams informed of new products and developments by progress boards, including Gantt charts.
- People: A skills matrix (or I L U O chart) indicates achievement from beginner to instructor. Operators on one vertical axis, tasks on the horizontal axis.
- Employee suggestions and employee of the month are more common in service than manufacturing - why?
- Have a mirror with a slogan such as 'You are looking at our most important source of ideas'.
- Methods: Keep those standards and methods at the workplace!
- Materials: Don't forget to keep footprinting up to date. Kanban remains the most visible effective means for production control.
- Maintenance: A maintenance 'red tag' board, showing all outstanding concerns.
- Money: There is a welcome tendency towards displaying company financial and sales data on the shop floor.
- Improvement: Keep a flipchart handy to note problems (which accumulate Pareto style) and suggestions with stage reached shown by filled in quadrant (suggested, investigated, being implemented, solved).
- Storyboards, showing the standard stages in kaizen events, recent successes and current progress.
- 5S: Display area responsibilities and 5-minute cleanup plans. Use shadow boards. Label and organise. Use kitchen organisation, not 'garage' organisation. Display audit results and winners.

Audio is also useful. At Toyota each robot cell has its own unique tune, which gets played over loudspeaker when there is a stoppage or problem. Maintenance engineers learn to listen out for the tunes they are responsible for. Tunes change with urgency.

Galsworth suggests appointing a 'visual co-ordinator', and a management champion, who encourages and monitors the status of visual management in a plant. She makes the important point that visual management extends beyond the shop floor to the design of forms, to the presentation of information, to office layout, and to the home, together saving countless hours of waste spent searching and clarifying.

Single Point Lessons

A useful and widely adopted procedure is the Single Point Lesson. These are found on display on the factory floor and:

- Focus on one single point for improvement
- Are highly visual – containing the steps, key points, and invariably a diagram or photograph
- Contain content that can be delivered in 15 minutes or less
- Address the main stages of learning – awareness, understanding, competence, ability to train others.

These stages are frequently shown in by coloured-in segments on a pie chart.

Note: See also the sections on Training within Industry (TWI) concepts that should be the foundation for Single Point Lessons.

Further Reading

Michel Grief, *The Visual Factory*, Productivity, Portland, OR, 1991

Gwendolyn Galsworth, *Visual Systems: Harnessing the Power of a Visual Workplace*, AmaCom, New York, 1997

6.7 Standard Work, Standard Operating Procedures, and Job Breakdown Analysis

Standard work is one of the pillars of the Toyota Production System. Standard work aims at creating processes and procedures that are repeatable, reliable, and capable. It is the basis for improvement. Standard work is the philosophy, Standard Operating Procedures (SOPs) are one mechanism of putting that into place.

The best standard, and the most sustainable, is a standard that is so good, so obvious that doing the task any other way is seen as just plain silly. This may have to be demonstrated. But that is the end state, the vision to strive for.

But beware of a 'standard' that becomes confused with a target that generates silly behaviour such as improving OEE by making bigger batches. Another poor use of a standard is where it becomes confused with an accounting or costing standard – especially with absorption costing where failing to meet the budgeted rate of work leads to under-recovery of overhead and hence pressures to over-produce to 'recover' the overhead. This is plain nonsense in the context of Lean. Likewise requiring office workers to tape standard locations for pens and folders on their desks is absurd.

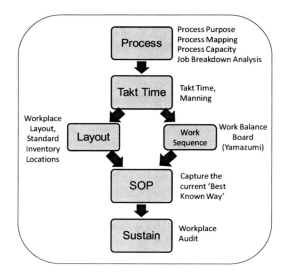

So, a balance needs to be sought. Three useful quotations set the scene:

'To standardise a method is to choose out of many methods the best one, and use it. What is the best way to do a thing? It is the sum of all the good ways we have discovered up to the present. It therefore becomes the standard. Today's standardisation is the necessary foundation on which tomorrow's improvement will be based. If you think of 'standardisation as the best we know today, but which is to be improved tomorrow' - you get somewhere. But if you think of standards as confining, then progress stops.' (Henry Ford, *Today and Tomorrow*, 1926)

'In a Western company the standard operation is the property of management or the engineering department. In a Japanese company it is the property of the people doing the job. They prepare it, work to it, and are responsible for improving it. Contrary to Taylor's teaching, the Japanese combine thinking and doing, and thus achieve a high level of involvement and commitment' (Peter Wickens, Former HR Director, Nissan UK)

'A proper (standard) procedure cannot be written from a desk. It must be tried and revised many times in the production plant. Furthermore, it must be a procedure that anybody can understand on sight. For production people to be able to write a standard work sheet that others can understand, they must be convinced of its importance.' (Taiichi Ohno)

Ohno believed that only by preparing their own job instructions could operators comprehend the details of their work and know why they should have to do things that way, and only then would they be capable of pondering other better ways to do the work. This is the basis of kaizen. In other words actually making operators and supervisors write work instructions means that they have to think about the way that the work is done.

From the first we learn that a standard is not static, but improves over time. From the second we learn that a standard is not imposed from on high. From the third we learn that standards are inherently practical. A good standard comes out of a bottom-up questioning culture – ever seeking a better, simpler, safer way to do a task – not a top-down imposed way of working. If you have no standard you cannot improve, by definition. So standards are part of PDCA.

At the outset it should be said that 'standards' does not mean the rigid, work-study imposed, job specification that is associated with classic mass production. Such standards have no place in the world of Lean. For such 'jobs' industrial sabotage and absenteeism are to be expected. Beware of the human-relations based reaction against 'work standards' which are often confused with work-study. Allowing standards that are too loose on critical aspects may lead to no standards, which in turn lead to decreased safety and productivity.

Also, beware of thinking that standards have no place in non-repetitive work such as maintenance, service, design, or senior management. Good, flexible maintenance and service work is built by combining various small standard work elements. Good design comes out of creativity combined with standard methods and materials, adhering to standard procedures and gateways. At Disney Florida, for example, visitors to Universal Studios walk through section by section. Within each section, the section time is divided into blocks and in each block certain loosely scripted material must be covered by the artist. Do visitors notice this? No, they enjoy the professional but personalised delivery.

Far from thinking of a standard as something that confines (as in adversarial work methods of old), one should think about a standard that enables and empowers. This is like the 'rules of engagement' in the military – where a modern soldier has to be free to make decisions, but under an umbrella of guidance as to what can and can't be done.

Spear and Bowen in a classic article discuss the apparent paradox of TPS that activities, communications and flows at Toyota are at once rigidly scripted yet enormously flexible and adaptable. They conclude that that it is the specification of standards and communications that gives the system the ability to make huge numbers of controlled changes using the scientific method. Without standards and the scientific method, change would amount to little more than trial and error.

Peter Wickens, former HR Director of Nissan UK explains that for an 'ascendant organisation' there must be both concern for the people and control of the processes. Without concern for people you have an 'alienated organisation'; without concern for the processes you have 'anarchy'. This is a good way of thinking about standards in manufacturing and service. Both dimensions are needed.

Hence, it is useful to think in terms of a 'double pareto' where the main activities to be standardised are identified, and then within those activities, the critical steps are identified.

Standards should not be there to 'catch you out', but to enable. This is like a tennis or golf lesson. You don't hide your weaknesses from the coach, you bring them out because you want to improve. That is the essential spirit that needs to be fostered.

Standard and Davies explain that there are 3 key aspects of standard work which need to be understood:

1. Standard work is not static, and when a better way is found the procedure is updated

2. Standard work supports stability and reduces variation because the work is performed the same way each time. Moreover variations (defects, deviations, discrepancies) are easily recognised.

3. Standard work is essential for continuous improvement – moving from one standard to a better standard without slipping back.

Management standards should exist for meetings, communications, budgets, and many other activities. Strike the right balance of detail.

Despite what some people think about Frederick Taylor, there remains 'one best way', with available technology, to do any task that will minimise time and effort, and maximise safety, quality and productivity. To some this may sound like boring repetition, but the 'new' standards are about participation in developing the best and safest way, mastery of several jobs, and the ability to adapt to changes in the short term.

Deming, in proposing the PDCA cycle, saw improvement moving from standard to standard. Juran emphasised the importance of 'holding the gains' by establishing standards following a process improvement, rather than allowing them to drift back to the old ways. Recently the 'Learning Organisation' has become fashionable, including 'knowledge harvesting' from everyone in the organisation. How is this to be achieved? By documenting experience; in other words by establishing standards from which others may learn. Supervisors should have prime responsibility for maintaining and improving standard work.

And in new product development, design, and product launch, do what Toyota does and accumulate checklists of things to look out for. The checklists become the accumulated knowledge of the company. Knowledge management. If you don't do this simple thing, you are doomed to always 'learn the hard way' and to 'start from scratch'. What a waste!

Leader-level standard work is very powerful. Establish a routine that requires different tasks, audits, or visits to be done at regular intervals. Perhaps, at director level once per quarter, manager level once per month, team leader level every day. Team members are expected to do a problem review every morning. That is the standard.

How do you build 'culture'? Though broad standards! By regularity. It becomes 'the way we do things around here'. We hold a problem review meeting every day. We discuss better ways to do tasks. We hold after action review meetings. We build checklists. We try to never let a customer go home dissatisfied. Culture emerges out of the standard practices. But standard practices are not left to chance – like many things in Lean – the what and the why (and the 'will do') comes down from the top, but the 'hows' are developed locally, sometimes with expert help. Be aware, if you are a manager, that standards begin at the top but are carried out at all levels. The day you walks past an unacceptable practice and don't enquire (note: do not impose), is the day that quality and culture begins to slip back.

6.7.1 Training on Standards

Please refer to the section on Training within Industry (TWI).

The TWI Job Instruction (JI) method, evolved during the Second World War, is probably the most thorough and well-proven approach to work standards and training. Ohno recognised the power of the approach and claimed that it was the breakthrough that enabled TPS to progress. Today, more than 60 years later, TWI JI methods remain a cornerstone of TPS and are virtually unchanged.

JI begins with the Job Breakdown Sheet where, for a task, the 'Important Steps', 'Key Points' and the reasons for them are identified.

Major Steps	Key Points	Reasons for Key Points
Prepare the patient	1. Set out central line kit 2. Check the lab reports 3. Lay patient on back 4. Place rolled up towel between patient's shoulder blades	1. Immediate access to materials 2. Prevents potential adverse effects of the procedure/check to see if procedure could potentially be harmful to the patient 3. Makes access to vena cava easier 4. Makes finding the clavicle easier
Apply anaesthetic	1. Swab chest with antiseptic 2. Inject 5cc of lidocaine	1. Prevents infection 2. Keeps the patient from feeling excessive pain
etc	...	...

Importantly, whilst the aim of the sheet is to document the 'one best way' (a phrase from FW Taylor), it does not attempt to document everything. Hence, an 'Important Step' is a work element where work is advanced. Steps such as 'open door' are not recorded because, whilst

necessary, it is not important. Often, there will be a number of 'key points' associated with each Important Step. Key Points reflect the 'knack' or special points to look out for or do. It may be a feel (like pressure), a sound (like a click), timing, or something to notice. Safety issues are always key points. The reasons for the key points should be known. If there are more than (say) seven important steps – depending on complexity – segment the task.

Then, actual JI can begin. JI has four steps (taken from a JI card):

1. Prepare the worker. This aims as setting the person at ease, finding out and recognising what skills are known already, explaining what the task is for, and setting the operator in the most comfortable, safe position.

2. Present the task. Important steps are shown and talked through one at a time. Then it is shown again, this time going through the key points. Patience and time is required. The worker should know all the important steps, and the number of key points in each. Questions?

3. Workers turn. At least three rounds are required. First, the worker goes through the whole task. The instructor patiently corrects errors. Second, the worker repeats the task, this time explaining the important steps. Third, go through again, explaining the key points. 'Continue until you know that they know'.

4. Follow up. The person then works on their own, but must know exactly who to speak to in case of problems. Checks are carried out regularly but with decreasing frequency.

Note that a job breakdown sheet is not a SOP. It is the basis of a SOP. The SOP is used for audit and recording, by the supervisor, not for training.

References

Patrick Graup and Robert Wrona, *The TWI Workbook*, Productivity, 2006

Productivity Press Team, *Standard Work for the Shopfloor*, Productivity, 2002

Much of the original TWI material is available free of charge. See http://chapters.sme.org/204/TWI_Materials/TWI_Manuals/TWIManuals.htm

Thanks for John O'Dwyer of Lake Region Manufacturing

6.7.2   Window Analysis and Standards

Window analysis is a framework used to confirm that standards are being followed, and to identify potential problems. It is particularly appropriate for assembly operations where following standard work is critical to quality. The method helps to establish the reason for the failure of a standard – whether caused by establishing the standard, communicating the standard or adhering to the standard. The method is used by, for example, Sony.

The method seeks to understand whether the issue identified is confined to one person or group, or is more widespread. This is done by referring to party 1 and party 2 , or group 1 or group 2.

The categories are: 'Known' or 'Unknown' – whether the correct methods are established and known – and 'Practised' or 'Unpractised' – whether the correct methods are practised 100% of the time.

There are then four conditions – refer to the figure:

A – only if methods are known and practised by all parties 100% of the time.

B – an adherence problem, when a method is established and understood, but not practised by all parties.

C – a communication problem, when the method is established, but some individuals are not informed about it.

D – a standardisation problem, where the right method is not established.

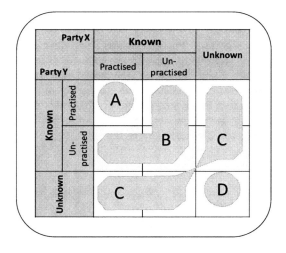

6.7.3   Variations on SOPs

Single Point (One Point) Lessons. This sheet or display is a focused single lesson that can be covered in 5 minutes or less. It is often used in TPM. The purpose is to reinforce areas where difficulties have been experienced in quality, safety, downtime, etc. It can be used as a reminder or a training aid. For training, progress – from trainee to coach - may be indicated, perhaps building a square '□' (I → L → U → □) to indicate progress.

Problem (Trouble shooting) Cards. These are 'what if' cards to cope with relatively rare but important contingencies. What to do if the chuck breaks.. Most air force pilots are used to the idea of consulting a card in an emergency – in order to avoid potentially disastrous mistakes in a time of crisis.

Test Cards. These include a small number of questions on the task – true or false, or multiple choice – used for standard ops where

- Tasks are done infrequently
- A new operator requires confirmation
- There is job rotation amongst the team.

The last two can be motivational for team members to devise – and also help them to think more about their tasks.

Today most SOPs are produced on computer using digital photos and laminated. This is good but can be inhibiting – consider having SOPs written in pencil but protected in a folder.

RACI (pronouced 'racey') charts are good practice for SOPs. They have wide application outside of SOPs. RACI denotes who is R = responsible, A = accountable, C = needs to be consulted, I = needs to be informed.

	Supervisor	Operator	Facilitator	Area Manager
Prepare SOP template	C	C	R	A
Write SOP	C	R	I	A
Approve SOP	R	I	C	A
Audit SOP	R and A	I	I	A

It is good practice to use PDCA for standards and develop a set of standard questions for each stage. For instance: PLAN: What characteristics are important? How stable is the process? Who are the stakeholders, and how should they be engaged? Who is likely to know the best known method (BKM)? How will it be piloted? How will be process be monitored? DO: Who will prepare, write, approve, audit (RACI). CHECK: What checks are needed to make sure the SOP has been written and implemented correctly? ACT: Who will train, implement, verify? How much time is needed?

Further Reading

Robert W Hall, 'Standard Work: Holding the Gains', *Target,* Fourth Quarter, 1998, pp 13- 19

Taiichi Ohno, *Toyota Production System*, Productivity Press, Portland, OR, 1988

Charles Standard and Dale Davis, *Running Today's Factory*, Hanser Gardner, 1999

Spear and Bowen, 'Decoding the DNA of the Toyota Production System', *Harvard Business Review*, Sept/Oct 1999

Steven Spear and Kent Bowen, 'Decoding of the DNA of the Toyota Production System', *Harvard Business Review*, Sept / Oct 1999

John Bicheno and Philip Catherwood, *Six Sigma and the Quality Toolbox*, PICSIE, 2005

6.8   Changeover Reduction (SMED)

Changeover reduction is a pillar of Lean manufacturing. The late Shigeo Shingo produced the classic work on 'SMED', and until recently very little has been added to what he said. However, a significant advance has now been made by a group at the University of Bath (McIntosh et al, 2001). Also Six Sigma methods have added useful dimensions on variation. Generally, for Lean, the reason to do changeover reduction is to allow for small batch flow and improved EPE performance.

Shingo, by the way, did not invent quick changeover. He codified it. It had already been known at Toyota before Shingo, and at Ford before that. In *Workplace Management*, Gemba Press, 2007, Ohno tells a story of how workers from Toyota Japan went to Toyota Brazil to learn about forge changeover which the Japanese did not believe could be done.

What is a Changeover?

There are three views. The first, narrow, view is the time that a machine is idle between batches (the 'internal' time). The second, and widely held, view is that it is the time from the last piece of the first batch to the first good piece of the second. A third view, is that it is the time from the standard rate of running of the first batch to the standard rate of running the second. This view therefore includes rundown and ramp-up times.

The classic Shingo methodology is:

- Identify and classify internal and external activities. Make a video?

- Separate 'internal' activities from 'external' activities. External or preparation activities should be maximised. Cut or reduce waste activities such as movement, fetching tools, filling in forms.

- Try to convert internal activities to external (for example by pre-heating a die).

- Use engineering on the remaining internal activities. There are many tricks, from quick release nuts, to constant platform shims, to multiple hole connections all in one. Both Shingo and McIntosh are an excellent source of ideas.

- Finally, minimise external activity time. Why? Because in small batch production there may be insufficient time to prepare for the changeover during a batch run.

The Formula1 pit-stop analogy is useful. But note this is a high cost changeover with minimum time virtually the only goal. It also involves slowing down and speeding up. Safety is important but again costs big money. The wheel is held in place by only one screw. Is this acceptable?

Another analogy is the standard seat belt, which has been safe, quick, reliable, adjustable and easy for all to use and widely acceptable. This is a design-led innovation – see below.

Mapping the changeover process is a standard approach. Begin by chunking the changeover into major stages – these will be, at least, preparation, actual changeover, and re-establishing speed and quality. Then divide each stage into steps. Use a brown paper chart with all the steps in the changeover going along the top of the chart. Then construct rows along the chart. The rows are as follows: Total time for the step. People involved, internal time, external time, and then three further rows – one to note whether the step can be reduced, simplified, eliminated, or done in parallel; another for other ideas and sketches, and the third row for photos. Use post-it stickers for the last three rows.

To standardise, the work combination chart and standard operating procedures are useful. See separate sections on these standard Lean tools.

Adjustment is an important consideration that can consume much time. Moreover, *adjustment is the root of many quality problems.*

List all adjustments on a separate sheet. Categorise into three. First, adjustments that should not be made – here think of ways in which the setting can be frozen or welded. Second, adjustments that have a limited number of standard settings – each of these should be indicated or marked, or ways thought of to achieve 'one touch' adjustment. And third, adjustments that truly need adjustment: for these try to devise a standard one best way. Failsafe if possible – it will pay big dividends.

Variation in changeover time is almost as important as the changeover time itself. If a changeover has large variation, then good scheduling practice is made difficult. Therefore, track the major elements of changeover and determine which stages have greatest variation. Then tackle variation as a separate exercise.

Changeover is part of OEE, so changeover reduction may be done under the auspices of TPM. The very thorough 9 step TPM approach will always be effective, but the effort to go through this level of detail is significant. TPM will also help highlight the importance of the changeover in relation to other losses, particularly when changeover is being done for capacity reasons.

To derive the target changeover time for mixed model flow, please refer to the Batch sizing section. Essentially, this involves working out how much time is available in a day (or week) for changeover, after making the standard set or 'campaign' of products. Then divide the available time by the number of changeovers required in that period. This will give the target changeover time. The process is iterative. If there is insufficient time to make all changeovers in a day, try two days, then three, and so on until a realistic changeover target time is found. Then, after a time, reduce the period. And so on.

It is useful to think strategy. Is the changeover to reduce time, to reduce cost, to improve quality, to reduce manpower, to limit maintenance, or a combination? Is the aim to increase capacity or to improve flow? Generally you can't have them all.

McIntosh et al say there are four elements to successful changeover. *Attitude,* including workplace culture and receptiveness to change. *Resources* including time, money, personnel, training, tools. *Awareness,* including the contribution of changeover to (flow), flexibility, inventory, capacity and awareness of different possibilities of achieving quick changeover. *Direction,* including leadership and vision, priority and ranking, (and presumably impact on the value stream). McIntosh usefully divides changeover reduction into three phases. The table below has been developed from their work. McIntosh et al make a strong case that there are two general approaches, organisation-led (i.e. SMED) and design-led. For each, there are four areas to address:

1. 'On line activities' – by internal and external task reallocation, or by designs that allow the sequence to be altered – for example simultaneous rather than sequential steps

2. Adjustment – by reducing trial and error by for example indicators and shims, or by design which allows 'snap-on' adjustment

3. Variety – by standardisation and standard operations or by design which reduces the possibilities of variation – pokayokes

4. Effort – by work simplification and preparation or by design which incorporates simplification – for example fixing multiple hoses by one fixture.

Phase	Tasks	Issues
Strategic	Identify opportunities Focus priorities Identify the approach to use (Organisation led or design led)	Internal team? Consultants? New equipment? Dedicated equipment? TPM / OEE approach?
Preparatory	Existing performance records Variation in times? Postponement? Fix targets (e.g. via VSM)	5S of tools and dies Sequence dependencies? Use blitz? Involve different shifts?
Implementation	Video SMED technology Engineering changes Pokayoke? Regularity and sequencing	Records Incentives? SOPS Sustainability

The choice between organisation-led and design-led, depends on objectives, on how much you are willing to spend, and on sustainability – design-led locks in improvements much more than organisation-led. McIntosh at al suggest that a 'Reference Changeover' be developed. This is what Lean practitioners would call a 'paper kaizen' activity. It involves collecting data (e.g. by video), identifying and cutting all waste, rearranging activities, and doing the changeover as efficiently as possible. This establishes the theoretical benchmark.

Six Sigma methodologies have also been attempted on changeover. It is now clear that the SMED methodology is much more effective at reducing time, and should always be done first. But then Six Sigma analysis can be useful to examine the causes of time variation. If the changeover is important enough, building the distribution of changeover times and examining for normality of distribution and then seeking root causes of, for example, bi-modal time distributions can be worthwhile. Design-led changeover involves a whole tranche of possibilities: breaking task interdependencies and automating adjustments (for example incorporating a measuring scale); making parts more robust or lighter; pokayoke, incorporating built-in tools (e.g. welded spanner), improving access, and mechanisation and robotics. Clearly there is overlap with TPM.

Finally a few tips:

- Measure and record changeover times. Many changeover times have fallen by doing this alone..

- Involve the team in analysis. Do not rely only on Industrial Engineers.

- Make a video, and get operators to record and critique. The video must remain their property. When making a video try if possible to use two or even three teams. One team to video the big picture, one team to record hand movements, and another to record the paperwork. Ideally, each team should have a cameraman, a recorder to make notes, and a commentator to speak about what is happening. Put their ideas up on a board at the workplace.

- Consider a financial incentive for consistent improvement in changeover, whilst discouraging incentives for more production.

- Remember the equation: Changeover time x no of batches = constant. In other words as changeover time comes down, this must be converted into smaller batches. Resist the temptation just to gain extra capacity.

- Q: 'How do you get to Carnegie Hall?' A: 'Practice, man, practice'. It's what grand prix teams do.

- Use trolleys onto which all tools and equipment are placed, and which can be wheeled to the changeover machine.

- Regularity in the schedule helps. If everyone knows that Machine A is changed over every day at 9 a.m., then everyone from forklift driver to setter will be on hand.

- Tool and die maintenance is a vital, but sometimes overlooked, part of setup reduction. Don't compromise.

- At bottlenecks, use a team for the changeover, bringing in operators from non-bottleneck machines.

- Use appropriate quality control (such as SPC and Pre-control) procedures.

- Be aware of the optimal sequence of changeover times.

Further Reading

Shigeo Shingo, *SMED*, Productivity Press, Portland, OR, 1985

R I McIntosh, S J Culley, A R Mileham, G W Owen, *Improving Changeover Performance*, Butterworth Heinemann, London, 2001

6.9 Small Machines, Avoiding Monuments and Thinking Small

The small machine concept is one of the least recognised Lean facilitators. The general principle is to use the smallest machine possible consistent with quality requirements. Having several smaller machines instead of one bigger, faster 'monument' allows flexibility in layouts, easier scheduling, reduction in material handling, less vulnerability to breakdown, less vulnerability to bottleneck problems, possibly reduced cost (through a mix of capability), and through phasing of machine acquisition, improved cash flow and more frequent technology updates. Do work improvement first, and only then do equipment improvement.

The related sunk cost principle means that the priority should be with minimising present and future costs, not with keeping machines working to 'pay off' a cost that has already been incurred. Therefore utilisation is less relevant unless it is a capacity constrained machine. Remember, however, a poorly utilised machine can become an effective constraint.

Old Machines. The small machine concept can be extended to older machines. The best machine may well be an old machine that is quality capable, that is permanently set up, located just where needed, and that is written off in the books so that no-one cares about utilisation. It is throughput and lead-time that count. Beware of scrapping old machines that are still quality capable for machines that are faster.

Self Developed Machines. Why should a machine be 'at its worst when new'? Because it may not yet have had pokayoke devices fitted, may not yet be quality capable, may not yet have had low cost automation devices integrated with it, and may not yet have been developed for multiple operations, and especially if variation has not been tackled.

Automation. The prime reason for automation in Lean is for quality. The principle is not to automate waste. So simplify first. Ask whether a low cost solution is possible – a gravity-feed rather than a robot. Good reasons for automation are dull, dirty, dangerous and hot, heavy, hazardous. Another good reason is reduction in variation. A bad reason is to reduce people. Remember, machines don't make improvement suggestions. Schonberger offers excellent advice in what he terms 'Frugal Manufacturing'. In essence:

- Get the most out of conventional equipment and present facilities before implementing large-scale automation projects.

- Keep control over manufacturing strategy rather than turn it over to newly hired engineers and computer technicians or to a turnkey automation company.

- Build up your capability to modify, customise and simplify your machines. Do not expect commercially available general-purpose equipment to be right for your products. The ability to modify continually is becoming increasingly important as materials, technologies, quality standards and products change and improve.

- Approach bigger, faster machines and production lines with caution. High capacity and cost tend to dictate production policies,

and immobility and inflexibility do not accommodate shortening product life cycles.

- Understand that big machines, separated equipment, and long conveyor systems disconnect people, obscure opportunities for merging processes, and result in divided accountability: automation has the potential to lower costs and minimise variations in quality, but it makes sense only when it solves clear-cut problems and when it costs less than simpler solutions introduced incrementally. Small Machines are a part of a wider Lean issue – the advantages of thinking small.

This goes back to Schumacher's classic work *Small is Beautiful*. Pil and Holweg discuss four advantages of small-scale operations.

1. Tapping into local networks, as decentralised R&D labs as opposed to large centralised facilities, are able to do in Cambridge (US and UK).

2. Responding to customers, as several manufacturers such as Nypro and Johnson Controls find when they set up facilities near customer sites.

3. Rethinking human resources – developing people more quickly by giving them greater responsibility in small operations. (South Africa has been a hotbed for the development of automotive CEO's, or training airline pilots in smaller airlines.)

4. Driving Innovation as per steel mini mills and discount airlines.

Further Reading

John Bicheno, *Fishbone Flow*, PICSIE Books, Buckingham, 2006

Richard Schonberger, 'Frugal Manufacturing', *Harvard Business Review*, 1987

Frits Pil and Matthias Holweg, 'Exploring Scale: the Advantages of Thinking Small', *MIT Sloan Management Review*, Winter 2003, p.33-39

7 Mapping, Assessments and Analysis

'It's not what you don't know that hurts you. It's what you do know that ain't so.'

(Will Rogers)

7.1 The Value Stream Implementation Cycle

Mapping and assessments are major analysis tools in Lean. Mapping is the 'Meta Tool' in the Lean Toolbox because the mapping tools should guide the use of all other tools.

Remember that all mapping and analysis is waste unless it leads to action. Doing mapping is not the same thing doing Lean. Do not fall for Paralysis by Analysis.

7.1.1 What is the Aim of Mapping?

The real purpose of mapping is to design the future state. It is a visualisation exercise – a vision of the current state and of the future state. This is done by establishing priorities for Lean implementation, both short and medium term. Mapping is also an excellent vehicle for involvement and participation. For many, participation on a mapping exercise is their first practical exposure to Lean outside of a classroom or after a 5S exercise. Mapping is also a great tool for idea generation.

Although maps should be on display, mapping is not for decoration, it is for action – obvious, but be wary of framed or laminated maps. Date all maps, and take them down after actions are taken. Having old, out-of date maps on the wall that no-one, except visitors, looks at any longer is poor practice.

Good mapping practice is to show 4 maps: Current State, Future State, Ideal State, and Action Plan. Maps should be developed by the area's people for the area's people. Maps should be signed off by all the participating mappers, but especially by the people from the area just mapped.

7.1.2 Before You begin Mapping...

Mapping is a powerful tool, but is not the full answer. First, clarify the aim, and the implementation period. 'If you don't know where you are going any road (map) will get you there'. What are the essential issues that a mapping exercise aims to achieve? Short term cash flow, longer term productivity, survival, or is it simply that 'mapping is such a usual thing to do in Lean that 'I suppose we better do an exercise ourselves'.

Second, is scope. Defining the value streams is discussed later, but there is an immediate issue of scope. Where does it start and where does it end? Is it to be a local exercise, a pilot, or is it be in-plant with an idea of extending out to the supply chain in the future. Why start in the factory? Why not with customer service? Many companies are far better with their internal operations than they are with field service, or delivery, or installation. Or, start with administration – the factory has been worked over many times but never the office. Often, the biggest problems are in the information flows that support physical operations. So, one might need to do some preliminary high-level maps to help answer these and similar questions.

Third is performance. A value steam map is also a snapshot in time, so it is not capable of picking up vital variation information. In other words, a value stream map is good for muda, less good for muri and mura. You will want to understand:

- The delivery performance, and the variation thereof – not one average figure!
- Demand management. (Read the section!)
- Is there a pattern to delivery performance – preferential customers, seasonality of performance.
- Customer satisfaction, and reasons for dissatisfaction.
- Lead time variation.

Remember that a mapping exercise should be at least as much about the information flows as about the physical flows. It is tying the two together that is the real benefit.

Before You Begin Mapping…Part 2: Two Essential Paretos.

Lined with scope is focus. What elements of the value stream should you focus on? Here there are two 'essential Paretos' – the Lead Time Pareto and the Cost Pareto. Both are important to help steer you away from irrelevant or fashionable exercises that will make very small contribution to either customer effectiveness or to the bottom line. For these exercises you will need the help of the production control and finance departments.

The Lead Time Pareto is an estimate of the length of time of each of the elements of total end-to-end lead time – order to delivery. Draw out a Gantt chart or critical path network from the time that an order is received to the time of delivery (for a make to order item) or the time from planning a new batch to its delivery into finished goods (for a make to stock item). As a preliminary exercise, estimates will do. You may well have to get finance and manufacturing planning in a room together to work this out.

The various elements typically comprise many of the following:

- Order entry time: paperwork. The time from receipt of order to entry into the manufacturing and planning system.

- Credit verification.

- Manufacturing planning time.

- Schedule assembly time – needed to consolidate orders into balanced assembly sequences.

- Configuration time: the time from entry into the system to completion of the configuration. In assemble to order or make to order this could involve design or CAD time and configuration checks. This time may be zero in make to stock environments, and may be zero or near zero in repetitive operations.

- Procurement or material acquisition time: the time taken to procure materials and components, to kit (if done) and to bring the materials from the store to the point of use. In repetitive operations some elements of procurement time could be regarded as zero where they are done routinely or in parallel with order entry time or configuration time.

- Non-specific manufacturing time: the time taken for 'variety as late as possible' manufacturing stages, where components or subassemblies are not specific to a final product or order. Note that this time element may overlap or be done in parallel with order entry time and configuration time.

- Order-specific manufacturing time: the time taken in those manufacturing stages which are order- or customer-specific. This time may be zero in make-to-stock.

- Order launch.

- The move, queue, changeover, and run times for each stage, although some of these elements may be zero if there is a cell. Also tackle the longest subassembly sequence, not minor non-critical path subassemblies.)

- Time spent as WIP or semi finished, between stages
    o Inspection / quality control time
    o Finished goods store time
    o Delivery time
    o Invoice time
    o Payment time.

Of course, some of these will overlap or be done in parallel. The important thing is – which of these or other stages takes the most time and is on the critical path? So, don't work on the physical flows if it is the office stages that take the most time, and vice versa.

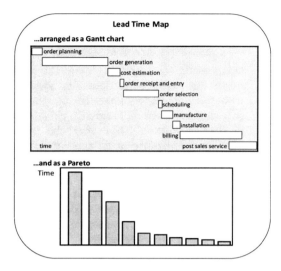

The Cost Pareto is simply a list of the major cash flow elements of the chosen value stream for a typical period – say a month or a quarter. Typical elements are:

- Sales (in the period following, usually)
- Material costs (careful here – show materials actually used in sales, and show material increases or decreases separately)
- Throughput (= sales – material costs)
- Direct wages
- Management costs
- Other overheads
- Energy costs
- Depreciation
- Consumables
- Sales and marketing
- Freight and distribution
- R&D
- Other costs

Arrange these in a histogram of descending order. The idea, of course, is to focus on those areas where the biggest hits are likely – material costs perhaps – and to de-focus on low impact areas – direct wages perhaps. A good idea is to plot these on a diagram showing money along the vertical axis, and estimated ease of achieving (say) a 10%

reduction along the other axis. See the figures for each of these Paretos.

7.1.3 A Note on Intervention Theory and Change

If you are a Lean expert or champion, do not fall into the trap of doing the mapping yourself. The results will seldom be sustained – however skilled you may be.

Gerard Egan's widely used and proven intervention theory (used in counselling) has three stages that are very similar to mapping stages: current picture, preferred picture, action plan. The first stage is about skilled, active listening, to help uncover blind spots. No advice or critique is given, but empathy is required. Use open, not closed, questions. Recognise past achievement and give respect. The task is to reframe the perspective, to uncover blind spots. Points of leverage are discovered. The second stage is jointly developing the destination. This is about exploration and developing commitment. It has three sub-stages – possibilities, change agenda, and developing commitment. Stage three also has three sub-stages: possible strategies, selection of best-fit strategies, and plan – what, when, where, how. Patience is required. Think of the tortoise and the hare!

7.2 Stages of Mapping

The main stages of mapping are Current State, Future State, and Action Plan. But this is seductively simple! It is very much the case that what you get out depends on the effort you put in. Beware of picking the 'obvious' solution or of using mapping to justify what you were going to do anyway! There are ALWAYS multiple solutions. So think about Future State**s** (plural).

Before you even begin mapping, start your exercise by brainstorming out what the team thinks are the major issues and problems. It is also useful to discuss what the team thinks success will look like. What would you like the future customer experience to be, and what would you like the future employee experience to be? Be bold! Mapping is a time for big picture transformation, not incremental change. Tim Hurson in 'Think Better' calls this productive, or breakthrough, or 'tenkaizen' thinking as opposed to reproductive, incremental, or kaizen thinking.

Then, when you have done your Future State map(s) you can return to the issue list and to the success list, and see if your concepts address these.

Russell Ackoff suggests that you don't work forward from the Current State, but that you work backwards from the Future (or Ideal) state. Start from the assumption that the factory burned down yesterday.

References

Tim Hurson, *Better Thinking*, McGraw Hill 2008

Russell Ackoff et al, *Idealized Design,* Wharton Publishing, 2007

7.2.1   *Organising for Mapping*

The following sections refer to the Figure 'Mapping and Implementation' below. What is usually needed is:

- A 'Lean Promotion Office' or some supporting organisation of expert facilitators. Read *The Facilitators Fieldbook*.
- Perhaps a Lean consultant or sensei
- A value stream manager for each major value stream
- Team leaders and supervisors from the areas – not just factory areas but administrative areas as well
- A few good operators or clerical staff – from different shifts
- Possibly other key staff – perhaps people from Quality, Maintenance, Accounts
- Consider if key customers and suppliers should be involved
- One or two outsiders to ask the 'silly' questions, and learn the concept for their turn later on

- Middle manager involvement is highly desirable – the wider the scope, the more necessary.

So perhaps a team of 10 for a bigger project. Less for smaller projects. Not all have to be there all the time. A core team of say 7 is good.

Materials are the obvious things – brown paper rolls, lots of post its, felt pens, pencils, erasers, etc. It is useful to have layout diagrams if these are available.

A dedicated mapping 'war room' is highly desirable.

There are computer packages for mapping, but if used at all should be at the last stage – it is much better to allow participation around a 'messy' wall chart than a neatly produced map done by one person. You can document the value stream for record and communication purposes – but not as the working document.

Remember, mapping is a 'gemba' activity – not an office activity!

7.2.2   *Pre-Mapping Workshop*

Hold an initial workshop on expectations. Review the essential Paretos discussed above. Outline the sequence. If necessary use the workshop for some initial training on mapping mechanics and wastes. Perhaps visit other areas that have already been mapped. Watch a video on mapping – there are several available. But the principle activity is to clarify the issues raised in the 'Before You Begin' section above. An important consideration is to define the 'system boundary' – what is to be included and what excluded. A SIPOC diagram is useful: Suppliers, Inputs, Process, Outputs, and Customers. Take time on this one. Be guided by areas of responsibility but also natural flows of work and buffers. In perhaps half of situations it will be necessary to identify the particular value streams to be mapped using Product Family Analysis. To concentrate on the most time-consuming part of the value stream, often begin with Overall Lead time Pareto.

Include a discussion about when to map. Plan to map on a representative day – midweek, probably. Try to avoid peak periods or leave periods.

Prepare the ground by briefing all staff in whose area mapping will take place. This is not only simple courtesy, but also allays fears. Consider having far wider representation at this initial meeting than just the mapping team. Show the Brown Paper Chart (see the section below) to clarify the area for everyone. Get the whole team to walk the route. Come back and brainstorm out the types of map that will be required. There are several, as outlined below, and not all are necessary.

7.2.3   *Determine Takt*

Takt time is often an important consideration and calculation. See the section on takt time and pitch time. But note that

- Takt can vary along a value stream – for instance where one section works a single shift and another works double shift. But always bear in mind the delivery or customer takt time interval.

- Takt can also vary with time – across a year.

- Takt can be stabilised by changing the working time.

- In some situations, takt has little or no relevance or meaning - in many admin functions, in design, in large vessel or batch manufacture, in situations of erratic demand. In these cases, do not bog down with ultimately meaningless discussion. In these situations, capacity management and flexibility is much more important.

7.2.4   *Shared Resources and Variation*

Of crucial importance is to take particular care of shared resources – those key machines or workstations, and sometimes key people that are shared between value streams. For these, capacity analysis is required. It is simply wrong to work through a value stream mapping exercise making the assumption that a resource is dedicated when it is not. Where there are dedicated resources, the

total load must be calculated. This involves tracking all the parts that cross the resource in a typical period, not just those parts that are from the value stream. The load is the summation of all parts that cross the workstation in the period x the unit run times, plus typical changeover times during the period. As noted under Muri and Mura, be particularly careful where the load exceeds say 80% of available (demonstrated) capacity. So, OEE data will also be useful.

Remember, variation is the enemy of Lean. See 'A Formula for Lean'. If you don't recognise variation you can't see your enemy – and it could prove fatal. So the team will want to know about variation at key resources – how much variation in

- Cycle time
- Changeover time
- Uptime
- Quality
- Demand placed on the resource.

It may be that case that, coming out of this sub-exercise alone, these turn out to the root cause of many delivery and lead times failures. Important future state actions may be decided on.

7.2.5   *Mapping Begins*

Split up into sub teams to do the individual basic maps. For the 'Learning to See' current state you may have sub teams for material flows and information flows. Map what you see today. Ignore the fact that 'today is not usual'. Go to 'Gemba' and collect the facts.   Experienced mappers like to map working upstream so that they can collect customer requirements at each stage before moving on. If you find this confusing, work downstream.

Basic Mapping will often comprise assembling the five basic maps – Value Stream Map, Quality Filter Map, Spaghetti Diagram, Demand Amplification Map, and Cash Flow map. Throughout, maintain a high level view – keep to the main stages. Drilling down comes later. Certainly have one sub-team look at demand. See the section on Demand Management. However, the team should arrange for the data on variation and performance, as mentioned above, to be researched.

You will usually use brown paper and different coloured post-its - colours for information flow, physical activities, and ideas.

7.3 Mapping and Implementation

Mapping is best thought of as a PDSA process. It never ends. It is a process of discovery, exposure and reduction of more and layers of waste, amplification, and variation.

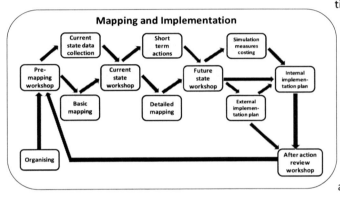

Current State Workshop

For a small exercise this may take place on the same day as the basic mapping. Otherwise schedule a specific day as near as possible to the mapping day. Assemble the basic maps. Talk them through with people from the area, to get validation. Calculate the time line, value adding ratio, and constraint utilisation. List the wastes that have been spotted. Together with the takt time, shared resource, and variation information,

There are three aims:

- to gain a clear 'big picture' overview appreciation
- to identify 'low hanging fruit' opportunities
- to identify areas for more detailed study.

Short Term Actions

These may be undertaken by mini kaizens or other quick kill initiatives – probably not by kaizen events at this stage because these require more detailed clarification. A definite time objective must be fixed.

The two financial maps – the cash flow map and the cost map, often grab the attention of senior managers.

Draw two time lines – one for physical flow, one for information flow. Use castellation. On the top of the castellation is written the value adding times, taken from the stage cycle times. On the lower castellation is the inter-operation time, estimated from the inter-operation inventory holdings multiplied by takt time, or the delay or queue for information flow. Adding these together gives an approximation of lead time. Note that if there are shared resources this estimation is incorrect because it fails to account for shared resource queue time.

Identify and draw on the loops or sub-systems that can be treated individually for analysis or implementation purposes.

Detailed Mapping

Here more detailed mapping is undertaken. Process activity maps may be made for selected areas. Also more detail on selected aspects of information flow or scheduling. The Information Value Stream Map can be used in selected areas where there are clear process flows and minimal branching resulting from decisions. Otherwise the Brown Paper Map can be used for selected processes. It is often useful to home in on variation of selected process or changeover times, defect rates, and demand data. Pay particular attention to mapping information flows. (This was relatively ignored in the *Learning to See* mapping text.) In some situations, mainly to bring home the point to management, preparing a Cost Time Profile may be useful. Detailed mapping and analysis for cell design and line design is described in a separate chapter. Cell layout diagrams and work combination charts will be used.

Future State Workshop

Schedule this day specifically to take place within days of the basic mapping. The aim is to develop (a) the next future state, (b) the action plan, and possibly (c) the ideal state. Use the guidelines of the future state questions mentioned later in this chapter and the Building Blocks discussed in the chapter on Scheduling.

The Future State map should not simply be the current state map with a few kaizen bursts added. This is too easy – and does not represent the potential of Lean. That is, it is Muda but not Mura or Muri – and all three are necessary. One must therefore do more extensive scheduling work, say using Lean repetitive scheduling principles or possibly drum-buffer-rope methods, and show the supermarkets, pull loops, and runner routes. Much of this is discussed in the Lean scheduling and Theory of Constraints chapters.

Draw the future state map and maybe the ideal state map. Feel free to annotate. Produce supporting diagrams and concepts, such as layouts, supermarket concepts, scheduling concepts, information flows, and people development. Try to keep all maps on display in the war room. Use the workshop as a workshop to present, to solicit ideas, and to gain buy-in.

Simulation, Metrics, and Costing

Simulation may be a possibility. It can take various forms: (a) physical simulation, such as full-size cardboard modelling, (b) paper simulation – whereby you work through a month's schedule on paper, or (c) more rarely, computer simulation using a modelling language. Toyota certainly uses 3D computer simulation to help design new facilities.

It is also essential to review the metrics that are in place for possible conflicts with the future state – for example a metric that encourages overproduction. 'You get what you measure' is an absolute truth. Ignore at your peril. Beware, for example, of OEE. Likewise review the costing system.

In particular, for a future state, it is important to work through the consequences on the financial statements of inventory reductions and other changes with your accountant. For these topics see the Chapter on Lean Accounting and Measures.

Internal Implementation Plan

The internal Action Plan covers what, who, when, and where. The team should present it to management. Explain it to the shop floor. Use Force Field Analysis as a means of both explaining and soliciting ideas. Gantt chart the activities. Develop a plan for the next 90 days – sometimes 180 days, but never more. Develop the Master Schedule for transformation. The Route Learning Map is beginning to be used as an effective implementation and communication device. Think in terms of bite size chunks. Do a little and return to do it again, rather than a big chunk that never gets there.

External Implementation Plan

The External Action Plan, covers upstream and downstream parties. It should include the reduction of amplification and variation. External implementation will eventually involve the whole supply chain. For this, the Supply Chain Structure map, and 'Seeing the Whole' (extended value stream map) will become essential.

After Action Review

To complete the current round of Plan, Do, Study, Act, it is vital that this workshop is scheduled after the review period of say 90 days. Everyone is expected to attend. It covers what was planned, what was achieved, why the difference, and what can be learned for the next cycle. Start and maintain a checklist for future mapping. Build in learning.

Further Reading

Peter Hines, Riccardo Silvi, Monica Bartolini, *Lean Profit Potential*, Cardiff Business School, 2002

Jeffrey Liker and David Meier, *The Toyota Way Fieldbook*, McGraw Hill, 2006, Chapter 3

Val Wosket, *Egan's Skilled Helper Model*, Routledge, 2006

Tom Justice and David Jamieson, *The Facilitator's Fieldbook*, 2nd edn., AmaCom, 2006

Thanks to Kate Mackle of Thinkflow for her inspiration.

7.4 Types of Mapping

7.4.1 Mapping Overview

Kate Mackle of Thinkflow has proposed a most useful way to begin mapping. See the figure and the table. This sets the approach and asks the essential questions.

The Questions	What is required to answer them and mapping the information flows)
1. How should capacity be offered to the market?	Understand the availability and quantity of delivered capacity
	Understand lead times and inventories by family
	The priorities for various types of demand
2. How are orders accepted and loaded	Order entry and capacity loading process
	How raw material is ordered from suppliers.
3. How is capacity and material availability planned?	Capacity management and material allocation.
	Determining and maintaining needed inventory levels.
4. How to schedule what should be made and when?	The scheduling process and procedures.
5. How to organise production and material handling between processes?	Production schedules and material movement methods and procedures
	Managing subcontractors.
	Synchronising support processes,
	Preparing and dispatching orders.

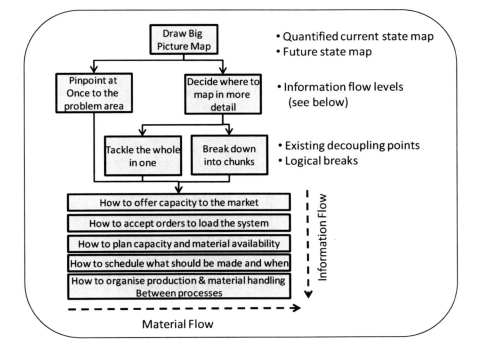

7.4.2 Brown Paper Chart

A Brown Paper Chart is a high level diagram showing the main product flow and stages. An example of a brown paper chart is shown below, taken from an automotive metal pressing company. It serves to clarify the overall logic of the plant. A supply chain version would show the main suppliers, service centres, supply routes, distribution routes, distribution centres, and main customers. Often products and percentages going through different channels would be shown.

The chart can become a focus in the 'war room' showing, by means of frequently updated photographs, graphics, and 'post it' notes, the progress and highlights. A 'Master Schedule' can go alongside – showing the Gantt chart of progress towards implementation. The team should gather around the Master Schedule at regular (weekly?) intervals to check progress.

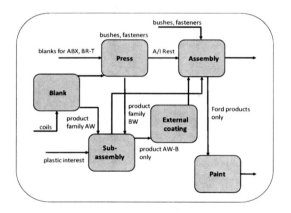

7.4.3 Product Family Analysis

Product family analysis is about breaking down the full product range into groups that can be managed together, or share a significant part of a value stream. It is the first step of value stream mapping and the basis of cellular manufacturing. There are strategic and technical considerations.

Strategic considerations

Strategic considerations involve, first, standing back and thinking about the ideal future state. 'Begin with the end in mind'', says Stephen Covey. This sentiment is echoed by Russell Ackoff in Idealised Design, where he urges the team to think that 'the system was destroyed last night'. Nadler urges team members to think about 'the plant after next'. In particular, Ackoff advises participants to

- 'focus on what you would like to have if you could have whatever you want ideally, not on what you do not want', and

- 'don't worry about the availability of resources'.

However, this thinking does not extend to fantasy – including as-yet-unavailable technologies.

The idea is to project forward, then work back and identify barriers and possible solutions. Don't work from current to future, work from Ideal to possible futures.

Strategic considerations would overlap with policy deployment considerations. See the chapter on this. It may, for example include

- dedicated customer cells (as in automotive first tier)

- synchronisation with major customers (as with Nypro and Dell)

- cells on customer sites (as with Jaguar, or spectacle suppliers located in shopping malls)

- outsourcing or insourcing

- setting up pulse or flow lines.

Some of these are discussed further in the chapter on Layout and Cells.

Reference

Russell Ackoff et al, Idealized Design, Wharton School Publishing, 2006

Technical considerations

After thought has been given to the strategic, technical considerations remain. The problem is how to group parts or products into cells or value streams. Briefly, there are three approaches:

- by inspection, or 'eyeball', or common knowledge – very frequently used
- the matrix approach – used occasionally
- more complex mathematical approaches – used rarely.

See also sections on cells and layout.

7.4.4 Lead Time Variation and Inventory Days of Cover

Lead time variation is important – perhaps as important as the overall lead time. We have already looked at the Lead Time Pareto, now look at variation.

Be sure to track the customer's end-to-end time, not the organisation's internal times. See the graph below. Note that control limits have been added. This is to identify 'special causes' or 'out of control' situations or times.

When mapping, it is important to look at 'end-to-end' performance, not just some minor subsystems. Measuring subsystems gives little indication of overall system performance. As Ohno was saying, 'all we are interested in doing is reducing the lead time from order to completion.' In the 3DayCar programme, the information flows were mapped for the entire Order-to-Delivery (OTD) process of major car companies.

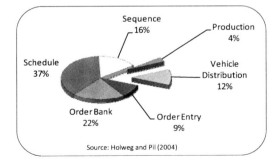

Source: Holweg and Pil (2004)

While most people in the car industry are focused on improving the vehicle assembly operation, it only accounts for 4% of the overall process delay the customer experiences!

When you map a system, look at end-to-end performance. There you look at outliers first, before you start improving the system.

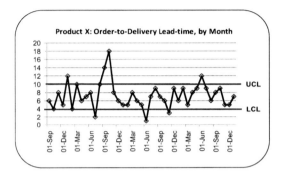

Throughout the following analysis it is frequently (always?) more useful to express inventory in terms of 'days of cover' rather than inventory units. Days of cover should relate to Sales units, not to money. You will need to get the total unit sales for a typical month and then to divide by the number of working days in the month – excluding weekends and holidays – to give the average daily sales rate. You may want to check this for several months, and also to get the normal range – upper and lower.

Then, in subsequent mapping, convert each accumulation of inventory to days of sales cover by dividing the inventory at that stage by the average daily sales rate. This is a more meaningful figure. It is generally more robust than calculating the time line by multiplying the inventory at the stage by the takt time - especially where there are shared resources or where takt is not very meaningful as in the case of some process industries.

Thereafter you will want to group the days of inventory into the appropriate main stages of raw materials, work in process, and finished goods. Of these, WIP is solely under the value stream's control whereas raw material may have to be held due to erratic supply and finished goods may have to be held to meet uncertain demands.

The next five maps are best regarded as the core set for plant mapping. They should be used together. Together they give a powerful picture of Lean status.

7.4.5 'Learning to See' Maps

The 'Learning to See' map has emerged as the most popular and clear way to illustrate the current and future state of a value stream. The method maps both material and information flows. It is quick to learn because it uses simple boxes to indicate stages, and other obvious symbols such as trucks, factories, and kanban cards. The tool is suitable for repetitive operations, especially where a single product or family is made. A powerful feature is that it 'closes the loop' from customer order to supply to manufacture ending with the delivery of the product. This closed loop is not shown on most detailed activity charts. An example is shown below, alongside the main mapping icons.

The point about these diagrams is that they are a clear overview that can be used for planning and participation meetings, from shop floor to top management. As a reference tool they can be placed on boards in meeting areas, and ideas can be added by Post-it stickers.

Progress can be charted. This creates the current state diagram. There are standard mapping symbols for the most common element. Some are shown. Many mappers invent their own supplementary symbols.

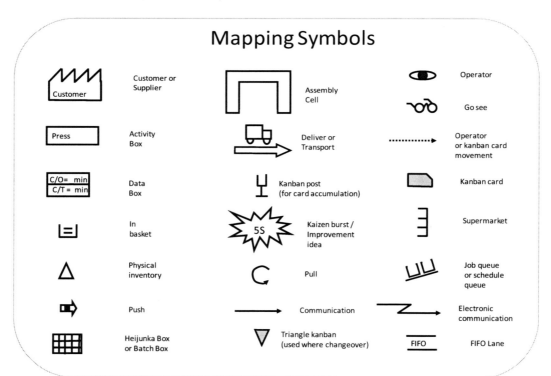

Mapping Symbols

Customer	Customer or Supplier
Press	Activity Box
C/O= min / C/T = min	Data Box
In basket	In basket
Physical inventory	Physical inventory
Push	Push
Heijunka Box	Heijunka Box or Batch Box
Assembly Cell	Assembly Cell
Deliver or Transport	Deliver or Transport
Kanban post	Kanban post (for card accumulation)
5S	Kaizen burst / Improvement idea
Pull	Pull
Communication	Communication
Triangle kanban	Triangle kanban (used where changeover)
Operator	Operator
Go see	Go see
Operator or kanban card movement	Operator or kanban card movement
Kanban card	Kanban card
Supermarket	Supermarket
Job queue	Job queue or schedule queue
Electronic communication	Electronic communication
FIFO	FIFO Lane

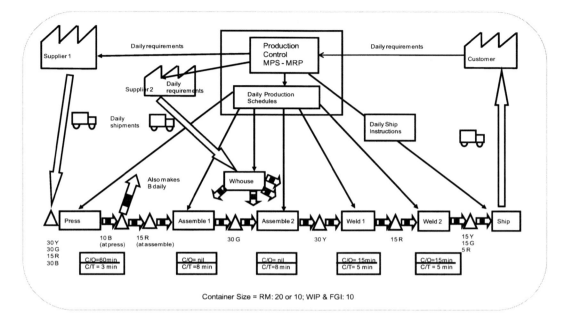

Container Size = RM: 20 or 10; WIP & FGI: 10

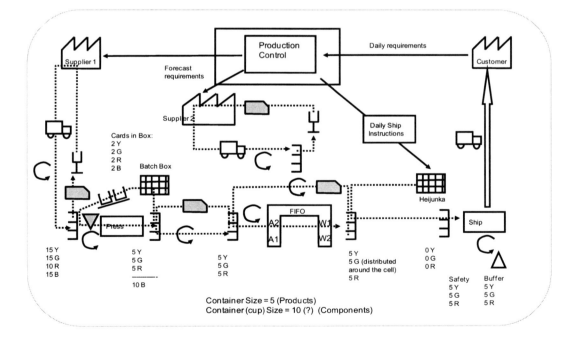

Container Size = 5 (Products)
Container (cup) Size = 10 (?) (Components)

To create the future state and ideal state diagrams requires two steps.

First, incorporate short-term improvements from the five basic maps. This includes waste reduction ideas. Show these as 'Kaizen bursts' on the diagram. The second step requires more in-depth knowledge of Lean possibilities. These are the subjects of Layout and Scheduling. However, it is useful to break up the value steam map into pull segments or loops, often separated by supermarkets. Then use these as building blocks for layout design.

Limitations

Of course, Learning to See has limitations that one needs to be aware of. Firstly, it considers only one product at the time, and does not consider any linkage on capacity on shared resources. Secondly, it is a static picture only. It is not able to capture variation at all. It therefore cannot make any statement about capacity or loading, which is a major shortcoming that the mapper needs to be aware of when analysing the result! See section above on Shared Resources.

Further Reading

Mike Rother and John Shook, *Learning to See*, The Lean Enterprise Institute, Brookline, MA, 1998

Kate Mackle and John Bicheno, *Lean Mapping*, APICS Lean Manufacturing Workshop series, 2003. (CD and book). www.apics.org

7.4.6   Spaghetti Diagram

The Spaghetti Diagram (or String Diagram) is a long established tool for more effective layout. It tracks the waste of transport and the waste of motion. It could not be simpler. Merely get a layout diagram of the plant and trace the physical flow of the product in question on the diagram. Mark on the diagram the locations of inventory storage points. Do not forget rework loops, inspection points, and weigh points. Calculate the total length of flow. Show component delivery flow paths in another colour. Again calculate the

length of travel. Wasteful movement and poor layout become clearly apparent. Do get the mapping team to walk the distance, rather than just to draw it. While the team is walking, get them to take note of variations in vertical movements – the more constant, the better.

A spaghetti diagram can also be used to map collection routes for parts, and external processing travel paths. Many plants have, for shock-tactics purposes, worked out the equivalent annual distance travelled in terms of, for instance, number of times around the world. Jim Womack once related the average speed of travel of an aerospace part to the speed of an ant!

The Learning to See map gives the logic of the main steps, the information flows, and the time line. The Spaghetti Diagram gives the geography. So they form a set. Strangely, this simple but powerful tool gets little or no mention in some mapping publications. At least two flows should be traced – the product flow and the regular (or irregular) material handling routes.

Lean layout groups inventory into supermarkets from which parts are pulled. Parts should not be scattered around in many locations. Parts are delivered to the line and products collected from the line by set material handler (water spider or runner) routes. The spaghetti diagram is the prime tool for establishing the best routes. The spaghetti diagram can also be used at the workplace level, for instance for changeover reduction analysis.

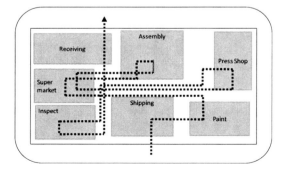

7.4.7   Quality Filter Mapping

Quality filter mapping aims to track the locations and sources of defects along a process route. The Quality Filter Map is a graph showing the parts per million (ppm) rate against process stage. Although this information may be collected and shown as part of a Learning to See current state map, a quality filter map adds emphasis. Two bars should be shown, Scrap and Rework.

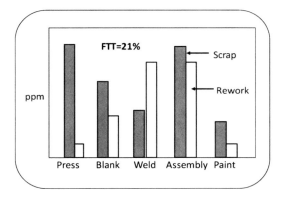

Note that scrap and rework should be recorded not only at points where the company records defects, but also at all operation steps. This is to ensure picking up what Juran refers to as 'chronic' wastes – the underlying defects, reworks, or inspections that have become so routine that they are not recognised as a problem. An example is the 100% manual touch-up welds done at the end of a robotic assembly line, which enjoyed zero priority for improvement but which, upon analysis, proved to be one of the most costly quality problems in the plant.

First time through (FTT) is often calculated as part of the Quality Filter Map. FTT is expressed as a percentage: 100 x (parts shipped - (parts reworked+parts scrapped)) parts shipped Note that if parts are reworked at several workstations the FTT figure can be negative.

An alternative to FTT is OTIF (on time in full) – the percentage of parts that are delivered both on time and in full. This percentage figure can be compared with the days of finished goods inventory. The more the days of finished goods inventory the higher the OTIF should be. For example, if there are 10 days of finished goods inventory why on earth is OTIF not 100%?

Quality filter mapping can highlight defects that are passed over long distances along a process route or supply chain only to be rejected beyond the point at which return for rework is not economic. Also, beware of parts that are passed onto constraint machines, thereby wasting capacity.

Alternatively, the **Yield** (output/input) percentage can be calculated at each stage. This can be useful in process industry and in offices.

Beware of accepting the official defect figures. 5 ppm at final dispatch may be the result of excellent process control, or of numerous inspections and reworks. In 1995 the story was told of a famous German car whose average time for rectification exceeded the total time required to build an entire new Toyota. The final build quality of the German car was, however, superb.

7.4.8   Demand Amplification Mapping

This tool maps what is termed the 'Forrester Effect' after Jay Forrester of MIT who first modelled the amplification of disturbances along the supply chain and illustrated the effect in supply chain games. It is also a form of the well-known run diagram used in quality management. Amplification happens in plant and in supply chains, but the latter has enjoyed more attention. Amplification is the enemy of linear production and Lean manufacturing, and results from batching and inventory control policies applied along the supply chain. For instance, fairly regular or linear customer demand is translated into batch orders by a retailer, then subject to further modification by a distributor adjusting safety stocks, then amplified further by a manufacturer who may have long changeovers and big batches, and then further modified by a supplier who orders in yet larger batches to get quantity discounts. The result is that, further along the chain, the pattern of demand in no way resembles the final customer demand.

An amplification map is plotted usually day-by-day across a month. There will be a line for each stage. For example, from purchasing, receiving dock, order entry, from completions at various stages, and from dispatch. In a supply chain, an amplification map shows orders, shipments, and inventory levels at each company in the chain over a period that matches the cumulative lead-time in the chain. It is quite a big job to get this data – but the results are often startling.

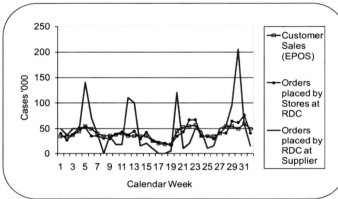

The figure shows an example from the grocery sector, which was collected by David Simons and Barry Evans from LERC. The chart shows how the EPOS (electronic point of sales) demand, which is what customers pay for at the till, is amplified as it is passed back to the supplier. Some distortion occurs when the store orders from the RDC (regional distribution centre), but then the manual intervention by purchasing at this supermarket chain causes major amplifications in the signal. This of course is not malicious, but is an effect that occurs when final demand is not transparent to the decision-makers, and forecasting takes over. Clearly the advantages of stable, regular orders from the customers are being destroyed. Life is being made very difficult for the supplier, and overall much more stock is held in the system. What is going wrong, and what should be addressed?

The amplification map is a great tool for getting at the heart of scheduling issues. It is also a good evaluation tool that forms part of a periodic report to management or as a tool for evaluating the process of Lean implementation. The amplification issue and its possible solutions in the supply chain context are discussed in the Supply Chain chapter.

Note that in order to create a meaningful demand amplification map it is important to pick a volume or runner product and a representative time horizon (generally 3+ months, avoiding Christmas and summer periods). Also make sure that the components and materials you map only go into the final product you are looking at so that you can show direct correspondence of the demand patterns.

Reference:

Jay Forrester, 1961, *Industrial Dynamics,* MIT Press, Cambridge MA

Lee, H. L., V. Padmanabhan, et al. , 1997. 'The Bullwhip Effect in Supply Chains' *Sloan Management Review* Vol. 38 No. 3, p. 93-102.

7.4.9   Financial Maps

It is possible to combine Learning to See maps with financial aspects to see where your money is tied up – how often are you turning the money around. This grabs senior manager's attention more than stock turns! Financial maps are also necessary to give the Lean enterprise viewpoint, rather than just Lean operations. Two maps are useful – the Cash Flow map and the Cost map.

The Cash Flow Map traces cash flows into and out of the company. This is important because this is 'real money'. Cash is king. Most are aware that a profitable company can still go out of business due to cash flow problems. How fast cash is turned over is the prime concern.

The Cash Flow Map highlights the gap in financing between paying for raw materials and components and receiving payment from customers. The gap has to be financed by the company. So, where are the opportunities to reduce the gap: payment lead time to suppliers, lead time due to internal operations, delivery, or waiting for payment?

The Cost Map is simply an extension of the Current State map, and shows a snapshot of the main direct costs relating to the value stream. Inventory costs are shown with respect to the raw material costs and do not reflect the cost accumulation of value added work as a part progresses. There are two reasons. First, the accumulated value is a matter of judgement and in any case takes time to calculate with little benefit. Second, it may be argued that a part-completed product is of no value to the customer until completed. The stage-by-stage costs show the direct costs – certainly labour, but maybe machine charge out rates or depreciation. (There is an argument on this, because machine cost is sunk. Perhaps show with and without. But, in any case, a consistent approach should be used.)

Note in the figure that a contribution profile is also shown. Contribution profiles are discussed in the section on the Essential Paretos. They are important because the contribution (sales price – direct costs) of all parts made in the value stream should be an important piece of information.

Note also that any shared resource should be highlighted. The cost of a shared resource should be apportioned between the sharing value streams, based on the time spent on the resource.

7.4.10   Human Resource 'Maps'

The future state map will carry with it requirements for the future required skill set. These skills should be determined. A human resource skill inventory or 'map' will begin at a high level and will list, for each level from manager to operator, the skill requirements and what will need to be known and standardised. The gaps between present and required skills will form the human resource development plan. Here, the TWI framework will be useful. See the separate section.

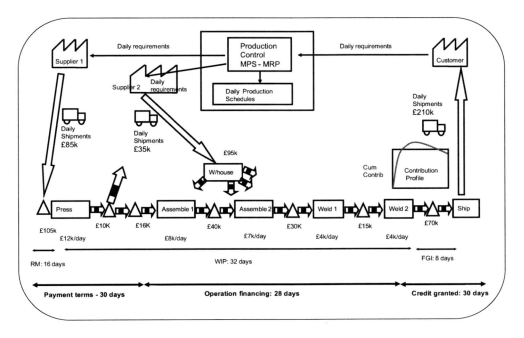

TWI has five categories of needs for supervisors – a 'supervisor' is widely defined as anyone who is in charge of people or who directs the work of others. The five are: Knowledge of the Work – specific to the company or process; Knowledge of Responsibilities – again specific to the company; and then the three TWI skills of Job Instruction (JI) – how to instruct; Job Methods (JM) – how to problem solve and improve; and Job Relations (JR) – how to work effectively with people. List these out as a matrix- similar to the skills matrix.

7.4.11   Cost Time Profile

A Cost Time Profile is simply a graph showing accumulated cost against accumulated time. Its beauty lies in its visual impact. Whenever value or cost is added, the graph moves upwards; a plateau indicates no value or cost being added for a period of time, for instance during delay or storage. The area under the graph represents the time that money is tied up for. The aim is to reduce the area under the graph by reducing time and/or cost. The technique is superior to a simple Pareto analysis of cost and time accumulations because one can immediately recognise where expensive inventory is lying idle and at what stage time delays occur. An example is shown in the figure. Notice two cumulative lines, one for total cost and the other for value. The difference between these two lines represents wasteful, non value adding activities and other cost accumulations such as the cost of money being tied up in inventory. Non value adding activities include inspections, transport, clerical activities, and rework. A cost time profile can be obtained directly from the process activity map, by multiplying by the costs of the various resources. If the process activity map is recorded on a spreadsheet, the calculations for the cost time profiles are easily done. When this is done, however, note that waste may still exist in a nominally value adding operation – for instance wasteful movements in an assembly activity, so that the lower profile does not represent the ultimate aim. The vertical distance between the two lines represents some

obvious wastes, but not all wastes. For example, the long plateaus also represent waste of unnecessary storage and inventory. The aim should be gradually to reduce the profile towards that shown in the bottom left hand corner of the figure.

Note that when bought in materials are added, this results in a vertical bar on the chart equal to the material cost. In practice, the time for most value adding operations is minuscule in comparison with the delay and queue times, so value adding operations also appear as a vertical jump.

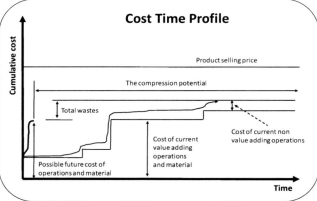

So a Cost Time Profile is a graphical method to identify when and where costs accumulate. They have been used extensively in conjunction with Lean manufacturing, supply chain analysis, business process reengineering and total quality. Look for long plateaus, especially later on in the process where costs have already accumulated. Attacking the long time plateaus will reduce cost, and improve responsiveness and quality.

The profiles are relevant to quality improvement because there is often a direct correlation between poor quality and wasted time. So for instance when there are delays due to rework, inspection or queuing, both costs and time accumulate. Many customers associate good quality with shorter response or delivery times.

The technique has been extensively used and developed by Westinghouse who used it as part of their Baldridge award-winning performance. It is equally applicable in manufacturing or office environments. Westinghouse made extensive use of cost time profile charts, but presumably this is not what caused the breakup of the company in 1997! The company used the profiles in a hierarchical fashion. That is the profile for each sub-process or product can be combined to form a profile for a whole section that in turn can be combined into a profile for a complete plant or division. Here total costs are used, so it is necessary to multiply the unit cost profiles by the average number of units in process. All processes must be considered, value adding as well as support activities and overheads. This therefore represents a total process view of the organisation, and may be used with process reengineering or Hoshin planning.

Note that conceptually the Cost-Time Profile is very powerful, and has top-management impact – which is its greatest advantage. However, in practice these charts are very difficult to draw, as the underlying data is generally not readily available.

Further Reading

Jack H Fooks, *Profiles for Performance: Total Quality Methods for Reducing Cycle Time*, Addison Wesley, Reading, MA, 1993

7.4.12 'Seeing the Whole' Supply Chain Mapping

Seeing the Whole Mapping is very similar to Value Stream Mapping described later. Only the scope is greater – the inter-company value stream rather than the in-plant value stream. A seeing the Whole map looks just like a Value Stream Map, except that plants replace process stages. Of course some intermediate stages may be warehouses or cross-docks. Information flows are in the top half, physical flows in the bottom half. The more detailed principles will be described under Value Stream Mapping; here the special features of supply chain mapping are mentioned.

The main benefit of Seeing the Whole mapping is to gain an understanding of the complete supply chain and to identify major co-ordination opportunities, rather than detailed kaizen implementation activities.

To this end, assembly of the mapping team from multiple companies is easily the most difficult and important issue. It has to be seen as a mutual benefit exercise – no hidden agendas. The team will be high level, because the issues are high level. The core of the mapping team is schedulers from participating companies.

What is a complete supply chain – how far upstream should you go? Answer: as far as possible, or pragmatically as far as cooperation from participating companies will allow. Even linking two companies is a worthwhile exercise.

Since the focus is on the complete value stream you can afford to aggregate most in-plant stages. In general, most plants will be drawn as single process boxes or stages. An exception is where there is a mix of shared and dedicated resources within a plant – for example a common press shop feeding a dedicated assembly line. These will often have separate scheduling systems. Ideally, each company in the complete supply chain will first have done an internal Learning to See map. But there are plenty of opportunities to go after, even if Learning to See maps have not been developed. For example, in the 3DayCar study it was found that by far most of the 6 week delay between placing an order and receiving a new car was due to information delays along the chain. Forget the internal physical changes and work on the information flows.

As with Value Stream maps, the main supporting maps are valuable – in particular the 'Demand Amplification Map', for which data should be collected alongside the 'Seeing the Whole' map. The focus should be more on the information flows rather than the physical flows. Do show the physical flows along the bottom of the map, but concentrate on the information flows. The physical flows can be 'black boxed', but real benefits accrue when getting into the details of the supply chain scheduling decisions, and the associated delays. When mapping the information flows, also record how often IT systems or databases are updated, or scheduling systems run: if a system runs only once per week, the average delay caused here is 3.5 days!

Keep the end customer in mind throughout the exercise. Intermediate customers (other companies) are important, but the supply chain exists for the end customer. Thus waste identification and opportunities are focused on the end customer.

A weakness of both Value Stream maps and Seeing the Whole maps is the way they deal with shared resources. (This aspect is generally ignored in *Learning to See*.) It is likely that there will be several shared resources in a complete supply chain. The way they are scheduled is key. What other supply chains are served, and how much capacity is devoted to the particular supply chain are important questions. For instance in one supply chain studied, a mid-chain participant could originally only devote one day a week to the particular supply chain. So, could changeover be reduced or buffer (supply chain supermarket) added to allow a more frequent EPE (every product every) – and who would pay for this? A detailed understanding of the scheduling assumptions and constraints is one of the great pay-offs. The scheduling building blocks (see separate chapter) are relevant in supply chains also.

Yet another weakness is ignoring variation – which is even more of a killer in a supply chain than in a plant. So do consider the vulnerability of the future state chain to disruptions, breakdowns, variation on delivery times, and quality problems. Consider the strategic location of supply chain supermarkets.

Further Reading

Dan Jones and Jim Womack, *Seeing the Whole: mapping the extended value stream*, Lean Enterprise Institute, Boston, 2002

Darren Dolcemascolo, *Improving the Extended Value Stream*, Productivity, 2006

Holweg and Pil, *The Second Century,* MIT Press, 2004

7.4.13   Order Tracking

Shapiro's 'Staple yourself to an order' article is an excellent guide for order tracking: follow through an order from receipt to dispatch. The aim is to decrease lead-time and improve customer service, and improve cash flow. Shapiro et al stress the importance of tracing the full 'order management cycle' from planning to post-sales service.

They state that there are typically 10 steps, some of which may overlap:

- Order planning
- Order generation
- Cost estimation and pricing
- Order receipt and entry
- Order selection and prioritisation
- Scheduling
- Fulfilment – including procurement, manufacture, assembly, testing, shipping, installation
- Billing
- Returns and claims
- Post sales service

Shapiro et al suggest drawing up a matrix chart with the 10 steps forming the rows and the various departments or functions forming the columns.

Trace the flow of an order, by arrows and activity boxes as it moves through the steps and departments. Identify who has prime responsibility for each step and who has a supporting role. Use the triangle symbol to indicate delays and queues. Estimate the length of these. When drawing the sketch, use different colours or shading for physical movement of paper and for flows by computer network, telephone and fax. The chart below shows a simplified chart of the order fulfilment process in the car industry, which takes (on average) 41 days to complete an order. Most of this delay happens in the information flow, at the various stages of scheduling:

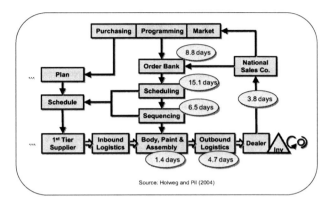

Source: Holweg and Pil (2004)

Now the questioning begins. The aim is to reduce time and waste. It is essentially a creative process. Preferably the people involved in the process should be used in its analysis and improvement. Bold thinking is a requirement, not piecemeal adjustment. The title of the classic article in *Harvard Business Review* by Michael Hammer gives the clue: 'Reengineering Work: Don't Automate, Obliterate!' That is the type of thinking that is required. Competitive benchmarking may be useful, as may the creativity encouraged by value engineering. The same *Harvard Business Review* article tells how Ford used to have 400 accounts payable clerks compared with just 7 people at Mazda.

The basic step is to examine the process chart and to split the activities into those that add immediate value for the customer and those that do not. Refer to the '7 wastes' for guidance. The concept is to achieve the added value of the product or service in as small a time as possible. Therefore try to make every value-adding step continuous with the last value-adding step, without interruptions for waiting, queuing, or for procedures that assist the company but not the customer. Stalk refers to this as the 'main sequence'. There are several guidelines:

- Can the non-value adding steps be eliminated, simplified, or reduced?

- Can any activities that delay a value adding activity be simplified or rescheduled?

- Are there any activities, particularly non-value adding activities, which can be done in parallel with the sequence of value adding activities?

- Can activities that have to be passed from department to department (and back!) be reorganised into a team activity? Better still, can one person do it? What training and backup would be required?

- Where are the bottlenecks? Can the capacity of the bottleneck be expanded? Do bottleneck operations keep working, or are they delayed for minor reasons? Are bottleneck operations delayed by non-bottleneck operations, whether value adding or not?

- What preparations can be made before the main sequence of value adding steps is initiated to avoid delays? E.g. preparing the paperwork, getting machines ready.

- Can the necessary customer variety or requirements be added at a later stage? (e.g. making a basic product or service but adding the 'colour and sunroof' as late as possible.)

- If jobs are done in batches, can the batches be split in order to move on to a second activity before the whole batch is complete at the first activity?

- Can staff flexibility be improved to allow several tasks to be done by one person, thus cutting handing-on delays?

- What are decision-making arrangements? Can decision making power be devolved to the point of use? Can the routine decisions be recognised so that they can be dealt with on the spot? Perhaps 'expert systems' can be used.

- Where is the best place, from a time point of view, to carry out each activity? Can the activity be carried out at the point of use or contact, or must it really be referred elsewhere?

- Do customers enjoy a 'one stop' process? If not, why not?
- If problems do develop, what will be the delays and how can these delays be minimised?
- What availability of information will make the value adding sequence smoother or more continuous? Is there more than one source of information, and if so can this be brought to one place? A common database perhaps? The established data processing principle is to capture information only once, and let everyone use the same data.
- As a second priority, can the time taken for value adding activities be reduced?
- Michael Hammer has some useful non - mechanical suggestions concerning assumptions. The following is based on his 'Out-of-the box thinking'.
- Are you assuming a specialist must do the work? (People)
- Are you assuming that purchasing will pay only after receiving an invoice? (Time).
- Are you assuming that record keeping must be done in the office? (Place)
- Are you assuming that inventory is required for better service? (Resources).
- Are you assuming that the customer should not be involved? (Customer)

Note: useful variations on this chart include adding Post-It notes to nodes (a data box or ideas), showing a time line below, labelling the flows with document types (paper, fax, etc.), adding supporting photographs, and adding rows for each activity type.

Further Reading

Benson Shapiro, Kasturi Rangan, John Sviokla, 'Staple Yourself to an Order', *Harvard Business Review*, July-August 1992, pp113-122

George Stalk and Thomas Hout, *Competing Against Time*, The Free Press, New York, 1990

John Bicheno, *The Lean Toolbox for Service Systems*, PICSIE, 2008. See Part 2: Service Mapping

Michal Hammer and James Champy, *Reengineering the Corporation*

7.4.14   Feedback (System Dynamics) Diagrams

Many of the maps discussed, with the exception of the Amplification map and Lead time profile, reflect a snapshot situation. However, it is frequently useful to attempt to capture the inherent dynamics and feedback loops. Merely drawing these out, even without quantification, can lead to much improved understanding. Remember Ohno's favourite word – understand.

An example is shown. This concerns a never-ending spiral that a company found itself in. Schedule instability drove short term line performance variation. This caused schedule over-runs that in turn reduced planned maintenance time. Reduction in planned maintenance produced quality problems that fed straight back to schedule over-runs. Short term line performance affected OEE that was also affected by long and variable changeover times. Poor OEE, in turn encouraged bigger batches. Bigger batches led to high inventory which produced shelf life issues – customers would not accept products with too short a shelf life. This waste and customer demands led directly to a fire-fighting schedule, a basic cause of schedule instability.

The example illustrates that 'root causes' are sometimes part of a feedback loop. Prioritising time for planned maintenance, even in the face of short-term customer demand failure, together with tackling changeover time turned out to be a good solution. See figure below.

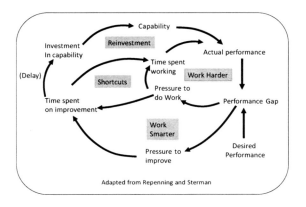

Adapted from Repenning and Sterman

References

Peter Senge, *The Fifth Discipline*, (revised edn.), Random House, 2006

Repenning and Sterman, 'Nobody ever gets credit for fixing problems that never happened', *California Management Review*, Summer 2001, pp 64 - 88

7.4.15   Process Activity Mapping

The process activity map is a far more detailed map than those described above, and should be used only on sub-processes where there are particular concerns. The earlier maps generally deal with hours and minutes; the process activity map works in seconds. But, it makes no sense to concentrate on the seconds before you have sorted out the hours and days. The map is a classic tool of the industrial engineer. The difference is that now they are used not just by work study officers or I.E.'s, but supervisors and operators as well. In fact, the first preference is for operators to learn how to use this most effective tool themselves.

The process chart is the prime tool for detailed analysis of value adding and non value adding activities at the micro level. The process chart lists every step that is involved in the manufacture of a product or the delivery of a service. Standard symbols are used to indicate 'operation', 'delay', 'move', 'store', and 'inspect'. (See figure below.)

The process chart helps identify wasteful actions, and documents the process completely. Good communication is an important reason to do this. The systematic record helps to reveal the possible sources of quality and productivity problems.

Many companies already have process charts. If they are available, beware! There are often differences between the 'official' process charts and the way things actually happen in practice. Are they up to date? The team or analyst should take the time to follow through a number of products, services or customers, documenting any 'horror stories' that occur. Often several actions and 'rework loops', unknown to management will be discovered. But it is not the purpose of the chart to use it for 'policing'. Often a team will draw up a chart for their own use in improvement and should not be obliged to turn it over to management.

Constructing the Process Map

Some process maps or charts can be long and complicated. If so, first break them up into sections of responsibility or physical areas. Then begin the detailed recording. See the figure. Preferably use a verb and a noun (e.g. select part, or verify document), and decide which of the standard categories (operation, transport / move, inspect, delay, store) the activity fits into. Note : the difference between delay and store is that store takes place in a specific store area or warehouse, whereas delay is a result of an in-process stoppage such as waiting for a container to fill up. You will differentiate between value adding and non-value adding operations.

While compiling the process map it is good to record distance, time, inventory, and the number of operators. If you are skilled at work study, also 'rate' the speed of work, but this is a secondary benefit. Also, record any wastes, comments, or interesting events. Many process charts are now supported with digital photographs.

While going around, take note of dates on inventory control cards, levels of dust, and container discipline – are containers moved in a first in first out, or a last in first out sequence? You may encounter some carousels where parts are moved or stored automatically. If so, mark them (unobtrusively if necessary) to get an idea of the length of storage or delay.

Step	Description	Symbol	Time	Inventory	People	Machine	Notes
1	unload truck	O	8 min	4000 pieces	2	forklift	
2	place at bay	D	3 min	2000 avg	0		temp store
3	move to line	⇨	1.6 min / trip	1000 per trip	1	forklift	4 trips
4	store at line	∇	330 min	4000	0		
5	press	O	0.06 / piece	1	1	AZ20 press	batch of 4000
6	store in container	D	12 min	2000			

A most useful exercise is to track one particular part, not just the type of part. However, this is often not practical due to the length of delays. So, mark a few parts and then come back and track their progress. Most manufactured parts have several branches or subassemblies flowing into the main sequence. It is a good idea to check on this first. Begin with the 'main sequence'. Then give attention to the major feeders, by time or cost, although it will seldom be necessary to track all feeders.

Also, you can 'black box' a particular sequence of operations. There is a special symbol to indicate this. Take particular care of inspection and rework points. Why is this necessary? Can it be done earlier? What happens to reject parts?

In work study there are more detailed charting or mapping techniques available, such as the two handed flow process chart, but don't go into these unless you have identified the particular step as an important bottleneck.

Analysing the Process Map

The first step is to classify the detailed activities into value adding, non value adding, and necessary non value adding activities. See the Basic Principles chapter. Note that some prefer simply to denote activities as value adding or non-

value adding, to avoid getting into long discussions about necessary. Some add green stickers to value adding steps, and red stickers for non value adding. (You will need many more red stickers than green.)

In general, the steps should then be analyzed using the 5 Whys (asking why 5 times over to get to the root cause - see the Kaizen section) and with the aid of Rudyard Kipling's 'Six Honest Serving Men' ('who taught me all I knew; their names are what and why and when, and where and how and who').

Thereafter, the creative or redesign phase begins. Some 'mechanical' considerations are:

- Can inspection steps be moved forward or eliminated?
- Can steps be done in parallel?
- Are there obvious candidates for automation?
- Is data being duplicated? Is it effectively shared?
- Can pre-preparation be done, before the event?
- Can steps be moved to another stage? For instance, can a supplier take over a non-core stage?

All the questions covered in the Order Tracking section are relevant here.

Selected Further Reading

Diane Galloway, *Mapping Work Processes*, ASQ Quality Press, Milwaukee, 1994

7.4.16 Value Stream Mapping and Policy Deployment: A TRIZ insight

When we draw a value steam map we are mainly concerned with the present or near future, and at a particular level of resolution – say the product value stream level, and not at the detailed process activity level. Similarly, in strategic Policy Deployment, we are sometimes concerned with SWOT analysis in the short term and at one level of resolution.

TRIZ, however, teaches that it is useful to look at the '9 Window' concept. The 9 widows are arranged in a 3 x 3 matrix. The 3 columns in the matrix relate to past, present and future, and the 3 rows to the wider system, the system, and the subsystem. This is a useful challenge and reminder. We are often concerned with the central cell (present, system), but there are 8 other boxes to consider.

Further Reading

Darrell Mann, *Hands-on Systematic Innovation for Business and Management*, IFR, 2004

7.5 Lean Assessments and Principles

In this section we review three assessment procedures by Kobayashi, Schonberger and Goodman. Such self-evaluation questionnaires or assessments are useful both for identifying areas of opportunity or weakness and for guidance on implementation. They can pick up areas that are not covered by maps – for example supplier performance, customer and employee motivation, or continuous improvement. Kobayashi's system is classic Japanese, concentrating on shop floor management. Schonberger's goes wider bringing in customers, bench-marking and perhaps a more Western view of employees. Note: The European Business Excellence model has become a well-used general assessment framework, but here we consider assessments more specific to Lean operations.

7.5.1 Schonberger's Principles

Richard Schonberger, author of the excellent *Japanese Manufacturing Techniques and World Class Manufacturing: The Next Decade*, and more recently *Best Practices in Lean Six Sigma* (2008), has developed 16 'Principles of World Class Manufacturing'. The 16 Principles are an excellent concise guide to Lean Operations. A company's progress on their Lean journey can be measured on a 1 to 5 scale for each principle. For 16 principles this means a maximum score of 80 points with 'adulthood' beginning at 53 points and 'maturity' at 67 points. A rating of 4 or 5 would be fairly challenging for many manufacturers.

The Principles are:

1. **Team-up with Customers**. Organise by Customer/Product Family. It is best is to be organised by, and focused on, customer families. Try to organise, end-to-end, by value stream.

2. **Capture and Use Customer, Competitive, and Best-Practice Information**. Aim: drive your improvement efforts with external data from customers (customer satisfaction/ needs surveys), competitive products (competitive analysis), and non-competitive best practices (benchmarking studies).

3. **Continual, Rapid Improvement in Universal. Customer Wants**. This is the achievement principle, measuring improvement in the eyes of customers. All customers, internal and external, want quality (Q), speedy response (S), flexibility (F), and value (V); these are universals of continuous improvement in the customer-focused organisation.

4. **Whole Work Force Involvement in Change and Strategic Planning**. This principle provides a framework for empowered, self-managed teams; 'front-line' includes professionals/technicians as well as operations/clerical people.

5. **Cut to the Few Best Components, Operations, and Suppliers**. This single design principle includes both product design and

design of the supply chain, which are related.

a. Both call for simplification and numerical reductions.

b. Reducing number of suppliers is unlikely to reach a lower limit, because new/ revised products, part numbers, operations, and outsourcing usually add new suppliers – requiring renewed efforts to simplify and reduce supply-chain breadth.

c. Diligence in holding down growth of part numbers/ operations at the same time holds down additions of suppliers.

6. **Cut Total Flow Time, Flow Distance, and Startup/ Changeover Times.** This principle focuses on three core Lean concepts.

7. **Operate Close to Customers' Rate of Use or Demand.** This principle concerns scheduling and synchronisation to the drumbeat of demand, takt time, monitoring schedule performance, seasonality, distribution centers/ distributors, and kaizen events.

8. **Continually Train Everybody for their New Roles**.

9. **Expand Variety of Recognition, Rewards, and Pay.** This principle 'closes the loop,' returning value to the work force for value created through continuous improvements – necessary to keep process improvement 'evergreen'.

10. **Continually Reduce Variation and Mishaps.** Everybody should know and use the mostly statistical (but simple) tools of capturing variation, mishaps, and unsafe or environmentally hazardous incidents, and isolating their causes.

11. **Front-Line Teams Record and Own Process Data at the Work Place**. Effective management of quality and process improvement requires that front-line employees, not just managers and technical experts, be in charge. Ownership: to be in charge, frontliners must be collectors and owner-users of the process data. Visual management: hidden data get less use than visual data; so visual management is a training topic in step 1, visual plotting of process data is a requirement in step 2.

12. **Control Root Causes and Cut Internal Transactions and Reporting**. The principle follows the notion of economy of control. Controls are needed most when processes are complex, incapable, failure-prone, variable; processes that are simple, capable, rarely nonconforming thus need few formal controls. Best control is no controls; instead fix the processes.

13. **Align Performance Measures with Universal Customer Wants**. This principle concerns the extent of use of QSFV (universal customer wants) as internal performance metrics (measures); Principle 3 is different, since it is devoted to measuring the extent of attainments on QSFV.

14. **Improve Present Capacity before New Equipment and Automation**. Improvement of present physical capacity (plant and equipment) via: (1) TPM (also called total productive maintenance); (2) simplifying operation, maintenance, setup, and process control; and (3) upgrading safety and health. Ownership: operators must acquire ownership of maintenance and safety (including safety from environmental hazards and degradation), just as they have with quality, and participate in equipment selection/improvement; maintenance/safety people must be teachers/ facilitators, just as quality people had to become under total quality management.

15. **Seek Simple, Movable, Scalable, Low-cost, Focused Equipment**. The ideal capacity for a family of high-volume standard products is a dedicated team with dedicated, minimal-setup capacity, run like a largely self-contained business unit. For low-volume, high-variety products, the ideal is a dedicated team with flexible skills and flexible, quick-change capacity. For example, a March snowstorm kept most Ford Romeo,

MI, Engine Plant associates home, but the Niche Line ran because it takes just a single, small team to build engines on that line. In either case, facilities should be as movable as possible – buildings equipped with solid floors, few impediments, all utilities available everywhere, rectangular shape; standardised equipment, often on wheels; modular tanks and piping, etc.

16. **Promote, Market, and Sell Every Improvement.** The organisation delivers QSFV so impressively that its customers are attached and remain.

Schonberger's Principles and comments are reproduced with kind permission from Richard Schonberger. Schonberger & Associates, Inc. offer a benchmarking service based on these principles. Address: 177 107th Ave. N.E., #2101Bellevue, WA 98004, www.wcm-wcp.com

7.5.2   Kobayashi's 20 Keys

Iwao Kobayashi's concept of the '20 Keys' has gained increasing acceptance as both a manufacturing assessment and an implementation guide to Lean manufacturing at shop floor level. The keys relate to 20 concepts fundamental to Lean operations, the majority of which are to be found in sections throughout this book. Kobayashi's concept is especially useful because:

- Guidance is given on the order of implementation of the keys

- A five point scale is provided for each key as an aid to internal evaluation, going from level 1 beginner to level 5 ideal (certainly beyond 'world class')

- Links between the keys are established. For instance, in order to reach higher level in the production scheduling key, actions are required in all other 19 keys. It is not possible therefore to reach higher levels in most of the keys without substantial progress in all keys.

- As an evaluation method, comparisons can be made with other organisations, although this applies to Schonberger and Fisher as well.

The sense of the 20 keys is given below showing the links to other sections in this book, and some comments on the range of characteristics required.

Note 1: For the full rating scale, from level 1 to level 5 for each key, as well as many useful tips, the full 285 page book is recommended.

Note 2: Some Western managers and critics of Lean manufacturing feel that Kobayashi's ideas on operators (for instance keys 3, 4, 6, 7, 10, 15) are much too regimented and that Western workers cannot be expected to become automatons. Lean enthusiasts tend to discount such views, saying that this represents a misunderstanding, and that standards and discipline are fundamental to continuing improvement.

Note 3: Some would say that the 20 Keys are essentially 20 individual, not-joined-up ideas, and as such are 'point kaizens' requiring some meta tool such as value stream mapping or policy deployment to link them together.

The 20 are:

1. Clean and Tidy (see section on 5S).

2. Participative management style or top-down bottom-up management. This seems to be basic Policy Deployment. From disorganised to interactive, crossfunctional involvement at all levels.

3. Teamworking on improvement. See section on Continual Improvement.

4. Overproduction, reduced inventory and reduced lead-time. See Muda and Time Based Competitiveness.

5. Changeover Reduction (see section of same title).

6. Continuous improvement at the workplace (see Gemba, Improvement).

7. 'Zero monitoring' (see Cell balancing, and 'autonomation').

8. Process, cellular manufacturing (see Cells, Kanban).

9. Maintenance (see TPM).

10. Disciplined, rhythmic working (see cell balance, and heijunka).

11. Defects (see Quality, pokayoke). From inspectors being responsible for defect detection, to process control, prevention, operator responsibility, and pokayoke.

12. Supplier Partnership (see Supplier Development, Supplier Associations).

13. Waste Identification and elimination (see Wastes, Mapping).

14. Worker empowerment and training (see People).

15. Cross-functional working, multiskilling (see People and implementation).

16. Scheduling (see Flow, Pull, Regularity).

17. Efficiency (see Kaizen and performance measurement).

18. Technology and microprocessors. (see Thinking Small).

19. Conserving energy and materials (see Wastes).

20. Appropriate Site Technology and Concurrent Engineering (see New Product Introduction).

According to Kobayashi, keys 1 to 4 are the basics, the starting point. These lead on to keys 5 to 20 which work together as a set. However key 4 is critical to time, key 11 to quality, and keys 6 and 19 to cost. Each of these latter 4 keys requires development in all the other keys to become fully effective. Kobayashi also uses the 'bean sprout analogy' which is that, in a field of bean sprouts no one sprout can grow much ahead of the others or else it will blow down. Similarly in Lean, development must take place in areas approximately equally.

Some companies, for example Arvin Meritor, use the 20 Keys not only as an assessment tool but also as a main platform for development. Their sites are continually striving to reach the next key stage in each category. In this regard Kobayashi's book is very useful.

7.5.3   *Rapid Plant Assessment*

R Eugene Goodson has developed a useful plant assessment tool that is aimed at more effective benchmarking and assessment of supplier plants. So when you go on a plant tour, have your team assess the visit immediately afterwards using the Goodson methodology. It can be used in your own plant, but that was not the original idea. It is very useful to use one of these assessment tools when you visit a plant on a benchmarking visit. Turn 'industrial tourism' into a much more focused activity!

7.5.4   *The Shingo Prize*

Established in 1988, the Shingo Prize is the probably the most widely accepted assessment methodology for Lean. Following a written submission, it is carried out over 2 days on site by a team of trained practitioners, not consultants. Many go in for Shingo not necessarily to gain recognition but for a detailed independent assessment. It is tough and comprehensive. Download the model and guidelines free of charge.

Further Reading on Assessments

Richard Schonberger, *World Class Manufacturing: The Next Decade*, Free Press, New York, 1996

Iwao Kobayashi, *20 Keys to Workplace Improvement*, Revised Edition, Productivity Press, Portland, OR, 1995

R Eugene Goodson, 'Read a Plant - Fast', *Harvard Business Review*, May 2002, pp 105-113

www.shingoprize.org

8 Layout and Cell Design

8.1 Layout, Cell and Line Design, Lean Plant Layout

Lean Layout sets the framework for any Lean transformation. It is important because you may have to live with the results of poor layout day in day out for years.

A general guide to transformation is given in the Lean Frameworks section. This gives the wider context and steps to be followed, including layout considerations. This section deals with specific layout detail issues.

Layout is usually approached as a hierarchy

- Plant Location.
- Area Layout
- Cell Layout
- Socio-Technical considerations
- Workstation Layout

In the table below cross-references to other relevant sections of the book are given. In this section the main focus will be on area layout cell design, with comments on socio-technical considerations of layout and on workstation layout.

Area	Book Sections
Plant Location	Value Stream Economics, The Essential Paretos, Creating the Lean Supply Chain
Area Layout	Essential Paretos, Value Stream Mapping, Scheduling Building Blocks
Cell Layout	The Wastes, Balancing
Socio Technical Considerations	People section
Workstation Layout	Ergonomics

8.2 Major Types of Layout: The Product Process Matrix

In the section on Manufacturing Strategy the Product Process Matrix was discussed. This is highly relevant to layout, the thesis being that a major influence on layout is volume and repetitiveness. The major types of layout are shown in the table on the next page.

Before you start....

An opportunity to do a major layout is an opportunity not to be missed. It will help or hinder your Lean direction for a long time. So, visit extensively. Read all you can. And think. Is this an opportunity to redefine the business? For example, you could compete on lead time instead of cost – like some in-mall opticians that deliver a pair of glasses in one hour. You might rationalise the product line. You may outsource processes that are not key to your business but be careful. See the section on Value Stream Economics. Or you may in-source for lead-time reasons. You may decide to trade inventory for machines – having excess machine capacity allowing quick response and make to order rather than make to stock with bigger inventories. Rajan Suri has pointed out that there are two approaches to cells – the technical and the managerial. The former involve calculations and waste elimination and the latter involve understanding the market, and where competitive advantage lies – for example in lead time, price, or responsiveness. You would be wise to consider both.

	Project	Job Shop	Cell	Line	Flow
Also known as		Process layout		Assembly line	
Examples	Civil engineering, large turbines	Custom manufacture	Component assembly, robotic welding	Car or electronic end item assembly	Chemical works, flows between vessels
Volume	One-off, low	Low, batches	Moderate	Moderate to high	Continuous or batch flow
Traditional Characteristic		Flexible not efficient		Efficient not flexible	
Scheduling	Critical path	MRP, Finite scheduling	Heijunka	Broadcast	Optimization software
Evolution	Lean construction	Modularity, pulse line	Longer cycle teams	Global body line	Smaller vessels; 'base' and 'flex' plants
Issues	Coordination, learning	De-skilling	Boring, repetitive tasks, acceptance from former job shop people	Boring, repetitive tasks	Becoming high tech, skill shortage
Lean challenges	Standard work for repetitive elements	Standard work	Pace of improvement, value streams or linked cells	Mixed model	Down-sizing

8.3 General Layout: Good and Not so Good at the Factory Level.

First, size is important. Schonberger suggests a general cut-off point at around 50k square meters or half million square feet. Why? Because plants above this size are in danger of becoming unfocused. The workforce becomes too large. Lines of communication are stretched. Gemba walk by managers becomes impractical. Of course, there are exceptions, like car plants. But then can the plant be broken down into sub-plants, preferably end-to-end, each with its own order entry, production control, dispatch, meeting areas, and so on? An outstanding example is Freudenberg-NOK.

Bad: Square Functional / Job Shop Layout. In the Lean world traditional functional layout is seldom justifiable. It invariably involves batch and queue, significant transport, and long lead times. Poor quality often accompanies this because of failure to detect problems quickly. Complex scheduling routes and floating bottlenecks are often a feature. Finite scheduling is not the answer- it is a bit like adding insult to injury. Even worse is where this type of layout occurs on multi-floors. Demolish and start again!

A Little Better: Rectangular end-to-end flow, with receiving at one end and dispatch at the other end. Although the main lines may flow well, invariably there are long transport distances to workstations far from the receiving dock.

Sometimes Better: Spine layout is a good choice for fast changing situations. HP are enthusiastic users. Here there is a central material handling spine with cell areas along the spine, sometimes on both sides. Automatic guided vehicles may run along the spine. A warehouse may be situated at one end of the spine. Cells along the spine can be added or subtracted. Two way flow is made possible. Even better is where there are also outside doors giving direct access to the cells. The bad news comes from locked-in material handling along the spine that often involves wasteful travel.

Much Better: Rectangular layout – perhaps on a 60:40 ratio and with numerous delivery doors along one of the longer sides and numerous dispatch areas along the other long side. Direct delivery to cells is made possible. This also allows parallel processing along short, dedicated value streams and the possibility of sharing labour resources between the parallel streams.

Innovative 1: Star-shaped or multiple E shaped building design. The arms of the star are sub-assembly lines but having numerous outside access points. One of the arms is final assembly.

Innovative 2: Assembly at Toyota's Tahara Plant follows a 'doubling back' or spiral concept – starting out and working in. It also incorporates breaks in the line that are there to facilitate 'line stop' creating a buffer between sections so that one section can stop without the whole plant stopping. (How many stops per day? How about over 1000! They stop and correct their problems;

competitors may ship theirs!) This shape allows immediate feedback of problems.

Innovative 3: Volvo's Kalmar plant is arranged in hexagonal areas, one of which is used for team based assembly.

All these cases are even better where suppliers are located on-site right around the parent site.

Modern design incorporates good height and big windows. Floors should be thick enough to support changing machine locations. But also think one-level ergonomic flow – this may mean having to locate the bases of some machines below the nominal floor level so that the level of the workpiece is maintained. On the other hand, tilting or changing the workpiece height for worker access and ergonomics, is also found.

8.4 Material Handing: Good and Not so Good at the Factory Level.

Bad: Long conveyors, especially powered conveyors. Why? They lock-in the waste of movement, and worse they get forgotten about as waste. Because they are barriers around which much travel has to take place. They work subtly against communication and quality. There is always the danger that they become another unofficial store.

Also bad: Forklifts. These may be inevitable depending on the size of the product. But they also subtly encourage material movement in big batches by stillage. They take up space and are dangerous.

Better: FIFO lanes, if short, can be effective in encouraging flow. If too long, they can become inventory traps. A long FIFO lane should always be questioned but, if found to be necessary, one long lane should be replaced by two short side-by-side lanes to reduce double handling. Good FIFO lanes are marked with warning colours to indicate if inventory is building up.

Better: Tugger trains. These go around a regular rout, calling at 'bus stops' at regular intervals. The best are low-bed on which human-movable wheeled containers are placed. Human-movable small containers are good if carried at ergonomically friendly heights. Standard stopping locations should be part of the concept. A balance needs to be struck between the size of the tugger and the frequency of routing.

Much Better: Gravity feed, short conveyors, linking closely located machines. Like FIFO lanes, they should include coloured inventory accumulation warnings or even pokayoke light-beam warnings.

Best: Generally, are hand trolleys. But, these need to be moved by material handler or runner, not by cell operators. They give maximum flexibility, at minimum cost with no risk of breakdown. They encourage small batch flow. Beware, however, of kitting trolleys. Sometimes justifiable, kitting is often double-handling waste.

8.5 Cells

8.5.1 Why Cells?

Cells have become almost universal so there is no need for long debate. Compared with a traditional job shop the advantages are massive reduction in lead time through one-piece flow, big reductions in inventory, simplified control, early identification of quality problems, improved possibilities for job rotation, identification with the item, and volume flexibility by adjusting the number of workers. The type of cell, however, remains an issue. Two basic types: long and thin, with short repetitive work content, or short but fat, with longer work content. Although the former may be more efficient (less inventory stocking locations, less tool variety, less training), it is also more boring so job turnover may be an issue.

8.5.2 A Cell is Not a Cell: The Cell Ideal.

Although cells are widespread, many do not conform to the cell ideal. The ideal cell has one piece flow, good visibility, minimal inventories between stations (not zero), an organisation that matches – cell supervisors and identification of workers with the cell, support functions – quality, maintenance, ideally scheduling – that are focused on the cell not the wider enterprise, supporting supermarkets and runner routes, and incorporated pokayoke devices.

Moving to one piece flow has huge advantages – most dramatically a cut in lead time. Consider 4 machines with a cycle time of 1 minute per part, each sequentially producing a batch of 10 parts. The batch emerges after 40 minutes. With one piece flow, the first part emerges after 4 minutes! The former case is called 'fake flow'. But lead time is only one aspect – dramatic changes in transport (moving away from forklifts?), space, and in early problem detection are also big advantages.

8.5.3 Starting Out on Layout

The starting point for area layout should be the Essential Paretos, as discussed in the section of that name. P-Q Analysis (or Runners, Repeaters, Strangers) gives an initial clue to organisation. When looking at the P-Q profile, routings are also relevant. Hence Hales and Anderson prefer P-Q-R. High volume suggests assembly lines or production line cells. A production line cell is used for one product family and its variants that share common routings. Moderate volumes suggest traditional cells that can cope with a variety of products that share common manufacturing characteristics and most routings. Low volumes suggest functional cells, or a job shop, with like machines grouped together, and some specialised routings or off-cell processes.

Contribution analysis (see the Essential Pareto section) is also important for possible product line rationalisation or design modification. Contribution per bottleneck minute is important for products that share constrained resources. This is particularly important for machine intensive products. You will want to take action on products which make low contribution and which tie up precious bottleneck capacity.

8.5.4 Area or Value Stream Analysis

Grouping parts into cells is a similar but more detailed procedure to Value Stream identification. There are several approaches.

Inspection (or 'Eyeball') Perhaps the most common procedure is to group by inspection or by customer. For example, two cells to make automotive components – one for Ford the other for Toyota. Such cells are often run according to the customer's methodology – FPS for Ford, TPS for Toyota. Alternatively group by product or family that 'everyone knows' has similar characteristics.

Matrix Approach Where routings are complex, draw up a table of products against process steps. Do not include minor process or processes that are visited by all products. Then rearrange into groupings. A simple example:

	1	2	3	4	5
A	X		X		
B		X		X	
C	X		X	X	
D	X	X		X	
E		X			X
F	X				X

This becomes:

	1	3	5	2	4
A	X	X			
E		X	X		
C	X	X			X
F	X		X		
B				X	X
D	X			X	X

Note the outliers, D1 and C4. Can these be re-routed? Common processes (paint lines? tool maintenance?) should then be located to enable flow to be as easy as possible. Likewise supermarkets, which should be grouped together to allow convenient runner or material handling routes.

An analytical approach can be used for really complex situations. One such method is the 'binary ordering algorithm'. For in depth explanation, see Lee Hales or Nicholas.

8.5.5   A Participative Approach

In the sections that follow, it is often useful to break into sub-teams, or at the very least require several alternative layouts to be developed. A variant of this is Toyota's '3P' (Production Preparation Process) where the group is required to come up with 7 alternatives for making a new part. This overlaps with the Design stage and more detail is given in the Design section of this book. Here we are assuming that the way in which the product will be made has been determined, and the remaining issue is layout.

The idea of using more than one team, or of challenging a team to come up with 7 alternatives, is simply to avoid fixation or 'groupthink' in the vital matter of layout. Do involve operators in these exercises – they are the ones who will have to live with the results every day. 'If they plan the battle, they will not battle the plan'.

The procedure involves: clarifying the needs, Developing and weighting the criteria, generating alternatives, comparing the alternatives, using Pugh Analysis, deciding the best alternative

Clarifying the needs involves collecting the customer(s) requirements and the projected volumes over the life of the product family. Flexibility and uncertainty would need consideration – what is the likely volume range, and life, the necessary delivery response time and projection of seasonality and other demand fluctuations. Target cost, quality, and material supply also need to be considered.

A need or issue could include a problem (competitiveness, quality), an opportunity (a new market, or improved visibility), a requirement (safety, legislation), an uncertainty (about demand or prices) or a controversy (internal viewpoints or politics).

Developing and weighting the criteria follows. The team would brainstorm and list the criteria or issues against which alternative layouts could be judged – typically cost, quality, delivery, safety,

flexibility, ergonomics, visibility, and possibly lead time, inventory, skill issues and others. Issues could also include points of annoyance or problems that the team has experienced with the current layout. Weighting the criteria can be done in several ways – simply voting, distributing points, or pairwise comparison where every combination is considered.

Generating the alternatives should be helped by the tools described in the following sections. It is better to do this in small teams of maximum 6 people rather than in one big team – particularly if operators and team leaders are likely to be overawed by industrial engineers or managers. This is an opportunity for involvement that should not be missed. However, perhaps precede the deliberations by a briefing by industrial engineers covering developments in technology and pointing out pitfalls.

Selecting and generating an improved design is by Pugh analysis which is described in a separate section.

8.5.6   Area Analysis by SLP

Systematic Layout Planning (SLP) is a robust procedure developed by Richard Muther Associates, and useful for area layout of offices and factories. It can also be used for micro workstation layout. It is quick and effective. For a more detailed explanation see Lee Hales, Hyer and Wemmerlov, or Tompkins et al. The procedure involves:

- Establishing the desirable closeness relationship between all major sections and departments, according to the vowel sequence AEIOUX – absolute, essential, important, ordinary, unimportant, and undesirable. All the relationships can be shown on a triangle matrix as below.

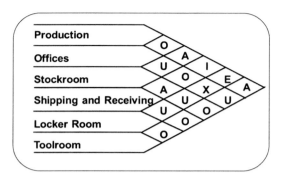

- Then, a 'space relationship diagram' is drawn with an initial layout or the current layout. The desirable closeness from the triangle matrix is drawn in using multiple lines: 4 lines for A, 3 for E, 2 for I, 1 for U, and a jagged line for X as shown below. Then, simply using an inspection approach, the departments are rearranged to minimise the total lengths of lines as shown. Alternatively, the lines can be coloured from red for A, through orange, yellow, black, blue – and rearranged with the idea of making the diagram 'cooler'.

- Lastly, the relative locations are fitted into the actual space available.

The Cell Flow Diagram: Incorporating quantities and routings: Vertical or Horizontal?

Once the broad value streams have been established, another technique derived from Richard Muther Associates is useful for more complex value streams where products or assemblies within the stream vary in quantity and routings. The issue is whether to split the value stream into a series of linked cells, or to have one discontinuous cell. The technique involves drawing a 'cell flow diagram'. This is simply a schematic of the sequence of processes, with the intensity of the volumes of assemblies or components shown by numbers of lines from 4 for high to 1 for low, as shown below.

This useful diagram is a provocative aid. You can see a main line and a support line – but the question remains should this be in one cell, two

parallel cells or a two sub cells leading into one operation.

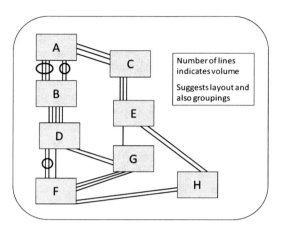

8.5.7   Cell Shape and Flow Direction

There is wide agreement on the virtues of U shaped cells. These include ease of balancing, improved communication and feedback for quality and other issues, improved visibility, and ease of control where one operator does the first and last operation.

But a cell need not be U shaped. Straight-through has flexibility and material handling advantages particularly for large artefacts. L shape may be chosen for storage considerations.

Most, but notably Toyota, think that flows around a cell should be anti-clockwise to facilitate ease of movement with the right hand. It is also thought to be the natural way – think of athletic tracks, dog racing, and horse racing. Even the planets (except Venus!) move anti-clockwise.

8.5.8   Other General Points

- The grouping of inventory into supermarkets is critical for Lean. Supermarkets provide the basic framework. They should be stable, while cells come and go. Try to have a few supermarkets, rather than inventory all over the place in little stores or in one central warehouse. See the Lean Scheduling Building Blocks section.

- Break up the value steam map into pull segments or loops, often separated by supermarkets. Then use these as building blocks for layout design.

- Avoid one big warehouse – especially an automated warehouse or automatic storage and retrieval system. There will always be a temptation to fill it. If you have an AS/RS already, set in place a plan to reduce usage then close it down.

- Establish a series of specific 'waterspider' routes – with material handlers making regular circuits. This powerful concept paces the work and flows information regularly via pull systems.

- Think three dimensions. Can deliveries be made from below or above? For example, can plastic be fed into injection moulding machines from the floor below thereby allowing a high state of cleanliness and separation of operators and forklifts?

- Don't get hooked on using old facilities. Better to demolish and move. Costs will quickly be recovered. Like Dell. Many plants grow like topsy, locating new work in any available space. They pay the price over many years.

- Locate design and engineering areas close to manufacturing. Make them share common break areas. Even better make engineers walk past production areas to get to their work area.

- Locate production control in the middle of the plant floor. If possible, managers' offices also. Don't make supervisors' offices too comfortable.

- Foster communication and visibility in the office by open plan layout, and common meeting areas

- Share information by visual displays. One idea is to have a data projector permanently on display and linked to current performance status and company news. Gather around Communication Boards. (See Manager Standard Work.)

8.5.9 Operator Considerations

There seems to be a slow but steady movement away from short cycle jobs towards longer, more interesting jobs. In other words there is a move away from long-thin layout to short-fat layout, or from one single line with short cycle tasks and takt time to several parallel lines with longer cycles and takt time. Job rotation is one solution for repetitive boring jobs, but is only a marginal improvement. Volvo tried parallel group assembly for staff turnover reasons. Staff turnover did reduce, but the additional problems were multiple inventory locations, hence more complex transport, multiple tooling, and group work norms some of which were high but others low. Training is also an issue. But this experience is no reason to reject longer cycle work. Longer cycle work is potentially more efficient because there is less balancing loss with fewer, longer cycles. So, what can be done to get the 'best of both worlds'?

- Group operators into tight groups, perhaps having several cells share a common supermarket. Group takt = takt time x number in the group.

- Give each operator a long cycle job (like assembling a complete photocopier) but get the group to work in parallel but start and stop together. They pace one another.

- This is often better than the whole group assembling one copier at a single location, with reduced cycle time, because they get in each other's way (called work interference).

- Get the operators to govern their own rate over (say) half a day, as long as they meet target by the end.

There is much to be said for combining operations rather than specialisation in a series. Think about a supermarket checkout. Would you rather progress through a series of checkouts, each one specialising: the first on fruit and vegetables, the second on drinks, the third on dairy etc. Life would be a pain!

8.6   Cell Balancing

Cell or line balancing is best done by operators themselves working with industrial engineers. Of course, for participation, there must be no threat of job loss. The steps are:

- Establish the takt time – derived from customer demand and work hours. Note that demand maybe able to be levelled, and work hours changed to stablise the takt time.

- Balance for different target rates. A good Lean cell should be able to change easily from one rate to another. Perhaps a rate change is done within a day (morning and afternoon?) or across a week.

- Establish the target cycle time. This will be less than takt time – normally 90% of takt, lower for high variation work, higher for very low variation work. If you balance to 100% of takt, one little problem and you will miss the target. See the Lead Time / utilisation graph, in the Principles section. It is a good idea to think in terms of several cycle times – to cope with different levels of demand.

- Remove obvious wastes and establish good practice. Do this without timing. If it is an existing cell, make a video and ask operators to critique. Look out for movement wastes, poor ergonomics, signage, and inventory footprinting. If there are multiple shifts, compare the same operation across all shifts with operators from each. Ask the operators to come up with the 'best of the best' or the ideal way.

- Consider the best layout as if one operator was running the cell. This will give the best layout.

- Then establish work element times. Ideally get operators to work out the times themselves. Use a video (?). Time at least 10 cycles. Get the operators to select an appropriate cycle time for each work element. This will not be the average, minimum, or longest time but at the pace for a good, steady rate. We are not talking work study here; that is anathema to many people. Generally synthetic time standards such as MTM are not satisfactory – not because they give the wrong answers but because they send out the wrong message. Don't add in an allowance factor. Rather factor breaks into the takt time calculation.

- Be particularly aware of non-repetitive work. For example:

  - Does an operator go to fetch parts every so often?

  - Are there interruptions? Why?

  - What documentation or records are needed?

  - Is there a break in the normal rhythm of work? Why?

  - What happens when material is delivered? Is there a break to sort out documentation, orientation, container placement, etc.

  - In all these cases, what can be done to make the work smoother and less wasteful?

- Inventory in a Cell. Remember the ideal is one-piece flow. A cell is not a cell if batches continue to be run. That is, if there is a batch between each stage. Merely changing the locations of machines to create a U shape does not constitute cell manufacture. If batches continue to be run, it is called 'fake flow'. However, although one-piece flow and not batches is the way to go, this does not mean having one piece per workstation! At least initially. Start with a few parts between stations as a buffer, and then remove them as flow becomes more established and problems are ironed out.

- Focus on detailed work station ergonomics and movement. Operators should stand and move, not sit, except for accurate work and hand assembly. Avoid the need to bend or reach. Use standard laws of ergonomics.

- A word on automation: Lean is not anti-automation in cells, but is cautious about it. Good reasons for automation are quality and 'dull, dirty, dangerous' or 'hot heavy hazardous'. A bad reason is to cut staff. Firstly robots don't improve. Don't lock in waste. Automation is not as flexible as people. Automating to allow machines to run unattended is good. So is auto-eject.

- An activity sample can be a useful supplementary exercise to collect data on wastes that may be occurring. See section below.

- After all the work time elements, value add and non value add, have been accumulated, the *approximate number of operators needed can be calculated from the sum of all the operator work elements divided by the target cycle time.* This is the calculation for manual assembly work. Where there are machines, a machine cycle may govern if the machine cycle plus load and unload times is greater than the required cell cycle time. In this case two parallel machines or even complete parallel cells may be required. Where a machine cycle is less than the required cell cycle time, normally, activities can be found for the operator to do while the machine is running. You want to avoid an operator standing and watching a machine run through a cycle. Sometimes this is unavoidable for part of the machine cycle and, of course, time allowance must then be made. This is where the Work Combination chart is particularly useful. See later.

- If there is a very long cycle work element in the process, say a batch process or where work needs to go out for a subcontracted stage, it is still possible to run a one-piece flow cell. Here, simply have an output and an input buffer at that stage but otherwise run the cell as usual.

- Think whether the cell or team leader is included in the required number or acts in a support role. On a complex inter-dependant assembly line it is a good idea to have the team leader as a floater – to assist with problems and to cover for short absenteeism (minutes for toilet) or longer term (days for leave). The size of each team will then be a consideration – Toyota uses around 6 on its lines, more in less complex areas. In a less tight cell, the team leader may well be one of the regular operators.

- Make up a Yamazumi or work balance board. Preferably use magnetic strips for the work elements cut to scale to represent the work element times. Green for value add work, red for all others. The strips are then fitted in and accumulated on the board up to the cycle time line to represent the work of each operator. See the figure on page 133.

- Repeat the cell balance exercise for the appropriate number of work rates, or takt times. As different numbers of operators will be used, it may be necessary to incorporate additional buffers between operator routes. Also, work out the changeover procedure from one line rate to another. Quick changeover principles are relevant here. A hint: when adding a new operator to a cell, add him in the middle, not at the beginning. The latter will cause increased variation in output.

- Decide on the standard inventory quantities, containers and footprinting.

- Incorporate 'pokayoke' failsafe methods where possible. This is not only for quality, but to maintain flow.

- Decide on the 'what if' or andon signals and communications that will be needed. Establish the procedure for what happens when there is any sort of problem.

- Establish the production control system for the cell. Maybe Heijunka (see separate section) or a work by the hour board. In any case, include a conveniently located problem board where the reasons for failures to attain the required rate are noted as they occur. This, of course, is not 'blame and punishment' but problem surfacing. Far from blame, operators should be commended for writing perceptive reasons on the board.

- Establish the start-up procedures and checks for the beginning of the shift. Remember to deduct this time from the takt time calculation. Likewise, if there are regular maintenance or check activities that need to be performed after every (say) 10,000 cycles, establish these procedures and allow for time for them.

- Prepare standard work charts – the Work Combination Chart and the Cell Layout chart, completes the design. See page 132. The best people to prepare these are the cell operators working in conjunction with industrial engineers. In fact, having operators participate in the preparation of standard work encourages them to question work methods and helps with sustainability. The work combination chart is a Gantt-type chart showing the sequence of activities and times that each operator follows. Note that movement activities are not recorded but shown as wavy lines connecting the activities. The cell layout chart shows the geography or plan view of the cell, the routes that each operator follows in the work sequence, and very importantly the locations and quantities of the standard inventories. You will notice that the example shows operators moving back and forth. This is not desirable if operators can walk around the cell in a circle, either in the direction of material flow, or in a direction opposite to flow.

- Then implement and verify the standard work charts. There are two main reasons why standard work is not being followed: it cannot be followed because it is wrong, or a better, easier way has been found. Both are reasons for putting a new standard in place.

- For a new cell, do a 'cardboard kaizen'. Here, use full scale cardboard boxes to represent the machines and have operators walk around as if they were running the cell. Make adjustments. This is PDCA.

8.6.1   Activity Sampling

Activity Sampling is a quick, effective way to establish data on wastes and how people, or machines, are spending their day. Many times,

people don't know this about themselves! The method involves taking perhaps 250 random observations of each operator in the area over a representative period of time, perhaps a week, certainly a full day. If days are the same, sample a day. If days are different, sample over a week or longer.

There are formulas for calculating the number of observations to yield a certain desired confidence level, but generally 250 observations is satisfactory to gain a good impression. A random observation is just that – of course the observations must be spread out over the day or week. Decide how many observations to take in a day, and spread the observations more or less (but not exactly) evenly across the day. One way of avoiding bias is for the observer to turn his back on the subject and count down from say 20. Then turn and observe in that instant. With each observation note down what that person is doing at that instant – either value adding or non value adding by type – for instance walking, recording, watching, talking, etc. Then simply calculate the percentage in each category.

However, this should not constitute 'spying' or subversive data collection but rather an analysis activity aimed at improvement. It is best done by operators themselves. It must be explained to those being observed.

Process Stage	Value Adding	Waste Type					
		Wait	*Move*	*Insp*			
Inject	┼┼┼ I	┼┼┼ II	┼┼┼ ┼┼┼ ┼┼┼ III	II			

8.6.2   A Note on Work Element Times and Variation

Most time studies of repetitive manual work reveal that the distribution of times for a work element is skewed to the right – that is with a short tail to the distribution on the left and a long tail on the right. It is also similar to the Poisson distribution observed in many service transactions – most transactions take a short time but a few take a very long time. It is the long tail on the right that is the 'killer' for line balancing. The long tail is usually caused by quality or part problems rather than operator problems. If variation can be reduced – to cut the tail, then (strangely?) the average time for a job can be increased. This is almost a feedback loop – take longer over a work element and thereby reduce problems in that or subsequent elements, which in turn means that time for those elements can be increased! Of course, there are limits to this. It is a consequence of the non linear queuing curve discussed under Muri and Mura. It is also common experience when sitting in a hold up on the highway and finding that the 'fast' lane has become the slow lane. The 'fast' lane has greater variation. Think about the tortoise and the hare.

8.6.3   Cell Balancing Alternatives

- **CONWIP.** Not actually a balancing method, but CONWIP (constant work in progress) can be highly effective. CONWIP methodology is simply to let in one piece (or container or hours-worth) of work for each one that is let out. This, of course, stabilises the lead time and helps surface problems. It is ideal where there is a U shape and where the same operator controls the entry and exit points. What is needed is first to calculate the required number of operators for the cell using the methods given above. Then allow operators to self balance the work. The inventory in the cell can ebb and flow at any particular workstation, but naturally tends to accumulate at the longest cycle operation. This highlights the issue. Start 'loose' with quite generous inventory and gradually reduce it.

- **Drum Buffer Rope (DBR).** Similar to CONWIP, but used where there is a clear 'bottleneck' (or longer cycle time) stage. Here what gets let out of the bottleneck is what is let in at the first operation – the rope. The buffer inventory protects the bottleneck from running out of work. The buffer may be a time buffer (x minutes of work) where different products are made. No other buffers are necessary. Again, operators can arrange themselves around the cell but need to know that the bottleneck is the critical resource that needs to keep working.

- **Bucket Brigade.** This method of balancing is suitable where all operators know all the tasks. First, calculate the required number of operators as outlined above. Then line them up at the beginning of the cell – say three people in front of the first three tasks. Operator 1 starts work at workstation 1, then passes the piece to operator 2, who passes it to operator 3 for the third task. Operator 3 then progresses the work through all the remaining tasks until she completes the last task. She then starts walking back. In the meantime Operators 1 and 2 have been working on the second piece. When Operator 2 completes task 2, she moves to task 3 and so on until she meets Operator 3 walking back. Operator 3 then takes over and completes the remaining tasks, and again walks back. Operator 2 meanwhile turns around and walks back until she meets Operator 1 who is progressing the third piece. They meet, and operator 2 takes over piece 3 until she meets Operator 3. Operator 1 walks back and then begins piece 4. This is carried on until the cell settles down to the natural sequence of work. No detailed balancing or Yamuzumi board is needed.

Standard Operations Combination Chart

Process:	Block assembly	Required output		Takt Time:		Operational Takt (90%):		Date:	Mar-05
Part Name:	R2D2-block	1,160 units/day			55 secs		49 secs	Updated By:	COR

TIME (secs) — MANUAL / AUTO. / WALK

(● ● ● Idle)   Operation time shown in one-second units

Scale markings: 10, 20, 30, 40, 49, 50, 55, 60, 70, 80, 90, 100, 110, 120

OPERATOR	Sequence	WORK ELEMENT OR DESCRIPTION	VA	NVA	AUTO.	WALK
1	1	Pick-up @ store "a"	2			
	2	Unload/oad "B"		10	35	2
	3	Unoad/Load "Test"		10	35	5
	4	Unload/Load "C"		10	35	2
	5	Drop-off @ store "b"				2
		Return to store "a"				6
		Total		30		17
2	1	Pick-up @ store "b"	5			
	2	Unload/Load "D"		10	35	2
	3	Unload/Load "E"		10	35	2
	4	Unload/Load "F"		10	35	2
	5	Drop-off @ store "c"				2
		Return to store "b"				6
		Total		30		14
3	1	Pick-up @ store "c"	5			
	2	Unload/Load "G"		10	35	2
	3	Drop-off @ end of cell				2
	4	Pick-up @ start of cell				8
	5	Unload/Load "A"		10	35	2
	6	Drop-off @ store "a"				2
		Return to store "c"				8
		Total		20		24

Standard Operations Chart

Line Name:
PG U-shaped cell

Standard WIP:
11

Symbols:
Standard WIP △
Quality checkpoint ☆
Safety checkpoint: ✚

Quality/Safety Points:
Operator 1:
Step 2: Check manifold
Step 5: Avoid touching rod (hot)
Operator 2:
Step 2: Check tolerance
Step 6: Inspect sleeve

FINISHED GOODS

RAW MATERIALS

8.6.4 Balancing Mixed Model and Multi Model Lines or Cells

Balancing complex mixed model lines can be a very complex business! Mixed model interchanges products on the same line. Multi model lines usually run in small batches on the same line but with changeover operations between the various products. Each type can be paced unbuffered (like a mechanised line) or unpaced buffered (with operators moving pieces between stations). Paced unbuffered balancing can be calculated deterministically, other typyes generally require simulation. Readers are referred to the 'bible' and 'guru' on this topic, Armin Scholl.

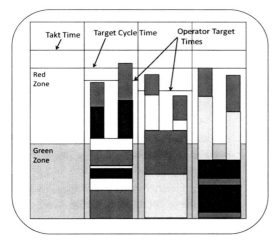

A simple case is illustrated in the figure, which shows a line balance board for a mixed model line with two products. Time is on the vertical axis. First, takt time is established. Then the target cycle time is set below the takt time to allow for general operator variation. A third line is the operator target times. These individual station times allow for relative complexity and uncertainty at each workstation. In the example, operator B has a much more difficult or variable task than the other operators. In the Green Zone activity times for the common elements are assembled. These activities are done for every product. Of course the Green Zone is variable for each workstation, so the zone limits are

approximate. In the Red Zone the work elements unique to each product are accumulated. In some cases, for instance A, a total product assembly time can exceed the operator target time, so long as the weighted average time reflecting the product mix, does not exceed the target time.

8.7 Chaku-Chaku Cell or Line

A chaku-chaku or load-load line, such as at Boeing, is the name given to a very compact and partly automated cell. It invariably includes one piece flow, automatic load and unload, and multiple pokayoke devices. The various machines are linked by gravity conveyors or chutes. Very often a chaku-chaku is used by one operator on an as-and-when needed basis, such as when feeding into a larger assembly.

8.8 Virtual Cells

Sometimes it is impossible to create a cell in one area due to size or environmental conditions such as a clean room. Stages may have to be separated. In this case a virtual cell may be a possibility. Instead of two areas with say four similar machines each being managed as a process job shop, consider changing to four virtual cells of two different machines each, managed as four distinct lines or cells. Operators move with the parts from area to area without intermediate buffers, so creating the effect of one piece or small-batch flow. The advantage is vastly reduced lead time and reduced scheduling complexity, against the penalty of greater transport and the need for cross training. This is really creating and running a value stream but in a traditional type of layout.

Operators identify with the line rather than with the process job shop. In the simple case cited, each operator would have to have the skills to run both types of machine and would move from one area to the other 'flowing' the product one piece at a time as far as possible. The old way would involve batch and queue; the new way would frequently involve setting up both machines as a flow line, albeit in separate locations.

8.9 Moving Lines and Pulse Lines

Henry Ford's original line was a 'pulse' line – in other words cars spent time at a fixed location before moving on to the next fixed location. This principle is now being rediscovered for a variety of large, slow moving, complex items such as aircraft engines, wings, aircraft and vehicle maintenance, remanufacturing, large transformers, electrical switching gear, earthmoving equipment, and ship and boat sections. Moreover, the principles may be applicable in areas as diverse as hospitals, construction, even education. The Lean principles of (relatively) fast, flexible flow fully apply. As with much of Lean a big issue is believing that a moving is possible in the first place – traditional ways (batch and queue, project management, complex scheduling, bottlenecks etc.) have been in place for decades.

As Henry Ford found a century ago, such lines are a revolution in productivity when compared with static build. As well as huge productivity and time gains, there is invariably a big reduction in space, a big improvement in quality (through improved standardisation and visibility), and big gains in training and apprenticeship. Historically, the American railroads were built with a type of moving line system, progressing 50 miles per day, and today track maintenance is just beginning to adopt moving line concepts.

for shorter station cycle times – say several hours. A moving line moves very slowly (perhaps in mm per minute) but continuously using a track or conveyor. One or several products are on the moving line at a time, depending on complexity. A pulse line uses a platform, such as 'hovercraft' cushion, to move between fixed stations at a regular takt time. Typically a small number of items are on the pulse line at a time.

Pulse or moving lines may be fed by supporting cells from which parts are pulled, or automotive style using a broadcast schedule to synchronise several lines.

The steps to set up a pulse or moving line are broadly the same as those used for a cell. Many of the steps are similar to the previous section. A few differences are given below.

1. Establish the product families. A line can be used for a class of products, like helicopters, even though there may be considerable customisation between individual units, but not for mixed products such as helicopters and aircraft.

2. Calculate takt time. This will determine the number of stations in a pulse line, and the total time in a moving line.

3. Develop standard work packages. Even in high variety lines there will be much fully standard work, and some semi-standard as for example in maintenance. Identify, time, and document. These are the essential 'Lego bricks' for such lines.

4. Accumulate and determine the standard work packages against the takt time. In moving lines, first determine the number of operators as for cells, but bearing in mind simultaneous operations, then balance the operators' work against the takt time. In pulse lines determine the number of operators per station – this will depend on simultaneous operations as well as technical considerations. In both cases, earlier stations could be more fully loaded whilst later stations are more lightly loaded to

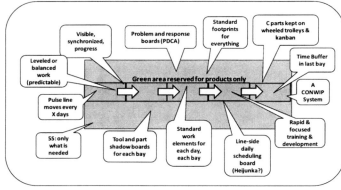

A pulse line is used where station cycle times are long – say several days, and a moving line is used

retain catch-up capability. Make allowance for uncertainty and complexity.

5. Establish standard locations and footprints for tools, and part trolleys. Each station should have its own shadow board. This is one of the great advantages of such lines, so pay attention to waste-free ergonomic micro-layouts. Get operators to participate in developing their own part and equipment handling systems. Keep frequently used tools and parts line-side. Get a 5S system established. Over time, work on rationalising tools and parts.

6. Establish pull systems for required parts. Try to pull as much as possible. Use the runners, repeaters, strangers / ABC classification system (see separate section). Establish priority kanban systems and supermarkets for supporting cells feeding the line. A and B parts should be stored on specifically designed wheeled trolleys, and moved to the exact location just-in-time.

7. Establish a progress signalling system. Visibility is another great advantage of lines, so capitalise on this aspect. In a moving line signalling is typically by marks on the floor corresponding to time and a light system so operators can report and display completions. In a pulse line, each day's work standard work elements are loaded via cards on a Heijunka-like board, and turned around when complete. In both types, it is necessary to pre-consider contingencies resulting in delays – transfer to next day or station (limited possibility in a moving line because downstream work location layouts will be unmatched), floating labour, stop the line, work overtime (not desirable?). Develop a board on which unforeseen problems are displayed, and an action sequence determined – perhaps like a TPM red card system.

8. Establish the planning system. For mixed model or variable time lines such as maintenance, the work packages in each cycle may vary. The work packages and manning will then have to be pre-planned subject to the constraint of the takt time. This planning is probably best done by a Heijunka-like manual capacity board that loads up the individual standard work elements.

Big and exciting opportunities lie in this concept being applied in maintenance, hospitals, and construction. 100 Years after Henry Ford, we are only just beginning.

8.10 Ergonomics

Good ergonomics – of both products and processes – should be essential for any manufacturer – Lean or otherwise. What makes Lean Ergonomics an extension to conventional ergonomics? This brief section does not deal with ergonomics per se (there are many excellent texts available), but comments on the Lean perspective on Ergonomics.

- Working to Takt or Rhythm. A regular rhythm can assist good blood circulation. By contrast, 'static' effort (for instance where a moderate rate of work persists for 1 minute or more, or slight effort lasts for 5 minutes or more – (Kroemer and Grandjean 1997) can obstruct the flow of blood. But beware of RSI.

- Lean favours standing rather than sitting (except for intricate work) – for flexibility to move between workstations, but also for posture and to help avoid lower back problems. Ergonomists recommend a combination with a predominance of sitting. Certainly standing or sitting without movement is poor practice. Sitting can lock-in an operator and inhibit movement. A good compromise is to have standing and moving operators but also frequent breaks with comfortable chairs in team areas. This fits in with 'standard work rate or stop' philosophy discussed under balancing. Some sitting can be accommodated in lines and cells – as per Toyota's 'Raku Raku' seats which swing inside a car to allow a sitting operator to assemble.

Some cell workstations are amenable to the best of all – allowing the operator to sit or stand.

- There is an inverse relationship between force exerted and duration of muscular contraction (see Kroemer and Grandjean, p 11). Toyota has developed an ergonomic evaluation system based on this relationship – a maximum force for each particular duration – exceeding the line calls for workstation redesign.

- The best workstations, both sitting and standing, allow for height adjustment both for the height of the operator and the type of work (higher for accurate, lower for heavier). Seats should be adjustable for height and backrest inclination. Look up the recommended heights of seats, work surfaces, and inspection surfaces for your own size operators in an ergonomic text. See readings.

- 5S. Take the opportunity to do 'Ergonomic 5S', not just 5S. Shadow boards for tools and parts need to be correctly located ergonomically. Try to maintain a natural posture at all times. The 5S principle of avoiding personal toolboxes can make sense both ergonomically (located at correct height and reach) and for standardisation reasons. Every course on 5S should at least say a few words on work heights and ergonomic workstation layout, lifting, lighting, controls, vibration and noise. Visibility principles should extend to ergonomics – for example seeing the progress of a moving line clearly marked by lights or time markings on the floor. Can a tool shuttle be used on a line to keep pace with the line?

- TPM and Quality related gages, dials and displays should also be designed using ergonomic principles. Operating ranges should be colour marked so normal conditions can be seen at a glance, lubrication levels made visible without bending, needle orientation aligned on dials, etc. (See Kroemer and Grandjean, chapter 8)

- There are many visual warning devices. An example is coloured stickers placed on all containers to indicate if they are human movable, human movable with care, or only machine movable.

Further Reading

Michael Baudin, *Working with Machines*, Productivity Press, 2007, and *Lean Assembly*, Productivity Press, 2002

Nancy Hyer and Urban Wemmerlov, *Reorganizing the Factory: Competing through Cellular Manufacturing*, Productivity, 2002

James Tompkins et al, *Facilities Planning*, Third edition, Wiley, 2003

H Lee Hales and Bruce Andersen, *Planning Manufacturing Cells*, SME, 2002

Jeff Schaller, 'Standard Work sustains Lean and continued success at Wiremold', *Target*, first quarter, 2002, 43-49

Mike Rother and Rick Harris, *Creating Continuous Flow*, LEI, 2001

Kevin Duggan, *Creating Mixed Model Value Streams*, Productivity, 2002

Armin Scholl and Christian Becker, 'State of the Art exact and heurustic solution procedures for simple assembly line balancing', *European Jnl of Operational Research*, 168 (3), 2006

Richard Schonberger, *Best Practices in Lean Six Sigma Process Improvement*, John Wiley, 2007.

K. Kroemer and E. Grandjean, *Fitting the Task to the Human*, 5th edn., Taylor and Francis, 1997

Jan Dul and Bernard Weerdmeester, *Ergonomics for Beginners*, Second edition, Taylor and Francis, 2001

John Nicholas, *Competitive Manufacturing Management*, McGraw Hill, 1998, Chapters 9 & 10

9 Scheduling

Scheduling is at the heart of Lean. All the other tools described in the book, can be seen as contributing to better schedule performance. Scheduling directly impacts lead time, delivery performance, cost and quality. Yet for many managers, incredibly, scheduling is not high priority. Nor is the position of scheduler or master scheduler a high-status job.

In some organisations, Lean concepts like 5S, kaizen, waste, standardisation, or Six Sigma, have become so prominent in the minds of senior management that manufacturing planning systems and procedures have been ignored. They are 'less sexy'. This is a serious mistake.

Some pointers or warnings:

- The **purpose** of the manufacturing planning and control function should evolve as Lean is implemented. There are three forms of evolution. First, the planning department should be less and less involved with execution and monitoring. Planning, though, remains important. Second, the role should evolve towards analysis and decision support rather than routine planning. Exactly the same comments apply to Accounting. Decision support means appropriate advice on current bottlenecks, appropriate buffer and safety stocks, changeover priorities, maintenance priorities, appropriate kaizen activity, and advice on level versus chase demand policy. Third, it should advise on appropriate supply chain decisions – where, when and how much to position inventory. All this means a much more high-level thoughtful role. Planning should in no way be threatened by Lean – quite the reverse. Only the best people will do.

- **Bills of Materials accuracy and structuring** is fundamental. Without good BOMs, material requirement planning – whether via an MRP system or not – is difficult or impossible. This information is needed for sizing of supermarkets and kanban quantities, and for planning runner routes. If BOMs are inaccurate, incorrect parts will be ordered and Lean will fail. BOMs should be restructured to reduce the number of levels as cells are introduced. Planning and modular bills, though emanating from classic MRP practice, are extremely useful in Lean

- **Safety Stocks.** When was the last time safety and buffer stocks were reviewed? Simply, whenever demand changes, or changeover is reduced, or lead times change, or OEE improves. In short, in a Lean programme, continually!

- **Batch Sizes.** Same comments as for safety stocks.

- **Routing files.** OK, you hopefully have reorganised into value streams and simplified your routings. But it would be comparatively rare to have done so throughout. Routings are essential for cell design, for capacity analysis, and for resource planning. If your routing file deteriorates, so will your Lean programme.

- **Organisation of Planning.** As Lean is introduced, planning should become more decentralised. Has this happened?

- **MRPII and ERP** have (often rightly) come in for some stick from Lean practitioners. But be careful of throwing the baby out with the bathwater. Although MRPII is bad news for execution, and sometimes bad news for planning – as Lean is introduced – it is very often an essential tool for analysis. If there are complex flows, shared resources, and changing demands and mix, how do you analyse where your bottlenecks are, how do you know your loadings (we know how critical this can be!), and how do you plan your future resources? The basic MRPII or ERP engine should be able to give you this information. In fact, this is probably its major purpose in a Lean implementation. Once you have got the information, only then can you begin analysis and improvement.

Sometimes the Production Planning office is not seen as having a part in Lean. 'Lean is for the factory floor but we need to get on with Planning' What a mistake to make! Visual management boards and improvement events, including cyclic reviews of batch sizes, safety inventories, planning lead time assumptions, lead time verification, and reviews of planning measures should be integral to Lean. In some enlightened companies, the best and brightest spend a period in manufacturing planning and control. It was not for nothing that George Plossl, manufacturing planning and control guru, made statements like 'The Master Schedule should be management's handle on the business' and 'lead times are what you say they are'.

Scheduling is the apex of 'System'. For a good, level schedule, many aspects have to come together in harmony – quality, standards, pull systems, delivery, material handling, and of course people aspects.

This chapter will begin with a section on the importance of the level schedule concept in Lean, and then progress to some frameworks for allowing a Lean schedule to work well. These are the building blocks and Lean scheduling concepts.

9.1   The Level Schedule

Several authors, but notably Richard Schonberger, have made a case for level scheduling to be the focus of any Lean implementation. This notion is directly compatible with Mackles' ideas on 'Creating Flow' as discussed in the Frameworks section of the book.

Why is a level schedule such a worthwhile aim, such a potent driving force?

- Muri and Mura are often cited as the root causes of Muda. Mura (unevenness) is directly related to level scheduling. Get the flow as even as possible – avoid waves and lumpy overproduction, followed by troughs of underproduction. Muri (overburden) causes instability and leads to instable schedules. Recall the exponential graph of queue or delay against utilisation: as utilisation or load on a resource increases

towards 100% of capacity, so delays build exponentially and the uncertainty schedule attainment increases.

- Level scheduling extends to suppliers and to customers – end to end along the supply chain. Suppliers appreciate regular orders and delivery schedules – this enables them to be more Lean themselves. Likewise if your customer places regular orders on you, your Lean programme is facilitated.

- The last point has a significant impact on inventory levels.

- The level scheduling principle is a challenge to marketing and distribution. What is causing the un-level schedule? Do customers really want large batches or would they really prefer small batch regularity? And, what is driving this behaviour – Quantity discounts? End of month reporting? MRP batching? Long changeovers? Get after the root cause of why the schedule is not level.

- Internally, delivery schedules are smoothed. A tugger route and bus stop system can be set up. Material handling is much more efficient. Perhaps you can ditch those fork lift trucks.

- Fast problem identification is facilitated by level schedules. As soon as a schedule deviates from plan it is noticed, and the root cause sought.

- If you have the level schedule ideal – what are the implications for layout, the small machine concept, value streams, and shared resources? You want to avoid bottlenecks and complex schedules, so set up clear value streams.

- The last point relates back to Spear and Bowen's 'DNA' points (see Philosophy section), especially the aim of having one clear communication path, and one clear unambiguous flow path.

- Internally, level scheduling is exemplified by mixed model scheduling. Small batches more frequently, rather than big batches less frequently.

- A basic concept is runners, repeaters, strangers analysis. Make the best use possible of regularity – repeaters. Make each day or week as similar as possible to the last one. Everyone works better with no surprises.

- And how about level schedules for management? This is the essence of what David Mann says about Lean Leadership – standard work for managers with meetings held at standard, regular times. In short, predictable level schedules.

- In Design, Intel and others level their Design process, bringing out new generations of chips on a regular cycle.

To summarise, the quality guru Phil Crosby talked about 'Ballet not Hockey'. Ballet style is the predictable, regular, level schedule. Getting it right and repeating it, as often as possible. Hockey (or football) style is the opposite. More fun, but the moves are not predictable. Every day is 'a complete new adventure'. And 'the same old problem is solved for the 500th time' as Crosby said. Not very good management or use of managers time. So, which are you?

9.1.1   *Achieving a Level Schedule*

Of course, a level schedule does not happen by accident. Demand must be worked on to allow a level schedule. This is discussed in later sections on Demand management and Sales, Operations and Purchasing Planning (SO&PP). But begin with the challenge that much underlying end-customer demand is level and / or predictable, not lumpy. For example the demand for many everyday items from tyres to toothpaste, many medicines, many staple foodstuffs, is level at the usage level. Why then does demand become so distorted? Yes, there are likely to be deliberate distortions – such as promotional activities, but there may also be causes that simply amount to unthinking actions, like 'getting around to it' type buying, or 'making a few more just in case'. These should be reduced – especially where there are cash advantages to so doing.

9.1.2   *Shared Resources and Lean Schedules*

Of course, many times a resource (machine or person) cannot be devoted to a value stream. But recognise that shared resources lead directly to complex schedules that in turn frequently lead to missed deliveries, longer lead times, and reduced quality. The management control system, including structures and coordination meetings is made more complex. So a shared resource is a 'monument'. There is a penalty that must always be paid by a shared resource. That penalty includes the factors just mentioned – but not least the cost of a more complex scheduling system and schedulers. Frequently these factors are not included when considering whether a resource should be shared between value streams. Untangle those spaghettis and seek smaller, dedicated machines. It may look expensive, but do the calculations – not forgetting the risk factor.

9.2   Constructing a Lean Scheduling System: Eight Building Blocks

In this section the intention is to provide a set of building blocks and scheduling concepts that can be slotted together Lego style to construct the framework for almost any Lean scheduling system. The eight building blocks provide (together with layout) the skeleton; the concepts are the circulation system. To this must be added eyes and brain (vision, strategy), the nervous system (Hoshin for deployment, and measurement system), and the muscular system (for quality and improvement), and the digestive system (to give energy and to get rid of waste) and the total system gives fast, flexible flowing human action. The eight scheduling concepts are takt, pacemaker, runners repeaters and strangers, kanban, batch sizing and 'every product every', material handling routes, and Heijunka for levelling and capacity management.

Goldratt uses VAT analysis as a guide for scheduling. See the section on Theory of Constraints and Factory Physics. This is very useful. However it is also high level. In other words there may well be, for example, a V type plant within an A type plant.

The blocks follow from the Goldratt section and cell principles. The idea is that the blocks can be combined to make up any factory. Note that the blocks identify the *stages* at which inventory buffers should be located, but the actual *locations* may differ. For instance, where a building block identifies that a buffer is required between stages A and B, the location may be immediately after A, mid way between A and B, or next to B.

Block 1: A is a constraint or bottleneck feeding B a non-constraint.

Q: Where should buffer be placed?

A: In front of A, (to ensure that it is able to keep working) but not in front of B (which can easily catch up). Beware, however, if B is sufficiently starved it may become the constraint.

Q: How much buffer in front of A?

A: Sufficient to ensure time coverage for frequent upstream disruption, but not for unusual events. May also include replenishment time for parts on a pull system, plus safety stock.

Block 2: B is a non-constraint feeding A which is a constraint or bottleneck.

Q: Where should buffer be placed?

A: As before in front of A, not in front of B. Except where B is at the beginning of a line, where buffer to protect against delivery fluctuations is required.

Block 3: A is a constraint; B and C are non-constraints

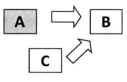

Q: Where to place buffer inventory?

A: In front of A and not in front of B on line A to B, (as before). But also in front of B on line C to B. Why? Because after the product has passed the valuable constraint, it should be delayed as little as possible, for instance waiting for a part from non-constraint C. If C goes down this will unnecessarily hold up the completion of B. Since A is the pacemaker the product will be delayed anyway – holding inventory between A and B will not help.

Think of an assembly line making cars – you would not want to delay the main line while waiting for a minor part. So definitely in front of B on line C to B if it is a 'C' part. However, if it is an A part it may be too expensive to hold a buffer in front of B, so a synchronised schedule must be arranged.

Block 4: A is a process with relatively long changeover, feeding two or more lines or cells, with little or no changeover.

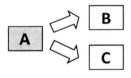

Q: Where to place an inventory 'supermarket'?

A: B and C probably would like to work whenever required. If so, a supermarket must be placed between A and B-C. If B and C work only periodically and one at a time, a supermarket can be avoided if A can be synchronised. Also, a supermarket will be needed in front of A so it can work on either part when called upon by B or C.

Q: How does A know what to work on?

A: You need a priority kanban (or accumulation kanban) system that indicates when buffers in front of B and C get too low. The target batch size is discussed in a section below.

Block 5: Two or more processes feeding a single constrained process.

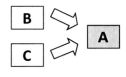

Q: How much buffer in front of A?

A: As with Block 1, sufficient to protect it from disruption. But you also need to know if B and C are both required to make a part A, or whether process A makes two products – one using a B and one using a C. The former requires a buffer of each, the latter requires a synchronised scheduling (pull) system, building inventory of B whilst A works on C, and vice versa. The priority (pull) system for B and C needs to be arranged accordingly.

Block 6: A, B, C, D are sequential operations.

Q: What are relevant questions here?

A: First, is there a constraint or near constraint in relation to overall takt time? If yes, split into pull loops, separated by supermarkets. Second, can the process be flowed, especially one-piece flow or pitch-time flowed. Pitch flow means you have one-piece flow, but can alternate different products every pitch increment. In other words is changeover time + time to make pitch quantity less than pitch time? If it can be flowed, then the processes can be treated as one cell and controlled by one pull signal, possibly with a supermarket or buffer only in front of A and after D.

Block 7 occurs in V plants, for example process plants such as steel. Resource A, of course, needs to be protected by inventory. But, changeovers and batches are inherent characteristics so resource C needs to make batches for A that will be sufficient to keep A running while changing over and making a batch for resource B, as well as changing over once again for resource A. In addition, inventory must be kept in front of A to protect against downtime failures by C. Here, of course, changeover times and reliability for the non-bottleneck resource C are very important in order to limit the otherwise large inventories that need to be kept.

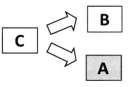

So, in this case, a long changeover at a non-bottleneck is as important as a bottleneck.

Block 8 occurs in situations where there are powered conveyors, such as bottling plants. Here, not only is sufficient inventory needed in front of bottleneck A (to protect against downtime by B), but also sufficient space for inventory build-up must be allowed after A to prevent A being clogged.

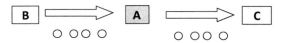

9.3   The Eleven Scheduling Concepts

These eleven concepts form a set which, together with the building blocks, enable most plants having any sort of repetitive production, and many with less regular flows, to implement a successful Lean scheduling system in a value stream. The eleven concepts make up a set. Of course, they can be used individually but in full Lean scheduling system most or all are used together.

1. Demand Smoothing
2. Takt and Pitch Time
3. The Pacemaker
4. Supermarkets and FIFO Lanes
5. Runners Repeaters and Strangers
6. Mixed Model Scheduling
7. Kanban and Pull
8. Lean batch sizing and 'every product every' (EPE)
9. Material handling routes (or 'Runner' or 'Waterspider')
10. Heijunka for levelling and capacity management.
11. Sales, Operation and Purchasing Planning (SO&PP)

9.3.1   Demand Smoothing

The smoother and more regular the demand can be made to be, the easier and better the schedule should be. Demand smoothing is a particular case of the Level Scheduling concept, the advantages of which were discussed in an earlier section.

This is discussed in the Preparing for Flow chapter.

Here note the earlier remarks about the inherent stability of demand for many types of product, at least at the point of use.

A good starting point is to draw a run diagram, or better a control chart with control limits to show normal variation. How much variation is abnormal or 'out of control' – and what are the causes of this? What are the causes of these special events? It could be just erratic ordering, or predictable events such as the start of a sport season. Where

the latter, of course, can advantage be taken of predictability to knock off the peaks and fill in the troughs of demand – assuming non time sensitive material? Can price be used to achieve smoothing as is already done in travel?

A run diagram may also reveal that large infrequent orders are being placed, instead of smaller more frequent orders. Why so? You can save your customer lots of money and yourself big shifts in your run rates by smaller more frequent orders. To ease worries on the part of your customer, you may agree to hold a 'permanent' buffer stock of his parts until the customer gets to believe that this 'magic' system – win/win for both – actually does work.

And, by no means least, can you help your customer in their ordering process? Maybe your product is a small proportion of your customer's business, but a large proportion of your own. So the customer may therefore simply never have looked into levelling the demand. Can you help him predict his own demand? Send out reminders that orders are overdue, or set up a vendor managed inventory system, or offer to take over (at no risk the customer) the ordering process – placing orders on yourself!

9.3.2   Takt Time and Pitch Time

The takt time is the drumbeat, and the pitch time is the repeating increment with which containers (or regular batch sizes) are moved. Takt is the available work time divided by demand over that period. Pitch is takt time x container quantity, and frequently used as the interval in a heijunka system. Takt and Pitch are discussed in greater detail in the Preparing for Flow section. They are not always relevant, so don't try to force it.

9.3.3   The Pacemaker

The single pacemaker is the stage around which the whole value stream within the plant is scheduled. One pacemaker per value stream. Having one pacemaker avoids amplification problems (see the mapping section) and creates synchronisation. If the pacemaker is the heart, the material handler is the circulation.

The pacemaker need not be a constraint or bottleneck, though it often is. It is usual to select a process well downstream as the pacemaker, so that upstream operations can be pulled. After the pacemaker, you would like flow to be first in first out (FIFO) or to go into a finished goods supermarket.

A pacemaker in a repetitive system works better with as smooth a demand. Also, the pacemaker will operate at or near the pitch increment as the drumbeat - typical would be 95% of the pitch time to allow for a little variation. This is really 'undercapacity scheduling'. It is common to use a Heijunka box at the pacemaker as the actual scheduling mechanism.

9.3.4   *Supermarkets and FIFO Lanes*

A supermarket is an inventory store where the runner 'goes shopping' to collect needed parts. Lean aims at flow - the ideal being one piece flow. Flow should take place between supermarkets. Supermarket areas should be grouped together to enable the material handler to visit on his or her regular routes.

With reference to value stream maps, often supermarkets are established at the boundary between loops of pull - say between a press and a group of cells, or where two value streams converge or diverge, or where two CONWIP loops meet.

It is permissible to have work in process inventory between workstations only when it is under visible kanban control (or CONWIP or drum buffer rope – see later). All other inventory should be located in relatively few supermarkets.

The finished goods supermarket is sometimes called the 'wall of shame' to indicate that demand management and schedule stability still require further development. In a finished goods supermarket the inventory should be continually reviewed. For example, use a marker system which shows if there is too much inventory in circulation because inventory below the marker is never called upon. A marker can be used for each container or location of a part in the supermarket, and then removed when the container is moved. If the container never moves for (say) a month the marker will remain, indicating possible excessive inventory.

FIFO Lanes are dynamic buffers of inventory between stages having different cycle times. Inventory accumulates in the FIFO lane while waiting for the next operation. The maximum inventory in a FIFO lane depends upon the time delay at the next stage – for example while waiting for a changeover. They may also used to link physically separated processes whilst maintaining the schedule. A FIFO lane should have a clearly marked maximum quantity. A FIFO lane should never be used as a work in process store, having permanent inventory. If the inventory in a FIFO lane never goes down to near zero, there is probably too much inventory in the lane. Good practice is to paint colours on the FIFO lane to indicate to the preceding operation when more parts are needed. Thus it is a form of kanban square. Where the green section of the lane is exposed replacement may take place. If the red is exposed, replacement should take place.

Merely setting up an in-process FIFO lane instead of an interim buffer point where the schedule can change, can bring large advantages for lead time, simplicity and lead time reduction. It is a way of achieving Spear and Bowen's third rule of the Toyota DNA. (Forgotten that? Read it again – it is highly relevant to Lean scheduling!)

A FIFO lane should have a definite maximum inventory. Typical would be where a second process is involved in a changeover whilst the first process continues to manufacture. Here, inventory queues in the FIFO lane between them. The minimum size of the FIFO lane needs to be able to hold the rate of production of the first process during the time the second process changes over. You would probably add some contingency. Another case is where the first operation works double shift whilst the second works single shift. The FIFO lane should then hold at least one shift of material from the first shift plus contingency. In both cases, the contingency should include the likelihood of breakdown in the next operation, and if the second operation is a bottleneck, the likelihood of breakdown in the first operation.

A FIFO lane should be filled from one end and emptied from the other end. A flag can be added where there is date sensitive material. For small part stores this can be a gravity rack for containers. For FIFO lanes used for large containers, two shorter side-by-side lanes is often preferable to one longer lane. This helps avoid double handling when the lane is emptied from the front necessitating moving up all the containers in the queue. With two shorter side-by-side lanes, pull takes place from one lane while the other is being filled. A signalling system (light? sign?) may be necessary for the material handler to indicate which lane to fill, thereby maintaining FIFO integrity.

Alternatively a FIFO lane can be controlled by a CONWIP signalling system – only letting in work at the start of a sequence of processes when work is let out at the end.

In mixed model production, a FIFO lane can be seen as a method to claw back time where the operation cycle time (on the next stage) for one or more products in a sequence is greater than the takt time. Time is lost on the next operation where the cycle time is longer than takt, so inventory accumulates in the lane. Inventory then decreases when products having a cycle shorter than takt go through the next operation. See calculation below.

In a supply chain a FIFO lane equivalent is the cross dock. The supply chain equivalent of a supermarket is a warehouse.

Sizing of supermarkets and their associated kanban loops is discussed on a following section. Sizing of FIFO lanes is discussed below.

A (by now hopefully) obvious point: do not use an automatic storage and retrieval system as a supermarket. They encourage more inventory, prevent visibility, and have slow response for a material handler or runner. And they may break down. However a small AS/RS (or carousel) can be used for consumables and slow response parts, if space and security are issues. Generally, though, they are things to be avoided!

Sizing a FIFO lane:

In the formulas below, LS is the lane size, but ignoring safety considerations due to breakdown etc. Note that if B has a shorter cycle time than A, no lane is necessary.

→ Size of a FIFO lane after a changeover operation A, feeding an operation B with longer cycle time than the changeover operation:

(c/over time) + LS (cycle time A) = LS (cycle time B). Solve for LS.

→ Size of a FIFO lane before a changeover operation A, being fed by an operation B:

LS = changeover time on A / cycle time of B

→ Size of FIFO lane needed where one or more cycle times are greater than takt, but average time for the mixed model sequence is less than takt. Consider only those products having a cycle time greater than takt in a repeating mixed model sequence.

LS = (accumulated times in the sequence that are greater than takt) / (takt time)

Example: A mixed model sequence is ABCDE repeating. Takt is 25 seconds. Cycle times are 15, 30, 40, 20, 10 respectively.

Average cycle time = 23 secs. OK, check.

Accumulated times greater than takt = 7 + 17 = 24

Lane size is 24/25 = 1, but rearrange so that sequence is A, B, D, C, E repeating. Probably have a stop trigger if there more than (say) 3 in the lane.

However if a batch of products, each exceeding the takt time, is run together, then the accumulated time must be for all of those products together. For example if 10 C's are frequently run together in a batch, then the accumulated times greater than takt is 10 x 15 =

150 seconds, and FIFO lane should accommodate 150 / 25 = 6 products.

Note: the example just given should be treated with caution. A rule of thumb is that when cycle times vary by more than about one third above the takt time, and particularly when there is high demand variation, there is a need to consider some alternatives. These are considered in the Mixed Model section below.

9.3.5   Runners, Repeaters and Strangers (RRS)

Runners, Repeaters and Strangers is a powerful idea for Lean scheduling, thought to have originated in Lucas Industries during the late 1980s. The concept is also known as Runners, Repeaters and Rogues.

A 'runner' is a product or product family having sufficient volume to justify dedicated facilities or manufacturing cells. This does not mean that such facilities need to be utilised all the time, merely that it is economic or strategically justifiable to operate such facilities on an as-and-when basis, and not to share them with other products.

A 'repeater' is a product or product family with intermediate volume, where dedicated facilities are not justifiable. Repeaters should be scheduled at regular slots. Even though the quantity may vary, the slot time should remain approximately constant. This brings advantages of order and discipline. For instance, maintenance and tooling know that a particular job requiring a particular die is needed each Tuesday morning, Suppliers get used to the regular order, setup resources are made ready, the forklift truck may be standing by, and so on. Regularity is the key: try for once per day at the same time; if this is not possible then (say) Monday, Wednesday, Friday at the same times; if this is not possible then (say) every week at the same time, and so on.

A 'stranger' is a product or family with a low or intermittent volume. Strangers should be fitted into the schedule around the regular repeater slots. They have lowest priority.

In constructing the production plan or schedule, begin by doing a Pareto analysis to split the products into runner, repeater, and stranger categories. Runners are of little concern so long as there is adequate capacity. They enjoy their own resources. Repeaters form the backbone of the schedule and should be slotted in at regular intervals as often as capacity will allow, maximising flow and minimising inventories. Make transfer batches smaller than production batches. Then fit the strangers around the repeaters in the schedule.

The distinction between repeaters and strangers is not always clear cut. For instance, you may have a very low volume product that nevertheless has steady demand – like 10 parts per month for an exotic car. It would not make sense to make, say, two parts per week especially if there are setup costs. So, here determine a sensible batch size and run it at off-peak demand periods.

This principle, of using low demand parts from the tail of the Pareto but grouped both to smooth demand and to achieve economic batching, can be used to great effect. Here perhaps is an application for the traditional economic batch size calculation rather than Lean type batch sizing explained elsewhere in this chapter. But remember to use Lean thinking in calculating the holding cost of inventory, setup costs, and the correct factor for '2' in the EOQ / EBQ formula – perhaps 1.5 would be more realistic.

The RRS principle is much like the way we run our lives. We have runners, for example heartbeat that goes on all the time, and we don't plan for these. But you may be conscious of keeping your heart in good condition through exercise. Then repeaters: we sleep every night perhaps not for the same length of time but every night. You know, without being told, not to telephone your friends at 3 a.m. Likewise you have breakfast every day. You use the opportunity to talk to the family, because they are all there without having to arrange a special meeting. What you don't do, even though it may appear more efficient, is to have one big breakfast lasting three whole days at the beginning of the month (one 'setup'). You organise your food inventories around these regular habits. Then strangers: you do different things each day, but these different activities are slotted in around the regular activities.

Once again, Quality Guru, Phil Crosby talks about running you business 'like ballet, not hockey'. In ballet you rehearse, adjust and do it the same for each performance. In hockey, each game is different. Runners, repeaters, and strangers allow ballet style management. Too often, it is hockey style - we collapse exhausted in our chair at the end of the week, feeling satisfied but having solved the same old problem for the 500th time.

	Runners	Repeaters	Strangers
A	Tight kanban	Tight kanban?	MRP / forecast
B	Loose kanban	Loose kanban	MRP
C	2 bin / ROP	2 bin	2 bin / 'go see'

Runners, Repeaters and Strangers, and ABC Classification

A useful way to think of Lean inventory control and control of parts is the table below. Note that here runners, repeaters and strangers refers to component parts not to end products. A component part may be used in several end items, whether the end items are runners, repeaters or strangers.

The columns are runners, repeaters and strangers The rows are the standard A, B, C inventory classification. A items are expensive, B intermediate, C are low cost commodity items.

The entries in the cells indicate the broad options. The table not intended to be applied to every company but is a broad guideline. Each company should develop its own matrix. An A class repeater is likely to be a candidate for tight kanban (that is, kanban with small safety stock). A class strangers are probably candidates for MRP or some forecast-based system; lack of repetition makes kanban less feasible. B class repeaters are probably candidates for kanban with more safety stock. Generally C class items should be managed by a simple procedure such as a two-bin system or reorder point system – perhaps periodic review for runners and continuous review for strangers.

A further dimension is lead-time, shown in the table below.

Here, VMI is vendor managed inventory and signal kanban is a launch activated signal system such as used for Ford engines or Johnson controls seats.

There should always be efforts to convert stranger parts into repeaters and repeater parts into runners, thereby reducing and eventually eliminating the need for MRP. This generally is a matter for design control.

	Runners	Repeaters	Strangers
A long LT	Loose kanban	MRP	MRP
A short LT	Tight kanban	Tight kanban	Signal kanban
B long LT	Loose kanban	Loose kanban	MRP
B short LT	Tight kanban	Tight kanban	Signal kanban
C long LT	2 bin / ROP	2 bin	2 bin
C short LT	VMI	VMI	Go see

Of course, demand for parts varies with time. With time, a runner may evolve into a stranger or vice versa. So it is necessary to keep tags on this. There is software on the market to do this automatically, and flag up when a part has shifted significantly between categories. Alternatively, the position of each part should be reviewed manually whenever a major change in demand mix of end products occurs.

Finally, please refer to the concepts of Demand Management and SO&PP. If demand is truly understood, many more products will be revealed to be repeaters. Infrequent moderate size batches should always be questioned – is that really what the customer wants? What the point of delivery usage – repeater or stranger?

9.3.6   Mixed Model Scheduling

Mixed model scheduling means scheduling ABC, ABC, ABC..... in a repeating sequence rather than in three large batches. There are several reasons for this: it is a powerful aid to cell balancing (by placing long cycle items next to short cycle items), it reduces WIP inventory and sometimes finished goods inventory, it may lead to better customer service, and (a big one) results in a constant rate of flow of parts to the line or cell by material handling, rather than at different rates for different products.

In practical terms, the degree of mixed model scheduling depends upon order sizing, shipment frequency and changeover. These in turn influence the EPEI (every product every interval). For example, if the usual customer pack or container size for A is 20, and B is 10, and demand is 40 and 20 per hour respectively, it will probably not make in a batch of one in a AABAAB repeating sequence but rather 20A, 10B, 20A, 10B or even 40A, 20B once per hour repeating. This pack size is in turn reflected in the pitch increment (takt x container size) used in setting up the heijunka box. See later. If the company ships twice per day, the best policy would be to, at least, try to make every product twice per day in the appropriate shipping quantities. 'At least' because there are balance and inventory risk factors which would favour an even more frequent sequence interval.

You also need to decide on a sensible period for the repeating mixed model sequence. This would in turn be related to the frequency with which a tugger vehicle or runner comes around.

In assembly operations with no changeover, mixed model operations should pose no problem. Where there are short changeovers, you can calculate the minimum feasible batch by changeover time + batch size x assembly time = batch size x takt time. Solving this equation for the batch size will give the desirable minimum number of a product that has to be kept together in a mixed model run sequence. For example if changeover time is 10 minutes, cycle time is 3 minutes and takt time is 5 minutes, then $10 + 3b = 5b$, or (batch) is 5.

Mixed model sequences are, of course, derived from product mix demand. So if you have two products, A with 66% demand, and B with 33%, then the best mixed model sequence is AABAABAAB. You work out the best sequence from the nearest lowest common denominator. Thus if demand for A, B and C is in the ratio 10, 5, 2 the lowest denominator is 2 and the approximate ratios are 5, 2, 1 translating to ABABACAA followed by ABABABACA. In a Heijunka box the mix model sequence is placed in the pitch increments. Some companies use standard pitch increments, and vary the length of the work day, others derive the pitch increment directly from takt.

Either way, first, the mixed model batch size or container size decides the pitch increment, and second, the number of pitch increments and the product mix decides the mixed model sequence. Thus if there are 48 10- minute pitch increments in the heijunka and the lowest common denominator demand mix is 6A, 3B, 2C, 1D then the day would be divided into 4 12-pitch increment repeating slots, each of ABABACABACAD.

That is the basic story. But sometimes not quite as simple. What happens when cycle times vary a lot (more than one third), or where there is high demand or mix variation, or all three?

- First, differentiate between repeater and stranger items. You will probably have to establish different safety stock policies for each category. Repeater items can be held in inventory as replenished in off demand periods. Strangers need to be built.

- For high cycle time variation between products, first ensure that long cycle products (or containers) are placed next to short cycle products. Then make use of FIFO lanes.

- If the mix or demand varies, ideally you will have balanced the cell for different rates. This will usually mean changing the labour requirements. Some companies are able to two or even three rates within a single shift.

- Lead time to make the product is a factor. If the lead time is long, use a supermarket to buffer (i.e. hold safety stock) for make-to-stock items (or repeaters), and then give priority to make-to-order items (or strangers). If lead time is short, you may be able to get away with just using FIFO lanes.

Essentially this is a two-by two policy matrix:

		Demand change	
		No	Yes
Mix Change	No	Standard operations	Change labour? Overtime?
	Yes	Employ safety stock? Use FIFO lanes?	Labour and inventory changes

Kevin Duggan suggests developing a 'Mix Logic Chart' to establish the policies in advance of the situation developing, rather than just working out what to do 'seat of the pants style' at the morning meeting. He gives several examples in his book. (It's back to Ballet not Hockey!). But there are other factors also including customer expectations and contracts, supplier relationships and responsiveness, the ability and willingness of the workforce to be more flexible, seasonality, trend and variation, and, not least, the cash available.

Mark Pyrah, who worked in an uncertain pharma environment (and building on DBO Services Ltd), says there are five categories to consider:

- **Normal:** where there is low incidence of zero demand for products – then make to stock
- **Erratic:** where the incidence of zero demand for products is 20% to 50% - so cost of safety stock would be high – try make to order.
- **Lumpy:** where there is high incidence of zero demand, accompanied by large and variable orders – make to order.
- **Slow:** where demand may be satisfactorily modelled by the Poisson distribution – make to stock.
- **Management Control:** where reliable calculation of a forecast is not possible and which does not fit any of the above.

9.3.7   Kanban and Pull

First of all it is important to distinguish kanban and pull. They are related but too often confounded. Pull systems are essentially those where production is based on consumption at the previous process, in other words there needs to be a demand 'pull' from a downstream process to initiate production (or stock replenishment) at the feeding process. This is different from push systems, where production is driven by a central schedule or forecast against which every process produces, irrespective of whether the preceding process demands it or not. Pull systems can be make-to-order systems, and generally are, but do not have to be. For example, you could supply products from finished goods inventory, and then replenish that inventory as it is depleted.

According to Hopp and Spearman, the main advantages of pull are reduced WIP and cycle time (because pull systems limit the amount of WIP since overproduction is not possible), smoother production flow, improved quality (as systems with short queues cannot tolerate high levels of yield loss), and ultimately, reduced cost due to all of the above.

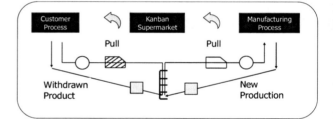

So how do pull and kanban link? Pull is the scheduling *principle*, kanban is one form of pull *mechanism,* albeit the most prominent one. Some other mechanisms are CONWIP, Drum Buffer Rope, 2-bin, faxban, and audio-based call. Remember also that pull is the fourth of Womack and Jones' Lean principles. That is deliberate. There is a lot to do before introducing kanban – reducing demand amplification, reducing changeover, creating more stable work through standard work, reducing the defect rate, and reducing disruptions through breakdowns. Do all of these first!

Kanban is an effective way of reducing Muda (waste) and Mura (unevenness), and to achieve Muri (overburden – overload).

Kanban is the classic signalling device for production pull systems. Nevertheless there remain uncertainties about types and quantities.

A basic classification is:

- **Production Kanbans**
    - Production (single or dual)
        - *Product*
        - *Capacity or Generic*
    - Signal or Triangle
- **Move or Withdrawal Kanbans**
    - In-plant
    - Supplier

Production Kanbans

Single Card Kanban

Traditional kanban is suitable in all stable manufacturing environments where there is repetitive production. In practice, the single card kanban category is by far the most popular type. It is easy to understand, easy to see, and reasonably easy to install. Single card kanban means that a single card (or pull signal) operates between each pair of workstations. Although there may be several single-card kanbans in a loop between a pair of workstations, each kanban is the authorisation both to make a part or container of parts and to move it to a specified location. Two major categories are product kanban and generic kanban.

Product Kanban

Product kanban is the simplest form of pull system. With this type, whenever a product is called from it, it is simply replaced. If there is no call, there is no authorisation so there is no production. In practice, the variations of this type include kanban squares (a vacant square is the authorisation to fill the square with another similar part), cards (which are returned to the feeding workstation to authorise it to make a replacement quantity as specified on the card), and other variations such as 'faxban' or 'e-ban'

(which operate in exactly the same way as cards, except that the pull signals are electronic not physical).

Product Kanban with Multiple Products

(a) Sequential Operations

In sequential operations having several different products, product kanban can be used between stations provided there are not too many products. Here, one partly completed product of each type is placed as buffer between each workstation. If product A is called for at the end of a line, triggers are activated sequentially along the line to make a replacement A. Other products do not move until they are called for. This system allows a quick response build from a limited selection of products, but has the penalty of holding intermediate buffers of part completed products of each type. Hence this system becomes impractical for more than a handful of products. The generic kanban type should then be used, as explained below.

(b) Assemble-to-Order Operations

A variation that is employed in several assemble-to-order operations (for instance, personal computer 'make to order') involves simply having shelves with at least one or two of all parts and subassemblies surrounding the final assembly area. When an order comes in, it is simply configured from the appropriate shelves. This then creates a blank space on the shelf, which is the signal for subassembly areas to replace that subassembly. Subassembly areas are themselves arranged into cells that pull parts from the shelf, and hence back to the store. In this way, literally millions of different configurations can be made under a pull system.

(c) Product Kanban with Synchronised Operations

Where there are several legs in a bill of material or assembly structure, synchronisation can be achieved by variations of so-called 'golf ball' kanban. Here, as the main build progresses, signals are sent to areas producing supporting assemblies to warn them to prepare the appropriate assemblies 'just in time' to meet up with the main build as it progresses along the line. Different colour 'golf balls' are moved (often blown by air or sent electronically) to the subassembly stations to signal them to prepare the exact required subassembly. This form of kanban can be used internally (say to prepare different windscreens or coloured bumpers) to go onto particular cars, or externally (for instance when sent to external seat suppliers to prepare the exact sequence of seats to meet up with a particular sequence of cars).

(d) Emergency Kanban

Emergency kanban is a 'special event' kanban that is inserted in kanban loops to compensate for unusual circumstances. Such kanban cards are of a different colour so that they can be distinguished easily. Such kanbans automatically go to the head of the queue so their requirements are dealt with as soon as possible. Having produced the additional quantity, emergency kanbans are withdrawn.

One variation is to have additional kanbans inserted to meet seasonal demand or to compensate for transport disruption, such as rail disruption or poor weather. These temporary cards are also withdrawn as soon as possible.

A Typical Kanban Sequence:

- When a product is picked from a container in Dispatch, a card is placed in a lane to await collection.
- The runner collects the cards from Dispatch and brings them to the value stream machine team leader's desk.
- The team leader raises the work order and hangs the card on the board next to the machine, in the appropriate row, and on the next available hook. Each product made by the machine has a row on the board. There are also columns painted green, yellow and red. Cards accumulate first in the green zone, then in the yellow, and finally in the red. There are usually several hooks under each colour for each product.

- The machine operator works down the board from top to bottom. This establishes the sequence of jobs in an EPE (every product every) cycle, that is determined bearing in mind the best changeover sequence. Before each product is started, the operator prints out the necessary paperwork.
- The batch quantity to be made is determined by rules. The normal situation will be for the machine operator to make a batch size as indicated by the number of cards. If however the only cards are in the green zone, he will skip over making that product. If there is any card in the red zone he will make this product next.
- As the batch is made, the cards are placed in the containers, the work order is back-flushed, and the container returned to Dispatch.
- Parts that are needed to make the product are also pulled by kanban. When the first component of a part is used, the operator detaches the kanban and places it for collection by the runner. The runner replenishes on his next route.

Let us apply this to an example process (see figure below), where a press operation replenishes the material for an assembly cell. These are the steps in a kanban-scheduling process.

1. The cell uses a container of product.

2. A Move card is sent to the post press supermarket.

3. The Production card is detached, and the Move card is attached to the container.

4. The container with the Move card is returned to the cell.

5. The Production card (triangle kanban) is sent to the Batch Board.

6. The number of Triangle kanbans accumulates and eventually reaches the target batch size line.

7. When the target line is reached, the kanbans for the product are placed in the press queue (file).

8. When the file reaches the beginning of the queue, the product is made.

9. The raw material kanban is detached and moved to the kanban post.

10. The triangle (Production kanban) is placed in the finished goods container, which is moved to the supermarket. The cycle is complete.

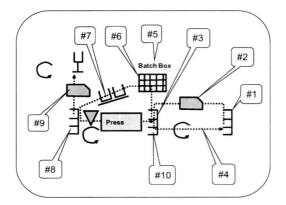

Capacity or Generic Kanban

Generic (or 'capacity') Kanban authorise feeding workcentres to make a part, but do not specify what part is to be made. The part to be made is specified via a manifest or a 'broadcast' system. It is therefore the preferable pull system where there are a large number of products, all of which have similar routings and fairly similar time requirements at each workstation. Generic kanban has less WIP than product kanban, but the response time is slower.

Signal Kanban

Where there is changeover, a signal (or triangle or priority) kanban is used. As parts are withdrawn, so kanbans are hung on the board under the appropriate product column. See figure. A target batch size is calculated for each product (see Batch Sizing section later), and the target is marked on the board. When a sufficient number

of kanbans has accumulated to reach the target, a batch is made. The operation has visible, up-to-date warning of an impending changeover. In normal circumstances the batch is made when the target level is reached. If there are problems kanbans may accumulate beyond the target level. This would indicate higher priority. Normally, a batch is made to cover all the kanbans in the product column. In very slack periods, a smaller batch may be made to cover only the cards on the board.

Dual Card Kanban

Dual card kanban, long established at Toyota and increasingly elsewhere, uses both production and move kanban cards. Production (or signal) kanbans stay at a particular workcentre and alternate from input board to finished goods container with production kanban attached. The workcentre operator uses them. Conveyance kanbans stay between a particular pair of workstations and alternate between move card mailbox and full container (with conveyance card attached). The material handler uses them.

A material handler collects up the conveyance kanbans from the mailbox (usually as part of his regular runner route – see also section on Takt and Pitch time) and takes them to the appropriate feeder workstation. There, the material handler detaches the production (or signal) kanban from the full container, and attaches the conveyance kanban to the container. The production kanbans are returned to the kanban board of the workstation. When works starts on the batch, as authorised by the production kanban, the operator detaches the conveyance kanban from the container and hangs it on the mailbox for the upstream conveyance.

The dual card and signal kanban systems, working together, mean that the move quantity does not have to equal the make quantity. This is good for linking several operations using a pacemaker or Heijunka system. Also, production kanbans have very short lead times because they stay at the workcentre. This means quick response and lower inventory.

Move (Withdrawal) Kanbans

In the single card system, these are simply kanbans that trigger parts delivery to the line, either from an internal supermarket or from an external supplier. In the dual card system, move kanbans work with production kanbans or signal kanbans as above.

The Ford Kanban System

Ford calls their kanban system SMART (for Synchronous Material Availability Request Ticket). There are several variants:

- SMART cards are for slow moving, small or inexpensive parts. These are collected by the material handler on his regular route and returned to the SMART office where they are scanned by bar-code reader. Flashing lights indicate to the material handler the priority for replenishment. This is a 'loose kanban' or slow response system.

- SMART call is for fast moving or heavy or expensive parts or where space is limited on the line. The operator presses a button lineside when a re-order point is reached. This is a 'tight kanban' or fast response system.

- E SMART gives pull signals directly from the line to external suppliers. Such parts bypass the warehouse / supermarket.

- SMART squares painted on the floor indicate the exact stopping point for the front wheel of a forklift or tugger vehicle, to optimise unloading.

Other Forms of Kanban

A kanban carousel is a storage rack on wheels, that is rotated. The back is filled while the front is being used. Good for kit parts. A post on a trolley has an adjustable height indicator that can be used for a variety of parts. The kanban quantity is adjusted according to demand. (Both from Arvin Meritor.) A sequenced in line storage (SILS) is a sloping gravity feed rack on wheels for mixed model heavy parts moved between next-door

supplier and consumer. A slowly rotating two-tier table stores parts below and has packing operations on top. The table rotates at takt time around a group of assemblers. There are load and unload stations. Some automotive plants use a shuttle that travels along with a car over a set of workstations, for custom products such as car seats.

Other forms of pull or kanban are CONWIP (constant work in process) and Drum Buffer Rope (DBR). These are multi stage kanban loops. CONWIP links the first process stage to the last. When a hour of work is let out at the last stage, an hour of work is let in at the first stage. DBR is similar but links the bottleneck with the first stage. These robust systems are less sensitive to process interruptions in the middle stages of the value stream, and are an attractive proposition where there is high variety, differing cycle times or product mix, or where there is good probability of disruption due to quality or breakdown. They are discussed in more detail in the section on Constraints and Factory Physics.

Rules of Kanban

- Downstream operations come to withdraw parts from upstream operations.

- Make only the exact quantity indicated on the kanban.

- Demands are placed on upstream operations by means of cards or other signals.

- Only active parts are allowed at the workplace. Active parts should have specific locations.

- Authorisation to produce is by card (or signal) only.

- Each kanban card circulates between a particular pair of workstations only.

- Quality at Source is a requirement. Only good items are sent downstream.

- The number of kanbans should be reduced as problems decrease.

Numbers of Kanban Cards

In line with Lean Manufacturing, the correct answer to the question of the number of kanban cards should generally be 'less than last time!' The well-known water and rocks analogy applies. That is, reduce the inventory levels by removing a kanban (or by reducing the kanban quantity) and 'expose the rocks'. Note that the philosophy of gradually reducing inventory by removing kanbans is 'win-win' approach: either nothing will happen in which case you have 'won' because you have found that you can run a little tighter or you 'hit a rock' in which case you have also 'won' because you have hit not just any old rock, but the most pressing rock or constraint. This is what Toyota has done for decades.

The general rule on kanbans is therefore to start 'loose', with a generous amount of safety stock, and to move towards 'tight kanban' gradually, but steadily. Probably in the majority of cases the number of kanbans is calculated on a safe assumption of having comfortably sufficient inventory in the replenishment loop. However, Ohno warned about an excessive number kanbans – thereby loosing the responsive 'feel' of a pull system. If you really want to use a formula, try what follows.

Calculating the Number of Cards: Introduction

In general, kanban works like the traditional two-bin system. In the two-bin system the reorder point ROP is calculated thus:

$$ROP = D \times LT + SS$$

where D = demand during the lead-time LT between placing and order and receiving delivery, and SS is the safety stock. This familiar formula is the basis for all kanban calculations. If the container or stillage quantity is Q, then the number of kanbans is simply

$$N = (D \times LT + SS) / Q$$

where N should be rounded up.

It is often better to think of a safety lead time instead of a safety stock. The safety lead time (ST) is a time buffer making allowance for the unplanned stoppages. In this case the formula is:

$$N = (D \times (LT + ST))/Q$$

Calculating the number of kanbans is usually not an independent calculation but is tied in with batch sizing (EPE) and supermarket considerations. These are all brought together in a following section.

(a) Number of Cards where there is Changeover

Lead-time must include changeover + batch run time + queue time + delivery time. In Lean we often think of the EPE cycle (every product every – see batch sizing section). The EPE cycle is the interval in days between running a product, so it will include changeover + run time + queue time but not delivery time or safety time. The delivery time will frequently be linked with the material handling (or waterspider) cycle or pitch time, and is in effect the response safety time between the point of consumption and the changeover operation. The formula then becomes

$$N = (D \times (EPE + delivery\ time + safety\ time))/Q$$

If demand for A is 30 parts per day and A is run three times per day, with a 1 hour delivery cycle. 2 hours is the safety time. The container quantity is 5 and the operation works an 8 hour day. Then N = (30 x (0.33 + 1/8 + 2/8))/5 = 4.25 kanbans. Of these, the safety lead time accounts for (30 x 2/8)/5 = 1.5 kanbans, so you would have to select 4 or 5 kanbans in the loop. Often these would be 'Triangle' kanbans to indicate that a changeover operation is part of the loop.

The batch size is 30/3 = 10 or 2 kanbans, which would be the trigger point on the kanban board next to the changeover operation.

(b) Number of Cards for Assembly Operations or from Suppliers

In repetitive assembly operations where there is no changeover, the demand is expressed in units per day, and the lead time LT is the time required to go through all the necessary steps between 'placing the order' (hanging the kanban on the board) and receiving it. This will normally include the usual lead time elements of run + wait + move. Note that run time should be the time to fill the container, wait time should include both pre- and post-waiting for movement and waiting on the kanban board or mailbox before the order is actioned. Where parts are obtained from an external supplier, the lead-time will be the expected lead-time for delivery as used in any inventory calculation. Demand and lead-time should always be expressed in compatible units; say demand per week and lead-time in weeks.

Safety lead-time should also be allowed. This will reflect any uncertainties in delivery, quality, breakdown or other disruption. Note two points. First, the principle of moving from 'loose' to 'tight' pull and second, the fact that safety stock has usually already been somewhat allowed for in the rounding up calculation to calculate the number of cards.

A Final Note on Kanban

In the preceding sections, traditional kanban has been explained. The weakness of traditional kanban is that it assumes repetitive production (even where generic kanban is used) and also a fairly level schedule. Where the schedule is not level, quite significant buffer inventories between the various stages may be idle for lengthy periods, waiting to be pulled. This is of course 'muda'. Further complications are routings that may vary significantly between products, and variation in processing times resulting in imbalanced lines and temporary 'bottlenecks'. In such circumstances traditional kanban systems can sometimes have more inventory than MRP push systems. Some variations have been developed to overcome these limitations. These include Drum Buffer Rope

and CONWIP, which are discussed in the Factory Physics section.

Notice also that the number of kanbans depends on demand. This means that when demand changes, the number of kanbans should change. In an unstable environment there could be quite a bit of adding and subtracting kanbans. When takt changes, kanbans will often have to change. This means the schedulers have to be vigilant. Some have suggested that an MRP system be used to generate the required number of kanbans. This does not sound to us like generally good advice. Apart from sounding like a reversion to job cards, the philosophy of MRP is inherently that of the job shop rather than the flow shop. MRP is fine for planning, but not for execution.

In the next section kanban is linked with batch sizing and supermarkets.

9.3.8   Batch Sizing

This section gives an introduction to batch sizing and scheduling in situations where changeover remains a significant factor. Of course, you should still continue to attack changeover times, since any reduction improves the flow and reduces batch sizes.

First, a few words on the economic batch quantity. From a Lean perspective, this approach should be totally rejected. Major criticisms include:

- no account is taken of takt time or flow rate
- classic 'batch and queue' thinking
- changeover cost has to be given as a cost per changeover, whereas changeover teams are usually a fixed resource
- inventory-holding costs are often understated
- capacity is assumed to be infinite, and
- demand is constant and uniform - a Lean ideal, but sometimes not a practical one.

The EOQ formula is theoretically sound but practically useless for batch sizing. The four variables in the formula are:

1. Demand which is uncertain.

2. Setup cost – Do we use the average, marginal, or zero if no additional resources are used?,

3. Inventory holding cost. We know that, in Lean, this is much higher than the traditional cost because of quality, space, and lead time considerations – but by how much?

4. The Cost of the item. Should this be at the point of changeover or the final cost? Is overhead included? What happens when a small order is received, then a large one?.

And finally the '2' in the formula derives from the unlikely situation of constant demand. So if you still want to use it.....

The general Theory of Constraints batch sizing guideline is to increase batch sizes on capacity constrained machines, while reducing batch sizes on non-constrained machines. If changeover teams are available they should be used to carry out more changeovers on non-constrained machines, with corresponding reduction in batch sizes, so that such machines become fully utilised in either changeover or running. The resulting reduction in WIP can be used to justify employing more resources on changeover. As a general guideline, this makes good sense, and is compatible with Lean thinking.

Every Product Every (EPE) Batch Sizing

The EPE concept is an important Lean idea that establishes a regular repeating cycle. A Lean ideal is to run every product every day. This would be excellent for service and inventory. EPE regularity has big advantages for standard work, quality, predictability, suppliers, changeover time, and regular time for improvement. 'A good Lean schedule is a boring schedule' is a good maxim. An EPE cycle is often referred to as a 'campaign'

The basis of batch EPE is to make the batch as small as possible by doing as many changeovers as possible in the available time.

→ Time for changeovers = Total available time – Total run time

→ Number of batches = Time for changeovers / changeover time

Example: ACME makes 6 products A,B,C, D, E, F in a press shop. All changeovers take 30 minutes. Demand for the products translates to daily actual run time of 3, 2, 0.5, 0.5, 0.5, 0.5 hours respectively. Net available working time per day (after breaks, routine maintenance, team meetings) is 8 hours per day.

Then total run time per week = 7 hours

Total changeover time for all products = 3 hours

EPE interval	Run time	c/over time	Total time	Available hours	Feasible?
1 day	7	3	10	8	No
2 days	14	3	17	16	No
3 days	21	3	24	24	Just!
4 days	28	3	31	32	Yes (1 spare)
5 days	35	3	38	40	Yes (2 spare)

The 4 day EPE cycle seems attractive. A schedule (showing run times) would be as follows:

	A	B	C	D	E	F	C/ov	Total
Mon			2	2	2		1.5 hr	7.5 hr
Tues	5					2	1.0	8
Wed	7						0.5	7.5
Thur		8					0	8
Fri								
Total	12	8	2	2	2	2	3	31 h

The 5 day EPE has 2 hours spare that could be used to do an extra 4 changeovers per week. Note that you could not run A every day because that would mean that product B would have to run on more than one day. This could lead to the following schedule (in run hours for each product).

	A	B	C	D	E	F	C/ov	Total
Mon			2.5	2.5	1.5		1.5 hr	8 hrs
Tues	3.5				1	2.5	1.5	8
Wed	4	2.5					1.0	7.5
Thur		7.5					0.5	8
Fri	7.5						0.5	8
Total	15	10	2.5	2.5	2.5	2.5	4.5	39.5 h

As can be seen it is reasonably complex, so it may be much more simple to run a 4 day cycle. The moral of the story is – go for the simplest!

An Improvement?

If the product range going through a machine has a strong Pareto shape – in other words some products having very high demand but there is a long tail of very low demand products – the following is worth considering. Simply run the big demand items more frequently and the smaller demand items less frequently – thereby reducing the total WIP inventory, but retaining the same number of changeovers. Start at the extreme ends and go on making the tradeoff until it is no longer worthwhile.

In the example, the demand ratios are 6:4:1:1: 1: 1. It would seem attractive to run A with an EPE of 2 days, and F, with an EPE of 8 days. B,C, D, E remain on a 4 day EPE. It is probably not worthwhile trading off B against E because the benefits are marginal. Thus, in the table below, some product batches are carried forward from one day to the next, without changeovers.

	A	B	C	D	E	F	C/ov	Total
Mon			2	2	2		2 hr	8 hr
Tues	6	1.5					0.5	8
Wed		2.5				4	1	7.5
Thur	6		1.5				0.5	8
Fri		4	0.5	2			1.5	8
Mon	6				1		1	8
Tues	2.5	4			0.5		1	8
Wed	3.5	4					0.5	8
Total	24	16	4	4	4	4	3	31.5 h

The EPE and Batch Size Calculation

We know that a better way to calculate batch sizes is to use the available time to do more changeovers and to drive down the batch size. Available time may be time on a machine or setters time. This should always be the starting point rather than some silly EOQ calculation. However, there is a situations where the approach needs modification.

- Where there is lots of time available for extra changeovers but where there is a definite marginal (not average) changeover cost in terms of materials used, or by creating a new bottleneck with some other resource such as a forklift truck.

- Where quality requirements are so onerous that scrap almost inevitably results.

In these cases the marginal cost of the changeover needs to be set against the reduction in inventory.

Another way of saying this is that batch size reduction (and changeover reduction) has diminishing returns. There will come a point where it is not worth doing an additional changeover. Beware, however, of using this as an excuse. Very quick SMED changeovers can be a huge benefit for flexibility – even if there is sufficient time to take longer for a changeover.

A Further Batch Size Consideration

The standard EPE calculation works well for constant demand. What happens when demand is not constant? The following is one good practical solution.

- First, look at demand over a year (say) and establish 'plateaus' – average levels of demand during periods of the year.

- Do the standard EPE calculation using average period demand, then calculate the normal period of coverage. In other words, what EPE cycle does this cover?

- Then make batches depending on the demand during the cycle – this will mostly work out OK because some products will be above, others below average demand. This method is similar to the 'period order quantity' rule used in MRP.

- Decide when a batch is too small to make economically. Do this batch early and carry over to the next period.

Deriving the Target Changeover Time

An alternative is to calculate the target changeover time that will allow an EPE of (say) one day. In the above example there is an average of 7 hours of demand run time per day, leaving 1 hour for changeover. 6 changeovers are required for an EPE of 1 day, so the target changeover time is 10 minutes. For a 2 day EPE it is 20 minutes per changeover. This is a very useful calculation since changeover times below 10 minutes will not yield further inventory or lead time reductions, but could give more free time for other improvement activities. But, can this time be effectively used?

A Formula for the EPEI (Every Product Every Interval)

It is necessary to calculate the EPEI (every product every interval) for

- every family of products
- on every machine

The formula is:

$$EPEI = \frac{\sum (\text{changeover time per campaign})}{\text{Total changeover time per day} - \sum(\text{run times per day})}$$

Example: Three parts are run on a machine. A campaign is the repeating sequence of all products. Assume that each product is run once in each campaign. The effective workday is 8 hours

Part	Daily Demand	Run time req'd per day (hrs)	Change-over time (hrs)	Batch size (hrs)	Batch Size (units)	Contain-er size	No of contain-ers/batch
A	300	3	1	7.5	750	100	8
B	400	2	2	5	1000	100	10
C	50	1	2	2.5	250	50	5
Total		6 hrs per day	5 hrs per campaign				

Then EPEI = 5 / (8-6) = 2.5 days. This means that there should be a repeating cycle every 2.5 days.

Batch sizes are A: 3 x 2.5 = 7.5 hours run time; B = 2 x 2.5 = 5 hours run time; C = 2.5 hours run time. The batches will however need to be rounded to match the container size.

Different machines will of course have different EPEI and batch sizes – necessitating a FIFO lane between them.

If however there is flow sequence of machines that all products pass through, and that are regarded as one value stream, then do this calculation for each machine and fix the batch as the largest batch calculated for any machine.

Remember that where there are *sequence dependent* changeover times, you should use the above calculation with average changeover times to get the approximate batch size. Then make the products in the correct sequence to minimise total changeover time. However, use 'constant sequence, variable quantity' batches – that is, when a batch is due to be made, make up to the target line. But monitor these quantities and adjust with time.

Buffer or Safety Inventory in Kanban Loops

We (and others) have tended to use the phrase 'buffer inventory' for any excess inventory. This is OK, but it is better to distinguish between two types.

Buffer inventory caters for uncertainty in customer demand – an external factor. Safety inventory caters for problems that occur in the process – an internal factor. It is good practise to separate the two types of inventory – for visibility and problem solving – even though this may involve a little extra inventory. Generally buffer inventory is held only for end items, or subassemblies that may be sold as spares. Safety inventory can be held anywhere to protect against internal disruptions.

Buffer stock protects against short-term customer variation, but not against trend, seasonality or promotions. For these, the supermarket needs to be resized and / or the line should be rebalanced. See the earlier section.

Buffer stock is often calculated from the service level, and associated number of standard deviations (z value), and from standard deviation of demand during the forecast horizon. So, buffer stock = average demand during lead time x z x standard deviation of demand.

Sizing Supermarkets and Kanban Loops

The previous sections can now be combined to calculate the necessary inventory in a supermarket and the associated kanban loop. The number of kanbans in a loop is integrally associated with the size of the supermarket.

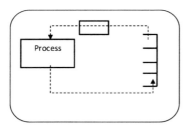

Notice that the supermarket symbol has an extra box. This is useful to remind one of the four inventory elements in a supermarket:

- buffer stock
- safety stock
- batch quantity (derived from EPEI)
- customer demand during the replenishment time.

The replenishment time includes the queue time at the process and the transport time to move the container back to the supermarket. Some of these four quantities may be zero. Buffer stock is held only for end items. Where a supermarket feeds a supermarket, the EPEI quantity would be zero. One point of caution: decide if a kanban is sent when the first or the last part is taken out of the container. It obviously affects the replenishment

time. The latter is a clearer indication; the former sends the message earlier.

Part	Daily Demand	EPEI Qty (conts)	Buffer Qty (conts)	Safety Qty (conts)	Replen time days	Replen Qty (conts)	Total Contain	Ave Contain
A	300	8	2	2	0.5	150/100 =2	14	7
B	400	10	0	2	1	400/100 =4	17	9
C	50	5	0	1	1	50/50=1	8	4

Consider the above EPEI calculation:

The total containers is the maximum size of the supermarket, and, where a kanban is attached to each container, the number of kanbans in the loop. The average containers assumes constant demand, so it is half the (EPEI + replenishment) plus the buffer and safety containers – rounded up.

The re-order point or 'run line' is the batch (EPEI) quantity. There are two ways in which this can work. Where there are kanbans attached to each container, a new batch is authorised when kanbans equivalent to the batch quantity have accumulated.. Alternatively, if the supermarket has clearly marked spaces, or has a FIFO system, when spaces are exposed equivalent to the batch size, a single batch authorising kanban is sent. A variant is to use a clearly marked container – when the first part is taken out of this special container a single batch-authorising kanban is sent.

Signal Kanban, EPE, and the Batch Box

A batch box is used where there are long changeover operations and several products requiring a priority to be determined. 'Long changeover' usually means a changeover taking longer than a pitch increment. Where a changeover is shorter than a pitch increment, a FIFO lane is preferable. A batch box is simply an accumulation device for kanbans. When sufficient kanbans for a particular product have accumulated the workcentre is authorised to make the batch. When production is initiated at the cell, the triangle kanban which is in each completed batch container is detached and moved to the batch box. When sufficient triangle kanbans have accumulated for a particular product and the batch quantity is reached, all the triangle kanbans for that product are moved to the batch queue.

Triangle (signal) kanbans accumulate on the board from the base card upwards until the target batch size line is reached. All the triangle kanbans are then placed in a batch queue box at the end of the batch queue. The batch queue should be a marked-up FIFO lane. In other words the accumulation of queue boxes should indicate the current backlog by colours. If the accumulation is in the red zone, special action or overtime is required.

Take the previous example of 3 products. The figure shows the accumulation of kanbans. Part B, having reached the run line, is authorised to run. The kanbans for B are detached and placed in the batch queue. Normally the parts are run in a strict sequence A,B,C .

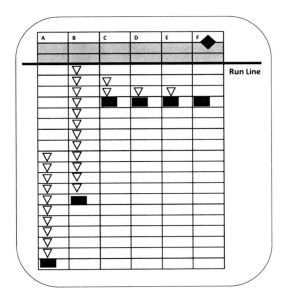

The dark bar in each column represents the base marker. Above the marker, kanbans accumulate from bottom to top. Base cards allow easy

adjustment when takt times change. The dark run line indicates the target batch size, determined from the EPE calculation above, expressed in number of kanbans above the base card. The board is of course dynamic, visible and up-to-date, and reflects changing priorities. A warning of the batch to be made can be clearly seen. In the example, B is due for manufacture and C is likely to be next. With such a board, a detailed schedule for the changeover is not required – it merely responds to the downstream pacemaker.

Minimum Batch Quantity with Changeover

The minimum practical batch quantity when there is changeover is sometimes governed by the *external* changeover time. In other words, unless the batch is greater than this minimum, the machine will be idle waiting for external changeover operations. This is a reminder that it is not just internal changeover operations that govern batch size. Effort may also have to be put into minimising external operations.

9.3.9 Material Handling (Runner) Routes

A runner (or waterspider or material handler or tugger) plays a central role in a Lean repetitive system. Far from being a dreary job, the runner acts like the information system holding the whole system together, noticing problems, and regulating flow.

The runner often starts and ends his regular cycle and the Heijunka box – perhaps every pitch increment. Certainly he or she will come around at regular intervals, visiting standard locations or 'bus stops'. During the regular cycle, the runner collects kanbans, picks parts, and delivers parts from the previous cycle. A constant route is followed. The runner may also collect finished products or issue the Heijunka card to the beginning of the cell.

The runner route should be carefully designed. It is a fixed time, variable quantity route. In other words although the interval is fixed the drop-off quantities may vary. It is therefore important to give consideration to the work balance. It would not be good if the runner was delivering lots of parts in the morning, but circulating with no drop-offs in the afternoon. Mixed model scheduling helps with this. So does using small containers.

You would like the runner to work at as constant a rate as possible.

If the runner experiences a shortage he draws on the buffer or safety stock as appropriate. He (or the team leader) should write up the reason for the occurrence immediately.

9.3.10 Heijunka (or Level) Scheduling

Heijunka is the classic method of Lean scheduling in a repetitive environment. It simultaneously achieves a level schedule or pacing, visibility of schedule, and early problem highlighting. It is usually used at the pacemaker process, and as such controls and paces the whole plant. Moreover it can be used as a form of '10 minute MRP' synchronisation tool. It can be used for production scheduling, for warehouse order picking, and in the office. Finally, it encourages schedules to be developed and controlled by supervisors at the Gemba.

Simply, Heijunka is a post-box system for kanban cards that authorises production in pitch increment-sized time slots. See separate section on takt and pitch time. A typical pitch increment is between 10 and 30 minutes. The box is loaded at the cell level by supervisors or team leaders. It is in effect a manual finite scheduler. As with kanban, a Heijunka system is always visible and up to date. You can see at a glance how far behind schedule you are.

A Heijunka Box has columns from left to right for each pitch increment, and rows for each product or family. For each pitch increment – except break increments – a Heijunka card is placed in one of the product rows to authorise production of one pitch increment's amount of work. A pitch increment normally fills a (small) container of parts, so deviations, when planned work is not completed, are clearly apparent. The cell is authorised, kanban style, to produce only that amount at the specific time. A variant is that a cell

may be allowed to produce the next pitch increment's quantity, but no more. Alternatively, a material handler is authorised to collect only that specific quantity at that time – not before. Therefore, loading up the Heijunka box levels the schedule and withdrawing the cards paces production during the shift. Should any item fail to be ready for collection, or the cell is unable to start work, this is immediately apparent. The worst case of undetected production failure is one pitch increment.

Heijunka is the pacemaker of the material handling system. A regular material handling route should be regarded as an integral part of Heijunka. The resulting regularity of flow of materials and of information is a major advantage.

Where Heijunka is used with a finished goods store, the material handler takes the Heijunka card for the slot as authorisation to withdraw from finished goods. He detaches the production kanban on the container and sends it to authorise making the next batch. The production kanban is later attached to the completed container of parts, which is placed in finished goods. The used Heijunka card is placed in a box to be used tomorrow.

If demand cannot be met as per the Heijunka schedule, i.e. there has been a stoppage, the material handler draws on buffer stock but raises a flag to show that he has done so. At the end of the shift, the buffer must be replaced. Some users differentiate between safety stock to cope with line stoppages, and buffer stock to cope with customer surges in demand.

Mixed model scheduling is usually, but not necessarily, incorporated in Heijunka. The Heijunka box is loaded mixed-model fashion – not AAAAAABBBCCC but ABACABACABAC. See the earlier section on Mixed Model.

Heijunka is not a tool for the job shop or for highly variable production. Having said that, Dell Computer uses a 'sort of' Heijunka (but not called that), by loading up work into two-hour increments which are issued to the factory floor. It can be adopted for maintenance and long cycles.

A basic issue is whether to maintain a constant pitch time increment and derive the batch or pitch quantity as the takt time changes or to maintain a standard container quantity and derive the pitch increment as the takt time changes. The former seems most popular, leading to stability of material handling routes and rate of work. In this case, when the takt changes the container quantity should change and the number of pitch increments changes to meet the demand. You may end the shift with idle time or overtime. In a warehouse you simply accumulate the required number of picks to fill the slot. On the other hand, if you change the pitch increment as the takt time changes, rebalancing is required and material handling routes may have to change to fit in with the new pitch increment. Container sizes, however, remain constant.

Where there are very long work cycles or takt times such as with large items (refer for example to the section on the Pulse Line), the pitch increment can be made a fraction of the pitch time – normally a convenient time increment such as 30 minutes or 1 hour. This is referred to as 'Mini Pitch' or 'Inverse Pitch'. The Heijunka is then built around 30 minute standard blocks of work. The great advantages of levelling and pacing remain.

Heijunka cards may also be 'piggybacked' to achieve the effect of a 'broadcast sheet' as found in automotive plants. In this case a slot may contain several cards, each of which goes to a separate subassembly, thereby synchronising several streams automatically. This avoids having a separate schedule for each.

Finally, Heijunka should be regarded as the final Lean tool. Why? Because so much must be in place for it to be a real success – cell design, mixed model, low defect levels, kanban loops and discipline, changeover reduction, and operator flexibility and authority. But Heijunka is the real 'cherry on the top' – it is the ultimate tool for stability, productivity, and quality.

9.3.11 Sales, Operations and Purchasing Planning (SO&PP)

Sales, Operations and Purchasing Planning (SO&PP) has deservedly become well established in the material management field. SO&PP is a procedure that institutionalises regular meetings between sales and operations so that neither side gets caught short of inventory or capacity, and so that these classically adversarial departments work together both for customer satisfaction and for reduced cost. There are good publications available in this area, for example by Tom Wallace, which can be of benefit to mean Lean schedulers. Recommended.

But the concept needs to go further for Lean. Purchasing needs to be included. If Purchasing does not understand Lean flow concepts, or acts independently, buying large batches of components at low cost – but putting the supply chain at risk through delivery or quality failure, the whole value stream is put at risk. Particularly severe is the case when problems are only discovered some time later when the product is made and tested, leading to the whole batch having to be sent back. By then the purchasing manager who has received reward for his low cost purchasing success may have moved on. Even worse is where large batches are obtained with long lead times from half way around the world.

And it is not just a matter of synchronising sales with operations. Sales needs to be far more proactive than in traditional practice. Sales needs to understand the huge flow advantages of frequent moderate size orders rather than occasional very large orders. Don't just accept orders – only the right size and frequency of orders should be acceptable – and the advantages to both company and customer should be actively promoted. This begins with a thorough understanding of customer requirements – to the extent of being able to place orders for the customer when he needs them. If current metrics discourage this behaviour, senior management needs to step in. Demand management, the first of the Lean scheduling ideas and discussed above, needs to thoroughly understood by sales and marketing.

Hence SO&PP. SO&PP needs to enjoy senior management participation and support. Senior management needs to understand the advantages and requirements. Only senior management has the cross functional power – across sales, operations, and purchasing – to enable end-to-end level flow.

Further Reading

Kevin Duggan, *Creating Mixed Model Value Streams*, Productivity, 2002

Mark Pyrah, MSc Dissertation, Lean Enterprise Research Centre, Cardiff, 2005

Yasuhiro Monden, *Toyota Production System*, (Second edition), Chapman and Hall, London, 1994

James Vatalaro and Robert Taylor, *Implementing a Mixed Model Kanban System*, Productivity, 2003

Steve Bell, *Lean Enterprise Systems: Using IT for Continuous Improvement*, Wiley, 2006

Art Smalley, *Creating Level Pull*, LEI, 2004

Don Tapping and Tom Fabrizio, *Value Stream Management*. (Video series), Productivity, 2001

10 Theory of Constraints & Factory Physics

This section deals with the remarkable contributions of Eli Goldratt on Theory of Constraints (TOC) and with Hopp and Spearman's seminal work on Factory Physics. Both Goldratt and Hopp and Spearman are physicists turned production experts who have derived their theories from first principles.

Sometimes the Goldratt ideas have been seen as being in conflict to Lean operations. In fact, there is remarkable synergy. Some points of difference are discussed later, but for many they add to, rather than subtract from, Lean. But even then Toyota apparently uses finite scheduling software for new products.

Likewise, Hopp and Spearman's Laws of Factory Physics are fundamental statements that should be understood by every Lean practitioner. Their CONWIP system has gained a large following. A limitation of Factory Physics is that the classic book is a little too mathematical for some practitioners; this should not be a turn off – just ignore the maths and soak up the wisdom.

10.1 A Drum Buffer Rope Illustration

Consider a 6 step manufacturing process, as shown. Products flow from process 1 to process 6. Process 3 is the constraint or 'drum'. By definition, other processes will not be as heavily utilised, and will be idle from time to time. To ensure that constraint always has work to work on, it should be protected by a buffer in front of the constraint. There may or may not be inventory between other processes, but there should always be inventory in this buffer. Note two points about the buffer. (1) The buffer is actually a time buffer not an inventory buffer. In other words, there may be two hours of product A in the buffer on Monday, and three hours of Product B in the buffer on

Tuesday. (2) The buffer is strictly all the work that is scheduled on the constraint in the current period, and it may not actually be there yet. If the buffer is maintained the line will be working at its maximum throughput.

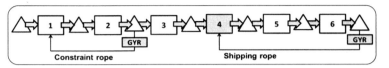

The buffer is coloured Red, Yellow, and Green. When there is sufficient work in the buffer, it is in the Green zone. With variation and possible upstream problems, the buffer might be depleted and be in the Yellow zone. This is a warning. If work is depleted such that it is likely to run out of work, the buffer is in the Red zone. If this situation is reached, the upstream workstations must work harder or longer (or 'sprint') to restore the buffer. A second buffer, at Finished Goods, is also coloured Green, Yellow and Red. These represent safety zones.

The 'ropes' are signals that connect the buffers with the workcentres, as shown. The shipping rope signals the constraint about what and how many products should be made. The constraint rope signals the gateway workcentre about which and how many new products need to be started.

If you have a morning production meeting, hold it at the constraint and do two things: Look at its performance, and look at the buffer inventory to see not only if it is of sufficient quantity but if it is of the right products.

This suggests that the schedule should be built around the drum. Forward schedule on the drum, and backward schedule on non-constraints. Then combine with the synchronous rules. This is not necessarily a simple thing to do. See the section on the Lean building blocks.

Note: The CONWIP (Constant work in process) system is similar, except that there is one rope connecting finished goods to the gateway. The nice thing about CONWIP is that inventory tends to accumulate at the constraint – just where you want it to be.

Both of these methods, DBR and CONWIP, can be used at the cell level, at the plant level, and at the supply chain level. As such they are major, versatile tools which should enjoy far greater acceptance, especially with manufacturing cells and in supply chains.

10.2 Throughput, Inventory, Operating Expense

Goldratt advocates the use of only three performance measures for operations. They are Throughput ('the rate at which the system generates money through sales'), Inventory ('the money invested in purchasing things that it intends to sell'), and Operating Expense ('the money that the system spends to turn inventory into throughput') as the most appropriate measures for the flow of material. We should note some important differences with more conventional usage of these words. Throughput is the volume of sales in monetary terms, not units. Products only become 'throughput' when sold. Some constraint or bottleneck, either internal or external governs this. Inventory is the basic cost of materials used and excludes value added for work in progress. Again, building inventory is of no use unless it is sold, so its value should not be recorded until it is sold. And Operating Expense makes no distinction between direct and indirect costs, which is seen as a meaningless distinction. The aim, of course, is to move throughput up, and inventory and operating expense down. Any investment should be judged on these criteria alone. This cuts decision making to the bone. All three have an impact on the essential financials of business (cash flow, profit, and return on investment), but in varying degrees.

Goldratt says throughput is the first priority, then inventory, then operating expense. Why so? Because throughput has an immediate positive impact on the three financials, whereas you can afford to *increase* inventory and operating expense provided throughput improves. Decreasing inventory has a one-off effect on cash, and can reduce lead-time but unless you are careful reducing inventory may reduce throughput. Likewise operating expense, but with the possibility of removing inherent skills of the business. Note that many businesses, when faced with a crisis, adopt just the opposite priorities – first cut people, then inventories, then perhaps try to improve throughput.

10.2 Dependent Events and Statistical Fluctuations

Goldratt believes that pure, uninterrupted flow in manufacturing is rare if not impossible. This is because of what he terms 'statistical fluctuation' - the minor changes in process speed, operator performance, quality of parts, and so on. Average flow rates are not good enough to calculate throughput. Goldratt has a dice game to illustrate this. Each round, five operators roll a dice to represent the possible production capacity in that time period. The average of each dice roll is 3.5, so one might expect that the average production over say 20 rounds would average 3.5 units per round. In fact, this does not happen because intermediate workers are from time to time starved of parts due to the rolls of previous operators. Try it. They are dependent events.

Even with very large buffer inventories between operators, part shortages sometimes develop. JIT, according to Goldratt, aims at attacking statistical fluctuation in order to enable flow – for instance by using SOPs. But Goldratt believes that this is both very difficult and a waste of resources - which would be better directed at bottlenecks. Hence the TOC rules in the next section.

This may sound like conflict. In fact, one should take the best of both. Yes, try to reduce statistical fluctuation, but also be aware of dependent events and bottlenecks. Also, Lean tends to ignore variation but should not. For instance in *Learning to See* there is little on variation. That may prove a disaster. (But see the section on Muda, Muri, and Mura.) Remember that TOC is particularly applicable in batch environments that are moving towards flow manufacturing. The more you reduce changeover times, the more you smooth demand, the more you reduce variation, the more you tackle waste, the better. Irrespective of TOC, Lean or any other concept

10.3 Constraints, Bottlenecks and Non-Bottleneck Resources: The Synchronous Rules

In this and the next section, the rules and 'laws' of Goldratt and of Hopp and Spearman are given. Together they make a powerful set.

A **constraint** is a resource with the highest load. A **bottleneck** is a resource that is unable to meet current demand. A **constrained critical resource** (CCR) is a resource that has the potential to become a bottleneck – by, for example, periodic overload or sufficient unreliability. There are four types of constraint: physical (a **bottleneck** in a plant is an example), logistical (say, response time), managerial (policy, rules), and behavioural (the activities of particular employees).

In plant scheduling, however, it is constraints which determine the throughput of the plant. For many this is a whole new, radical idea. There is generally only one constraint, like the weakest link in a chain. A 'balanced' plant should not be the concern of management, but rather the continuing identification, exposure, and elimination of a series of constraints. Eventually outside constraints may be the determining factor. Examples are a market constraint or a behavioural constraint. While working on the alleviation of the constraint, the schedule should be organised around this constraint. The principles (often referred to as the TOC or Synchronous principles) follow. Note that here the word 'bottleneck' is used, but it is often a constraint rather than a bottleneck.

1. *Balance flow, not capacity*. For too long, according to Goldratt, the emphasis has been on trying to equate the capacity of the work centres through which a product passes during manufacture. This is futile, because there will inevitably be faster and slower processes. So, instead, effort should be made to achieve a continuous flow of materials. This means, for example, eliminating unnecessary queues of work in froof non-bottleneck workcentres, and by splitting batches so that products can be moved ahead to the next workstation without waiting for the whole batch to be complete.

2. *The utilisation of a non-bottleneck is determined not by its own capacity but by some other constraint in the system*. A non-bottleneck should not be used all the time or overproduction will result, and therefore the capacity and utilisation of non-bottleneck resources is mostly irrelevant. Traditional accountants have choked on this one! It is the bottlenecks that should govern flow. However, shorter runs (hence more changeovers) are desirable on non-bottlenecks in order to smooth the flow to the bottleneck.

3. *Utilisation and Activation are not synonymous.* This emphasises the point that a non-bottleneck machine should not be 'activated' all the time because overproduction will result. Activation is only effective if the machine is producing at a balanced rate; and this is called utilisation. Notice that this differs from the conventional definition of utilisation, which ignores capacity of the bottleneck.

4. *An hour lost at a bottleneck is an hour lost for the whole system.* Since a bottleneck governs the amount of throughput in a factory, if the bottleneck stops it is equivalent to stopping the entire factory. The implications of this for maintenance, scheduling, safety stocks, and selection of equipment are profound! If you think about it, it also has deeply significant implications for cost accounting.

5. *An hour saved at a non-bottleneck is merely a mirage*. In effect it is worthless. This also has implications for many Lean and Six Sigma improvement activities.

6. *Bottlenecks govern both throughput and inventory in the system*. A plant's output is the same as the bottleneck's output, and inventory should only be let into a factory at a rate that the bottleneck is capable of handling.

7. *The transfer batch may not, and many times should not, equal the process batch*. A transfer batch is the amount of work in process inventory that is moved along between workstations. Goldratt says that this quantity should not

necessarily equal the production batch quantity that is made all together. Instead batch splitting should be adopted to maintain flow and minimise inventory cost. This applies particularly to products that have been already processed on bottleneck machines - they are then too valuable to have to wait for the whole batch to be complete. Note that MRP makes an assumption at odds with this principle, in assuming that batches will always be kept together.

8. *The process batch should be variable, not fixed*. The optimal schedule cannot, or should not, be constrained by the artificial requirement that a product must be made in one large batch. It will often be preferable to split batches into sub-batches. On true bottleneck machines, and on CCR's, batches should be made as large as possible between setup (changeover) operations (thereby minimising setup time), but on non-bottlenecks, batches should be made as small as possible by setting up machines as often as possible in order to use the time available. So the batch size may change from stage to stage. This statement is also a rejection of formulas such as EOQ (economic order quantity).

9. *Lead times are the result of a schedule, and cannot be predetermined.* Here Goldratt disagrees with the use of standard pre-specified lead times such as you usually find in MRP.

10. *Schedules should be assembled by looking at all constraints simultaneously.* In scheduling, constraints may be machines, labour, material. Look at them all together. In a typical factory, some products will be constrained by production capacity, others by marketing, and yet others perhaps by management inaction.

10.4 The Laws of Factory Physics

Hopp and Spearman, in seminal work, have developed and mathematically proven a series of fundamental relationships in manufacturing. They term these the 'Laws of Factory Physics'. The Laws are of prime importance to a deeper understanding of scheduling. The dice games (see earlier section) are a fun way to experience some of these basic laws. Here, some of Hopp and

Spearman's Laws are summarised in relation to the dice games. Of course, the dice game has exaggerated variability, but nevertheless illustrates the principles.

1. *In steady state, all plants will release work at an average rate that is strictly less than average capacity.* The dice game has an average capacity of 3.5. Over 20 rounds, the game inevitably produces less than 70 units. Goldratt calls this statistical fluctuations and dependent events – each workstation is dependent on earlier workstations.

2. *Little's 'Law'. Work in Process = Throughput x Lead Time.* (Derived by John Little, and the basis of Factory Physics.) This relationship applies to a machine, a line or a factory. So if throughput is 100 units per week and lead time is 2 weeks, then WIP is 200. Although this 'obvious' relationship is strictly only a good approximation, it is useful because it gives the 'gist' for a host of situations. Thus (a) reducing lead time implies reducing WIP or increasing throughput, or some combination, (b) it gives a way of estimating lead-time from WIP and throughput which are generally easier to determine, (c) it gives a way of estimating utilisation (an important part of OEE, but difficult to determine. If you know the throughput of a machine in jobs per hour and the cycle time in hours, you get the theoretical WIP at the machine in fractions of a piece. This is the utilisation, e.g. 10 jobs per hour, cycle time 2 minutes or 0.033 hour, utilisation is 33%.

3. *In an unconstrained system, inventory builds relentlessly.* (Not a Law of Factory Physics as such.) The longer you play the game, the greater the amount of WIP and the longer the lead time.

4. *Accumulation of inventory is not necessarily an indication of a bottleneck (or constraint).* (Not a Law of Factory Physics.) In the dice game, all stations are equally balanced, but accumulations build and decline, sometimes quite steeply, at particular workstations.

5. *Increasing variability always degrades the performance of a production system.* This can be illustrated by re-playing the game but requiring re-rolls for dice throws of 1 and 6, then another

game requiring re-rolls for throws of 1, 2, 5, and 6. Average capacity is retained at 3.5 but throughput steadily increases.

6. In a line where releases are independent of completions, variability early in a routing increases cycle time more than equivalent variability later in the routing. A dice game can restrict variation to between 4 and 5 in later workstations. In the next game, reverse the variation restrictions. The former has greater output. The implication is to make improvements and standardisation starting downstream.

7. In a stable system, over the long run, the rate out of a system will equal the rate in, less any yield loss plus any parts production within the system. If the dice game is played long enough, output will stabilise at an average of 3.5 pieces – but it takes a very long time, and huge quantities of inventory.

8. If a workstation increases utilisation without making any other changes, average WIP and lead time will increase in a highly non-linear fashion. Another demonstration is to increase utilisation by increasing the WIP levels – say start with 6 pieces, not 4 (a 50% increase). Output in no way increases by 50%.

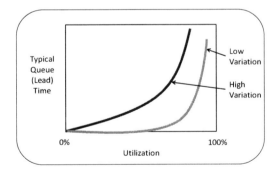

This is predicted by queuing theory, but also by everyone's experience of highway driving – adding a few cars makes no difference at low levels of utilisation, but eventually adding a few cars leads to jams. This phenomenon is the fatal flaw in MRP that assumes constant lead time at any level of utilisation below 100%. Queuing theory predicts sharply increasing waits above a utilisation of around 70% or 80% depending on variation. See the figure.

9. Cycle times over a segment of a routing are roughly proportional to the transfer batch sizes used over that segment, providing there is no waiting for a conveyance device. This Law is a variation on Goldratt's view that transfer batches should not be the same as and generally should be less than process batches.

10. Variability in a production system will be buffered by some combination of inventory, capacity or time. In other words if you do not attack variation, it will attack you in one or more of these three ways. Clearly a dice game where everyone rolls a constant number needs no buffers.

For Lean (and Six Sigma), the implications of these Laws are profound. Variation is the enemy. Planning and control systems must recognise the Laws. Execution systems must be adaptive.

Note that several of these laws and relationships can be demonstrated with the dice game discussed in the first (Principles) section of this book.

Further Reading

Wallace Hopp and Mark Spearman, *Factory Physics*, (Second edition), McGrawHill, 2000

10.5 A Conflict between Lean Thinking and the Theory of Constraints?

For the most part, in our opinion, there is little conflict and much to be gained, by treating TOC and Lean as fully compatible. This is particularly the case with regard to the Product Process Matrix presented in the Strategy section. Lean scheduling is more applicable to higher volume repetitive, and TOC to less repetitive situations particularly where there are shared resources between value streams.

Synergy and similarities. In an excellent paper, Moore and Schienkopf explain that TOC is able to identify the constraints that Lean is then able to target for waste reduction. Both are logical and pragmatic. Both share the goal of flow and throughput. TOC is better at identifying that handful of potential improvements that will make a real difference. In fact, the main criticism of Lean they identify is that Lean is less able to prioritise where to start with the improvements, although Lean's mapping techniques enable you to understand the system and its dependencies more clearly. TOC helps to identify and quantify the opportunities without taking the 'leap of faith' sometimes associated with some Lean implementations. Both encourage pull rather than push. However there are some philosophical differences.

First, **Shared Resources:** Lean, in general, tries to set up clear value streams with no shared resources and adequate capacity. Bottlenecks should be avoided. Spear and Bowen's third rule, ('The pathway for every product and service must be simple and direct'), discussed in the Value and Waste section, makes this clear. TOC attempts to accept bottlenecks and shared resources, and to schedule around them.

In real systems, even highly repetitive, there are often shared resources such as paint lines and press shops. There may even be bottlenecks. So the reality is that, often, TOC concepts will be very useful in many Lean environments, but the goal of clear, simple, unshared value streams should be ideal, desirable state. To help with this, the small machine principle (use the smallest possible capable machine) should always be used.

In general, a shared resource means more buffer inventory and increased lead time. So, wherever you have a shared resource, you should always calculate the cost-benefit of the alternative of dedicated machines set against reduced inventory, lead time and flexibility. Recall that decisions should be based upon the impact on throughput, inventory, and operating expense.

Second, **Utilisation, Local and Global Optima:** Here, hopefully, there is no great conflict. But there are still those who think that maximum utilisation at every resource is not only a good thing, but it leads to a global optimum. TOC teaches the opposite. Maximising utilisation at all but bottlenecks will lead to excessive inventory, and waste. Traditional accountants, beware!

Third, **Inventory between processes:** The classic Lean way is to have inventory between stages and to pull, stage-by-stage, with Kanban. TOC rejects this, in favour of Drum Buffer Rope (DBR). This links the drum (or constraint) with the point of entry, only allowing into the system the equivalent of what is let through the bottleneck. The Factory Physics version is CONWIP (or Constant Work in Progress) that links the end to the beginning. They are therefore multi-stage pull systems. Between workstations, inventory is allowed to fluctuate. It is therefore more robust to variation. It can easily be demonstrated with a dice game that, with high variation, stage-by-stage kanban results in reduced flow volume. Lean, TOC and Factory Physics all favour kanban to deliver parts to the line, but differ on in-process inventory control. Classic kanban, however, can highlight line imbalances and problems at intermediate stages faster than DBR or CONWIP.

Fourth, **Line balancing:** You don't balance a line in TOC. You control through the bottleneck. In fact, balancing to equalise work is considered positively harmful because 'statistical fluctuations and dependent events' together lead to a fall in output well below the balance rate. This does not happen with DBR, where the constraint is protected by buffer.

Fifth, the **Question of Waste:** It is important to know which constraints are affecting performance in any part of an enterprise. If, for example, you have a marketing constraint, it would be foolish to expend more effort on production. The thought that a constraint governs throughput of a plant has massive implications for investment, costing, and continuous improvement. Essentially, an investment which only affects a non-constraint is waste. Likewise, many continuous improvement efforts are waste. Waste walks may themselves be waste. This could be in conflict with standard Lean Thinking, and Six Sigma, but in fact exposes a weakness in both. Hopp and Spearman talk about 'free waste' or bad waste and 'tradeoff waste' or potentially good waste. Removal of good waste involves no penalty. Space and most transport savings are examples. Removal of tradeoff waste may have consequences for money or leadtime. Examples are some inventories, some over-processing, some transport. This leads to the following table:

	Free Waste	Tradeoff Waste
Affects Bottleneck	Do it now!	Calculate
Does not affect Bottleneck	Do it, but with low priority	Maybe not. Do only if money or flow improve

Sixth, **Costing** has also had a shake-up. 'Throughput accounting' uses the equation Revenue – direct materials - operating expenses = Profit. Here, there is no 'variable overhead'. Direct labour is treated as a fixed (or temporarily fixed) cost, and inventories and products are not re-valued on their path through the plant. More is said on this topic in the Measures and Accounting section.

10.6 The Theory of Constraints Improvement Cycle

Theory of Constraints (TOC) and the related Thinking Process (TP) was and is being developed by Eli Goldratt as an extension to his classic work *The Goal*. Goldratt claims wide applicability for TOC, not limited to manufacturing management. At the heart of TOC is the realisation that if a company had no constraints it would make an infinite profit. Most companies have a very small number of true constraints. From this follows Goldratt's five step TOC process of ongoing improvement.

The TOC Improvement Cycle has similarities to PDCA, but is more focused. It is an exceptionally powerful cycle for Lean, but often ignored by Lean practitioners.

1. Identify the constraint or constraints.

2. Decide how to 'exploit' the constraint. A constraint is precious, so don't waste it. Make sure you keep it going, protect it with a time buffer, seek alternative routings, don't process defectives on it, make it quality capable, ensure it has good maintenance attention, ensure that only parts for which there is a confirmed market in the near future are made on it.

Here the batch size should be maximised consistent with demand requirements. Have supermarkets that facilitate flow into and from the constraint.

3. 'Subordinate' all other resources to the constraint. This means giving priority to the constraint over all other resources and policies. For example policies with regard to overtime and meetings may have to change. Measures may have to change. Make everyone aware of the constraint's importance. For instance move inventory as fast as possible after processing on the constraint, reduce changeover time on non-constraints in order to reduce batch size and improve flow to the constraint, make sure that the constraint is not delayed by a non-constraint (a non constraint can become a constraint if it is mismanaged). The right batch size at a non-constraint is derived from doing the maximum number of changeovers that time will allow – in other words minimise the batch size and maximise flow.

4. 'Elevate' the constraint. Break it, but only after doing steps 2 and 3. Buy an additional machine or work overtime on the constraint. If it were a true bottleneck, this would be worthwhile. Be careful. It is seldom necessary to break a constraint because this is only necessary where the constraint is a bottleneck. Knowing the constraint is often a valuable piece of information around which planning and control can take place. If you break the constraint, it will move – possibly to a hard-to determine location.

5. Finally, if the constraint has been broken, go to step 1. Otherwise continue. Be careful that you do not make inertia the new constraint, by doing nothing.

Hutchins makes the useful point that there are five stages for each of these five steps. They are

- gain consensus on the problem

- gain consensus on the direction of the solution
- gain consensus on the benefits of the solution
- overcome reservations
- make it happen.

Further Reading

H. William Dettner, *Goldratt's Theory of Constraints: A Systems Approach to Continuous Improvement*, ASQC Quality Press, Milwaukee WI, 1997

Lisa Scheinkopf, *Thinking for a Change: Putting the TOC Thinking Processes to Use*, St. Lucie, APICS, Boca Raton, 1999

Eli Goldratt, *The Theory of Constraints*, North River Press, New York, 1990.

Robert E Stein, *The Theory of Constraints: Applications in Quality and Manufacturing*, (Second edition, Revised and expanded), Marcel Dekker, New York, 1997

Wallace Hopp and Mark Spearman, *Factory Physics* (second edition) McGraw Hill, Boston, 2000

Ted Hutchin, *Constraint Management in Manufacturing*, Taylor and Francis, 2002

Richard Moore and Lisa Schienkopf, *Theory of Constraints and Lean Manufacturing Friends or Foes?* Chesapeake Consulting Inc., 1998 (Supplied by Goldratt Institute).

Web site: www.goldratt.com

11 Quality

The goal of Perfection, the last of the Five Lean Principles, covers quality, delivery, flexibility, and safety. The Toyota Temple of Lean has two pillars, JIT and Jidoka (Jidoka being closely associated with quality, especially pokayoke). The two pillars are mutually supportive. For instance, improving quality improves Just in Time performance through less disruption and smoother flow. And improving JIT improves quality. Reduced batch sizes allow faster detection and less rework. Pull systems could be regarded as a quality tool. Layout influences quality through improved communication. Postponement reduces variation. Jidoka is a major way of exposing waste and improving quality through surfacing problems. Quality is one of a family of five inter-related concepts which together make up a foundation stone for Lean stability. The others are standard work, TPM, 5S, and visual management.

The prevention, detection and elimination of errors and mistakes forms part of Jidoka, one of the pillars of the Lean temple.

	Men / People	Machine	Method	Materials	Measurement / information	Mother Nature
Variation	Training, Experience	Tool wear, Vibration	Execution methods, standard work	Material variation	Gage accuracy	Temperature, humidity
Mistakes	Omission	Incorrect setup, software	Wrong method	Wrong material or part	Wrong instructions	Ignoring this type
Complexity	Individual differences, motivation	Difficult setup or adjustment	Difficult task, assembly complexity	Difficult to work or assemble	Unclear information	Interaction effects

For each of these three approaches, complexity, variation and mistakes, there are six possible sources of problems – man / people, machine, material, methods, and measures / information , mother nature. This gives a table, with examples, as shown. All of these need to tackled for a comprehensive attack on poor quality.

Hinckley makes the point that as variation is tackled via SPC and Six Sigma, and complexity via design simplification, the relative fraction of defects due to mistakes increases. For that reason Pokayoke is increasingly important. Hinckley states that the most effective order to tackle quality problems is first to address the product, then the process, and finally the related tools and equipment. Within each category, first simplify, then mistake proof, then convert adjustments to settings (i.e. one stop rather than fiddling back and forth), and finally control variation.

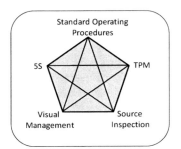

11.1 A Framework for Lean Quality

Perfection in Quality, according to Hinckley, should be approached in three ways:

- a reduction in **complexity** in product design and in process design.
- a reduction in **variation**.
- the prevention and reduction of **mistakes**.

	Product	Process
Complexity	GT, DFM, DFSS, QCC tools, Kano Model	DFM, Layout, SOPS, 5S, SMED, Mapping
Variation	Six Sigma, Shainin tools	Six Sigma, visibility, SPC, TPM, 7 quality tools, 5S, SOPS, Shainin tools, Successive inspection
Mistakes	Pokayoke, DFM	Pokayoke, 5S, SOPS, visibility

11.2 Complexity

Complexity is an interesting concept – although everyone understands it, as well as its negative impacts on any type of operation, virtually no one can define it properly!

What complexity does is to increase the need for management control – the more complex a system becomes, the more effort is needed to control it. Nobel laureate Herbert Simon distinguishes between *static* and *dynamic* complexity. Static complexity refers to those the elements or 'nodes' in the system that add to the complexity by being there, i.e. the more suppliers or product variants, the more complex the system becomes. Dynamic complexity on the other hand refers to the dynamic interaction between nodes, for example, the more volatile demand patterns are, the more complex managing the supply chain becomes (see bullwhip effect in chapter 16).

Complexity can also refer to both product and process. Product complexity refers to both the number of components and the difficulty of assembly. Process complexity refers to both the number of operations and the difficulty of each operation. Hinckley, following on from Boothroyd and Dewhurst, has shown that product defect rates are strongly related to assembly complexity.

11.2.1 Product Complexity

Quality Control of Complexity (QCC): Hinckley has developed a method called Quality Control of Complexity. The frequency of mistakes increases with increasing assembly complexity. The QCC method begins by constructing a tree diagram for assembling a product. Then the time required to complete the assembly is estimated from a set of tables covering alignment, orientation, size, thickness, insertion directions, insertion conditions, fasteners, fastening process, and handling. Alternative designs can then be evaluated based on the ratio of times to the power 1.3 (a value which has been found to be widely applicable). The results can be dramatic both for quality and for cost – perhaps a 50% reduction over the full lifetime. The power and simplicity of this technique should not be ignored.

Design for Six Sigma (DFSS): DFSS uses a defined set of steps called IDDOV (Identify, Define, Develop, Optimise, Verify) similar to the DMAIC steps in Six Sigma. It also uses a similar project organisation with Champions, Master Black Belts, Black Belts, and Green Belts. DFSS is discussed in more detail in the New Product Introduction chapter. The Identify and Define stages aim to clarify the customer and his or her needs. Typical tools are the Kano model and Quality Function Deployment. The Develop stage involves brainstorming and identification of alternatives, and their evaluation. Techniques include TRIZ, Pugh analysis (Concept Screening), and FMEA (Failure modes and effect analysis). The Optimise stage uses the Taguchi methodology for design optimisation and then for tolerance optimisation. In particular, parameter design uses a design of experiments approach to reduce overall variation by identifying and concentrating on the vital few critical parameters. The difference from standard Six Sigma is that here the concentration is on prevention and maximisation of benefit, rather than on detection and reduction of effects. Likewise tolerance optimisation concentrates on the vital few tolerances. Finally, the verify stage involves looking at the capability of the manufacturing process, and examining how the product will perform in the field by conducting experiments on prototypes and pilot tests. Capability studies, SPC and pokayoke are all relevant. Notice that, as with Six Sigma, a lot of attention is given at the front end. It is appropriate to take care at early stages to define the customer, the purpose, the environment, and the usage.

Group Technology (GT): Group Technology is a set of procedures aimed at simplifying products without compromising customer choice. It identifies similarities in function to reduce product and process proliferation. Thus a part designer would not start from a blank CAD screen, but would first search a database for products with similar functions. Similarly, for instance in selecting fastenings, she would make the selection from a pre-defined set rather than from an unlimited choice. The impact on part proliferation, on inventory, on manufacturing routings, and of course on quality can be dramatic.

Various GT coding and classification systems exist to assist both product designers and process designers. A part is described by stringing together a set of digits that cover for example material, usage, shape, size, machining, and forming. For cell design, particularly for complex machining cells, GT may be an early port of call to examine alternative methods and routings. Generalised classification systems can be complex, but frequently companies develop their own much more simple version which serves adequately.

Design for Assembly: Design for assembly, design for manufacture, and more generally DFx, are a key set of techniques for Lean processing simplicity. They impact time, cost, inventory, and quality. Design for manufacture is discussed in the Design and New Product Introduction section of this book.

11.2.2   Process Complexity

Process complexity may be independent of product complexity. A range of tools can reduce process complexity. The tools include:

- part presentation
- dividing work into tasks that can be completed in one or two minutes
- using standard operating procedures
- 5S

- simplified material flows and layout
- TPM
- SMED
- visual controls.

11.3 Variation

A principal approach for the reduction of variation is Six Sigma. But foundation tools for the limitation of variation include TPM, 5S, Standard work and changeover reduction. Tools for the control of variation include SPC and Pre Control.

Before starting out on a sophisticated Six Sigma programme, a Lean company should ensure that they have made reasonable progress with 5S, visibility and standard work and, in many environments, with TPM. This is akin to sending in the public health engineers before the medical specialists. The medics will have point impact, but it is unlikely to be sustained. The public health engineer working to achieve clean water and pollution from sewage is likely to have far greater and lasting impact. Then the medics take over to do their valuable work. This is not to say that a 5S program needs to be fully implemented – it a question of picking off the 'low hanging fruit'.

Statistical Process Control (SPC) is a good technique for variation monitoring and control, provided that its limitations are recognised. SPC is concerned with monitoring the process, not the product. If the process is good, and capable, then the products that are produced by the process will be good. However, SPC is probably not reliable for monitoring or controlling at levels of five or Six Sigma – below perhaps 1000 parts per million (0.1%).

11.4 Mistakes

The toolbox for the control of mistakes includes 5S, Standard operating procedures, pokayoke, self inspection and successive inspection.

Self Inspection and Successive Inspection. Self inspection is where an operator performs an inspection immediately after the manufacturing step is made. Successive inspection is where the next operator checks the previous step or steps. Such inspections are sometimes ridiculed because they are error prone, waste time (because every part is checked rather than a sample), and are usually unsophisticated. But do not be misled – these types of inspection are worthy of consideration because they provide immediate or short-term feedback and (in the case of successive checks) are capable of a high degree of reliability. For instance, if an inspection has 90% reliability, 100 out of 1000 defects would remain after the first inspection, 10 would remain after the second, and 1 (or 0.1%) after the third. This is a higher reliability and often faster than SPC. They do involve non value adding time, however.

Of course self inspection and successive inspection require good motivation and participation. This can be a failing. Relevant points are discussed in the People and Sustainability chapter.

Note that Shingo distinguished these types of inspection from 'judgement' inspection and acceptance sampling which involves both longer delay and increased risk of error.

11.5 Six Sigma

Although Lean and Six Sigma sometimes compete, in more enlightened companies they are seen as partners. Phrases such as 'Lean Sigma', 'Lean Six Sigma' have emerged. This is both good news and bad news. Good news because Lean has often tended to ignore variation, and because it is less strong at detailed problem solving (as opposed to problem surfacing). Together they make for a powerful combination. But also bad if each is defined too narrowly – as the first section of this chapter sets out there are complexity, variation, and mistake issues in a comprehensive approach to quality. Narrowly defined Six Sigma is not much about complexity or mistakes, and may downplay the role of foundation Lean techniques.

The term 'Six Sigma' derives from the spread or variation inherent in any process. Essentially, the Sigma level will tell you how many defects you can expect, on average, for that process. This is a powerful way of describing performance, as it can

be applied to any type of process – irrespective of whether it is in manufacturing or services. It is also limiting, as it only captures 'defects' as in deviation from the prescribed bounds of tolerance, and it assumes a normal distribution (the Central Limit Theorem is the foundation for this assumption). The following table puts the Sigma level into context by translating it into a percentage (how many results of your process will be within tolerance), and finally, how many parts per million do they Sigma levels relate to.

So why choose such a stringent performance level of *six* Sigma, if *four* sigma already only returns 0.6% defects? Because of 'process drift'. The originators of Six Sigma, Motorola, allowed for a 1.5 sigma drift. Thus, in a 3s process the result will in the long term become a 1,5 Sigma process corresponding to 93,32% within tolerance, or 6,68% outside (66,800 PPM).

Sigma Level	% Good	PPM	Illustration: Sigma levels as measure of spelling mistakes
1 σ	37%	632,120	170 spelling errors per page
2 σ	69.1%	308,537	25 spelling errors per page
3 σ	93.3%	66,803	1.5 spelling errors per page
4 σ	99.4%	6,210	1 spelling error each 30 pages
5 σ	99.98%	233	1 spelling error in one encyclopaedia
6 σ	99.9997%	3.4	1 spelling error in all books in a small library

For a 6s process however, a 1.5s shift will only result in 3.4 defects per million opportunities!

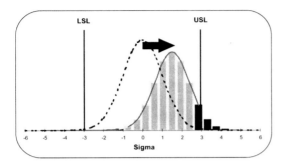

So, even with a 1.5 Sigma process drift, a Six Sigma process would be close to perfection. Of course, for some process such as airline flights, Six Sigma is not good enough.

Whether or not a process can achieve 3.4 defects per million, in a sense is not the point. The point is the rigorous process that moves one towards the goal. It is probably true that today most manufacturing firms are achieving 3 or 4 sigma performance, and most service firms achieve around 2 sigma. So Six Sigma is better thought of as a structured problem solving methodology rather than to do with product quality.

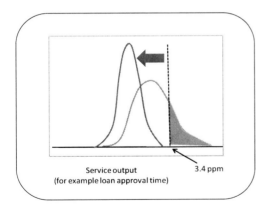

Six Sigma is concerned not only with reducing the number of defects but also with reducing the variation or spread. Part (or product) variation and

Process variation are both of concern. In the figure opposite loan approval time is the attribute most important to the customer. Six Sigma would aim not only to reduce the number of applications that are not dealt with in the specified time, but also to reduce the spread of approval times. Thereafter the target time could be reduced to form a new goal.

A starting point for Six Sigma is the belief in process. An organisation is characterised by processes, frequently cross-functional. The SIPOC model makes clear that a process has suppliers, inputs, the process itself, outputs, and customers. It is useful to go through these systematically.

Six Sigma has a specific methodology: Define, Measure, Analyse, Improve, Control (DMAIC) - in essence similar to the Deming or Shewhart 'Plan Do Check Act' (PDCA) cycle. Six Sigma progresses on a 'project by project basis', and is process oriented. These projects are generally fairly narrow and have definite begin and end points. It takes customer requirements into account at an early stage. A strong feature of Six Sigma is its bias towards data – measuring the variation of the process, and trying both to narrow the variation and to shift it within customer requirements – with 3.4 parts per million being the (frequently unattained) Six Sigma goal. Another feature is its strong financial bias – the benefits of every project are expected to show up in the financials, and are certainly costed. Many a Six Sigma Black Belt will say that Six Sigma is not about defect reduction, but about making or saving money.

Six Sigma is strongly based on statistics. Insistence on hard data is indeed a great strength. But Shingo warns, 'When I first heard about inductive statistics in 1951, I firmly believed it to be the best technique around, and it took me 26 years to break completely free of its spell'. Shingo's journey away from a statistics-based approach to quality should be required reading for every black belt and helps explain Toyota's lack of enthusiasm for Six Sigma (see below).

GE's version of Six Sigma revolves around six key principles. These are:

1. **Critical to Quality.** The starting point is the customer, and those attributes most important to the customer must be determined.

2. **Defect.** A defect is anything that fails to deliver exactly what the customer requires.

3. **Process Capability.** Processes must be made capable of delivering customer requirements.

4. **Variation** As experienced by the customer. What the customer sees and feels.

5. **Stable Operations.** The aim is to ensure consistent, predictable processes to improve the customer's experience.

6. **Design for Six Sigma.** Design must meet customer needs and process capability.

Six Sigma is driven by people qualified in the methodology. A useful innovation has been to recognise Six Sigma expertise by judo-type belts. A black belt typically requires four weeks of training held over four months and requires practical results. The four weeks correspond to Measure, Analyse, Improve, Control in the DMAIC cycle. Black belts often work full time on Six Sigma projects and typically aim at savings exceeding $200k per year. Master black belts are more experienced black belts who act as mentors. There are also Six Sigma 'Champions' who define the WHAT (a very important role, requiring cross functional and cross process knowledge), whereas Black Belts are concerned with the HOW. Some companies retain their Black Belts for between two and three years, and thereafter move them into line management positions or Champion positions. Green belts go through less rigorous training. Some companies, for instance Allied Signal / Honeywell have set goals of 90% of the workforce becoming green belts within 5 years. In a Six Sigma project there is typically a process owner, team leaders, a black belt, perhaps several green belts, and team members. Implementation and human issues are considered to be as important as the Six Sigma tools themselves.

Design for Six Sigma (DFSS) addresses design of product issues. This aspect is discussed in the New Product Introduction chapter.

11.6 How to calculate the Sigma Level of a Process

The sigma level of a process refers to the average percentage you would expect to fall outside the specific tolerance. So in order to calculate the Sigma level of a process, you first need to define the process and its tolerance levels, then you measure (with n>>30 measurements). The calculation of a Sigma level, is based on the number of defects per million opportunities (DPMO). An 'opportunity' is essentially every time you run a process, so DPMO tells you how many times (on average) you let the customer down due to poor performance, measured over 1,000,000 repetitions of that process.

Sigma level - tenths										
	0.0	0.1	0.2	0.3	0.4	0.5	0.6	0.7	0.8	0.9
1	-	-	-	-	-	500,000	460,200	420,700	382,100	344,600
2	308,500	274,300	242,000	211,900	184,100	158,700	135,700	115,100	96,800	80,760
3	66,810	54,800	44,570	35,930	28,720	22,750	17,860	13,900	10,720	8,198
4	6,210	4,661	3,467	2,555	1,866	1,350	968	687	484	337
5	233	159	108	72	48	32	21	13.4	8.6	5.4
6	3.4	-	-	-	-	-	-	-	-	-

Sigma level – whole number

In order to calculate the DPMO, two distinct pieces of information are required: the number of units produced, and the actual number of defects uncovered. DPMO = (Number of Defects / Number of units produced) X 1,000,000. Once you have the DPMO figure, use the table below to convert that into a Sigma level. Use the lower number of the interval, i.e. if your DPMO falls between 4.2 and 4.3, then use 4.2 as Sigma level.

To give an example: a manufacturer of computer hard drives wants to measure their Six Sigma level. Over a given period of time, the manufacturer creates 180,000 hard drives. The manufacturer performs 8 individual checks to test quality of the drives. During testing 4,302 are rejected. Overall, there were n=4,302 defects in 8 x 180,000 opportunities, which gives a DPMO = 2,987.5. Using the table above, this gives a Sigma level of 4.2.

2000.) But they also recognised key issues with Lean implementation such as the difference in the speed of 'kaizens' and Six Sigma projects, and the new roles of Black belts and Greenbelts. Effectively they have added to the Six Sigma practitioners' roles aspects of Lean principles, Values stream mapping and Kaizen methodology. Their internal case studies suggest that Lean Sigma projects produce results 2-3 times faster than normal Six Sigma projects.

Combining Lean and Six Sigma						
	Men / People	Machine	Method	Material / Product	Measures	Mother Nature
Variation	Lean (teams involvement, policy deployment) Kaizen	Six Sigma (CpK) Lean (SMED)	Lean (5S, SOPS) Six Sigma (SPC, DOE, DMAIC)	Lean Supply Six Sigma (SPC, DOE)	Lean (policy deployment) Six Sigma (DPMO, Gage R&R)	Six Sigma DOE
Mistakes	Lean pokayoke	Lean pokayoke	Lean pokayoke	Lean pokayoke	Six Sigma Lean	Six Sigma (DOE?)
Complexity	Lean (cross training, waste removal)	Lean (TPM, 5S)	Lean (waste removal)	DFSS Lean (GT, design)	Lean (policy deployment)	Six Sigma (DOE?)

11.7 Integrating Lean and Six Sigma

It may be argued that Lean and Six Sigma both have strong Deming connections. Deming placed emphasis on two main themes during his life – removal of waste and reduction of variation (see Deming, 1982). Waste reduction is central to Lean, and variation reduction is central to Six Sigma. Several large multinational manufacturing companies, for example Ford and Honeywell, have had separate Lean and Six Sigma programmes. Inevitably, these two powerful and widely used approaches have clashed and merged with titles to indicate this fact. The authors have identified Lean Sigma, Fit Sigma, Six Sigma plus, Power Lean, Lean Six Sigma, and Quick Sigma. Some of these are trademarked. There are likely to be others.

The integration of Lean and Six Sigma has become fashionable. To cite but a few examples: the TBM institute has made bold claims for its programmes by declaring 'LeanSigma utilises Six Sigma and Lean principles to reduce both defects and lead time with the speed of kaizen.' (Dean & Smith,

Mike Wader (2000), of Air Academy Associates, suggests that many world class companies are now following Maytag's lead and implementing top down Lean Sigma programmes where Lean is used to remove the waste and non-value adding activities while Six Sigma is used to control the variation within the value adding portion of the process, combining their tools and data sets to produce a comprehensive improvement programme.

One of the key reasons he notes for integration is to avoid the battle for funding between Lean and Six Sigma that occurs when the two programmes are run independently.

Area	Lean	Six Sigma
Objectives	Reduce waste Improve value	Reduce Variation Shift distribution inside customer requirements
Framework	5 Principles (not always followed)	DMAIC (always followed)
Focus	Value Stream	Project / Process
Improvement	Many small improvements, a few 'low kaizens'. Everywhere, simultaneous	A small number of larger projects - $0.25m cut-off? One at a time
Typical Goals	Cost, Quality, Delivery, Lead Time Financials often not quantified. Vague?	Improved Sigma Level (Attempt six Sigma, 3.4 DPMO) Money saving
People involved in improvement	Team led by (perhaps) Lean Expert. Often wide involvement on different levels.	Black Belts supported by Green Belts
Time Horizon	Long term. Continuous, but also short-term Kaizen	Short Term. Project by project.
Tools	Often simple but complex to integrate	Sometimes complex statistical
Typical Early steps	Map the value stream	Collect data on process variation
Impact	Can be large, system-wide	Individual projects may have large savings
Problem root Causes	Via 5 Whys (weak)	Via e.g. DOE (strong)

Drickhamer (2002) discusses how the adoption of Lean techniques prior to the application of Six Sigma projects can provide real benefits, removing the elitist strain from Six Sigma through teamwork while tackling the low hanging fruit with Lean.

This has a double benefit of removing much of the process noise that is the bug bear of Six Sigma projects. In two key insights he notes firstly from the Six Sigma perspective on blitz events, 'The Solution to many complex and long-standing problems can't be resolved using intuitive methods in less than a week', and secondly from the Lean perspective 'If you go and make everything a Six Sigma problem, you are going to constipate your system and waste a lot of resources'.

With this background, a comparative table is shown.

Once again, Martin Hinckley's useful framework for comprehensive quality improvement has been extended, below, by two more columns and used as a framework for suggesting the most effective approach – Lean or Six Sigma – and with some tool examples.

Toyota and Six Sigma

The classic Lean Company is Toyota. But, thus far, Toyota appears not to have employed Six Sigma in any way. Why? Frankly, the authors do not know but have held informal discussions with a number of Toyota staff. There appear to be at least six reasons, possibly more. These are:

1. The preference for pokayoke. See the earlier Shingo quote. That spirit seems to persist at Toyota where there are reported to be between 5 and 10 pokayoke devices for each process step!

2. The idea that problems and defects need to 'surfaced' immediately, not studied at length. TPS is packed with concepts designed to highlight problems as soon as possible. The systems include line stop, andon board, music when a machine stops, Heijunka (which can highlight non attainment of schedule within minutes), and of course a wide awareness of 'muda'. The thought seems to be to 'enforce' short term and continuous problem solving. Moreover, when problems are identified, the '5 whys' are employed to try to get to the root cause.

3. A worry about the elitism of Six Sigma, especially the 'black belt' image. The TPS way is for everyone to be involved in improvement, and hence a great reluctance to identify specialist problem solvers – however good. This is also reflected in policy deployment.

4. A 'Systems Approach'. Although Six Sigma would claim to use a systems approach, Toyota certainly uses it through value stream mapping and policy deployment. Hence, it avoids the sub-optimisation that is a risk in Six Sigma projects.

5. A belief that many quality problems lie in design.

6. Toyota has a significant improvement organisation in place, that undoubtedly extends the Six Sigma master black belt / black belt / green belt organisation. Refer to the improvement section of the book.

Further Reading

Keki Bhote, *The Ultimate Six Sigma*, AmaCom, New York, 2002

Keki Bhote, *The Power of Ultimate Six Sigma*, AmaCom, 2003

Paul Pande, et al, *The Six Sigma Way*, McGraw Hill, New York, 2000

Howard Gitlow and David Levine, *Six Sigma for Green Belts and Champions*, FT / Prentice Hall, 2005

Frank Gryna, Richard Chua, Joseph DeFeo, *Juran's Quality Planning and Analysis*, Fifth Edition, McGraw Hill, 2007

John Bicheno and Philip Catherwood, *Six Sigma and the Quality Toolbox*, PICSIE Books, 2005

Ron Basu, 'Six Sigma to Fit Sigma', *IIE Solutions*; July 2001

David Drickhamer, 'Where Lean Meets Six Sigma', *Industry Week*, 1 May 2002

Kaufman Consulting Group White Paper, *Implementing Lean Manufacturing*, 2001

C Martin Hinckley, *Make No Mistake*, Productivity Press, Portland, 2001

The authors would like to acknowledge the significant contribution of Brian Johns, MSc in Lean Operations and Six Sigma Master Black Belt, to this section.

11.8 Mistake-proofing (Pokayoke)

Defects due to mistakes may be more significant than defects due to variation, especially at higher Sigma levels. Moreover mistakes are less predictable in as far as they are not part of normal process capability. Mistakes caused in, for example, adjustment may have a severe impact on a whole batch even though the process is capable and 'in control'. As this is realised, mistake proofing is set to gain increased prominence.

The late Shigeo Shingo did not invent mistake-proofing ('pokayoke' in Japanese, literally mistake proofing), but developed and classified the concept, particularly in manufacturing. Recently mistake-proofing in services has developed. Shingo's book *Zero Quality Control: Source Inspection and the Pokayoke System* is the classic work. More lately C Martin Hinckley has made a significant new contribution through his work *Make No Mistake!*

A mistake-proofing device is a simple, often inexpensive, device that literally prevents defects from being made. The characteristics of a mistake-proofing device are that it undertakes 100% automatic inspection (a true pokayoke would not rely on human memory or action), and either stops or gives warning when a defect is discovered. Note that a pokayoke is not a control device like a thermostat or toilet control valve that takes action every time, but rather a device that senses abnormalities and takes action only when an abnormality is identified.

Shingo distinguishes between 'mistakes' (which are inevitable) and 'defects' (which result when a mistake reaches a customer). The aim of pokayoke is to design devices that prevent mistakes becoming defects. Shingo also saw quality control as a hierarchy of effectiveness from 'judgement inspection' (where inspectors inspect), to 'informative inspection' where information is used to control the process as in SPC, and 'source inspection' which aims at checking operating conditions 'before the fact'. Good pokayokes fall into this last category. According to Shingo there are three types of mistake-proofing device: 'contact', 'fixed value', and 'motion step'. This means that there are six categories, as shown in the next figure with service examples. The contact type makes contact with every product or has a physical shape that inhibits mistakes. An example is a fixed diameter hole through which all products must fall; an oversize product does not fall through and a defect is registered. The fixed value method is a design that makes it clear when a part is missing or not used. An example is an 'egg tray' used for the supply of parts. Sometimes this type can be combined with the contact type, where parts not only have to be present in the egg tray but also are automatically correctly aligned. The motion step type automatically ensures that the correct number of steps has been taken. For example, an operator is required to step on a pressure-sensitive pad during every assembly cycle, or a medicine bottle has a press-down-and-turn feature for safety. Other examples are a checklist, or a correct sequence for switches that do not work unless the order is correct. Shingo further developed failsafe classification by saying that there are five areas (in manufacturing) that have potential for mistake-proofing: the operator (Me), the Material, the Machine, the Method, and the Information (4 M plus I). An alternative is the process control model comprising input, process, output, feedback, and result. All are candidates for mistake-proofing.

According to Grout, areas where pokayoke should be considered include areas where worker vigilance is required, where mispositioning is likely, where SPC is difficult, where external failure costs dramatically exceed internal failure costs, and in mixed model and JIT production.

Shingo says that pokayoke should be thought of as having both a short action cycle (where immediate shut down or warning is given), but also a long action cycle where the reasons for the defect occurring in the first place are investigated. John Grout makes the useful point that one

drawback of pokayoke devices is that potentially valuable information about process variance may be lost, thereby inhibiting improvement.

Hinckley has developed an excellent approach to mistake proofing. He has developed a classification scheme comprising five categories of mistake (defective material, information, misadjustment, omission, and selection errors).

The last four have several sub-categories. For each category, typical mistake proofing solutions have been developed.

Thus, having identified the type of mistake, you can look through the set of possible solutions and adapt or select the most suitable one.

Hinckley quotes Hirano in listing the five most useful mistake-proofing devices. They are

1. Guide pins, to assure that parts can only be assembled in the correct way

2. Limit switches, that sense the presence or absence of a part

3. Mistake-proofing jigs, detect defects immediately upstream of the process ensuring that only the correct parts reach the process

4. Counters, that verify that the correct number of parts or steps have been taken

5. Checklists, that remind operators to do certain actions.

In their book, *Nudge*, Thaler and Sunstein remind us of the power of habit and repetition in avoiding mistakes. These are called 'automatic systems', where you are 'programmed' by habit. (Try your reaction time to a 'Go' light shown in red.) Thus in oral contraceptives, the best pattern is one per day (not one every other day), preferably taken at the same time each day. The pills for days 22 to 28 are placebos. So there is a double pokayoke – the one per day packaging, and the regularity even when not needed.

Pokayoke Types

	Control	Warning
Contact	Parking height bars Armrests on seats	Staff mirrors Shop entrance bell
Fixed Value	French fry scoop Pre-dosed medication	Tray with indentations
Motion step	Airline lavatory doors	Spellcheckers Beepers on ATMs

Adapted from : "Failsafe Services" by Richard Chase and Douglas Stewart, OMA Conference, 1993

Further Reading

Shigeo Shingo, *Zero Quality Control: Source Inspection and the Pokayoke System*, Productivity Press, 1986

C Martin Hinckley, *Make No Mistake*, Productivity Press, Portland, 2001

A wonderful award-winning web site on Pokayoke containing numerous examples and pictures is at: http://csob.berry.edu/faculty/jgrout/pokayoke.shtml

12 Improvement

The essence of Lean is improvement. Without improvement, any organisation will fail. To be pervasive, improvement needs to reach all levels and involve all value streams or processes, internally and along the supply chain. For improvement there needs to be a problem – and seeing the problem is the first problem. Problems are not necessarily big things. Since virtually nothing is perfect there is almost always an opportunity – a problem – to which an improvement cycle can be applied. People are often astounded when they hear (as in the Harvard Jack Smith case) that even at Toyota Kamingo plant which has been doing Lean for 50 years, a new manager was tasked with finding, on average, one improvement every 20 minutes. These are mainly small adjustments. But each one counts and they accumulate. Peter Willmott, UK TPM guru, does a 'spot the rot' exercise where participants can easily spot 100 potential improvements within an hour at even the most developed plant. It is a matter of 'learning to see'.

12.1 Improvement Cycles: PDCA, DMAIC, 8D, IDEA, and TWI

A recognised and understood improvement cycle gives a disciplined framework for the process of improvement. It is of great value to have a standardised approach to improvement in any organisation. There are several variants, but all basically similar. The cycles can be used on various levels – from Hoshin or Policy Deployment at the strategic level to value steam implementation, to organisational change (the 'unfreezing, changing, refreezing' cycle is a variation), to training, and to the smallest process change.

Whatever cycle is used, it should be thought of as an overarching approach often used with supplementary tools such as 5 why, root cause problem solving, and force field analysis.

12.1.1 Plan, Do, Check Act (PDCA)

PDCA or the 'Plan, Do, Check, Act' or 'Plan, Do, Check, Adjust' cycle is without doubt the most widely used improvement cycle – but may just be the least understood. In the West many organisations are apt to just 'do' and neglect the P-C-A. PDCA is the scientific method. Deming said that each of the four stages should be balanced. So not a 'quick and dirty' on plan, lots of attention on do, zero on check, and little on adjust or standardise. To truly learn PDCA requires mentoring, particularly in the hypothesis and check stages.

The predecessor to PDCA was 'the Shewhart Cycle' (or PDSA – plan, do, study, act) after Deming's mentor Walter Shewhart but the cycle has come to be named after Deming himself. PDSA is still preferred by some. PDCA sounds simple and is easily glossed over, but if well done is a powerhouse for improvement. PDCA is considered a foundation of the Toyota Production System. Deming taught that you should think about change and improvement like a scientific experiment – predicting, setting up a hypothesis, observing, trying to refute it, and attempting to learn what was wrong with the original hypothesis. We are not talking about a statistical hypothesis test to verify results – but rather making a prediction as a way of testing, developing your understanding of the process, and thinking through the improvement.

Plan or Hypothesis is the supposed first step, but how do you plan when you don't yet know the facts or situation. Remember that PDCA is an ongoing cycle. So Check may will often be the first step, involving perhaps demand analysis, mapping, variation and delivery performance studies. If you start with check, then the cycle is identify problems, propose countermeasures, identify solutions, implement and sustain.

Plan (Hypothesis, Propose countermeasures). Plan is not just about planning what to do, but about communication, 'scoping', discussion, consensus gaining and deployment. Begin with the customer - seek to understand their requirements. The idea of hypothesis is important: make a prediction of the desired outcome and later review to see if the hypothesis was correct. This is to help understand – one of Ohno's favourite words. The plan stage should also establish the time plan. It is claimed that leading Japanese companies take much longer to plan, but then implement far faster and more smoothly. You need to be clear what the goals are, and how to get there. Attempt to identify constraints beforehand, so force field analysis is a good idea. Try to identify root causes.

Do (Try, Identify solutions). This should be an easy stage if you have planned well. It is about carrying out the improvement, often in a test phase. But to 'Do Good' requires interpersonal implementation skills. Force field analysis is most useful.

Check (Observe, reflect, learn). A vital learning stage, but too frequently an opportunity lost. Is it working as you predicted? Did it work out as planned? If not, why not, and what can we learn for next time? The US Marines call this 'after action review' or AAR. Time needs to be set aside to Check, for example at the end of a meeting, or at the end of a 180 day future state implementation plan. Keki Bhote refers to B vs. C (Better vs. Current) analysis. This is to see if the improvement is sustained or as a result of the 'Hawthorne effect' which ceases when observation ceases. Six Sigma black belts would check the statistical significance – assessing the alpha and beta risks (accepting what should have been rejected, and vice versa). Once again ask about root causes. Also check if there are any outstanding issues.

Act (Adjust or Standardise). Often adjustment will be required. If it has worked well, then standardise the important stages. As Juran says, 'Hold the gains'. A standard reflects the current best and safest known way, but is not fixed in stone forever. Without this step all previous steps are wasted. Think about improvement as moving from standard to improved standard. A deviation from standard procedures indicates that something is amiss. See the section on standardisation. Consider if the new way can be incorporated elsewhere. Communicate the requirements to everyone concerned - this includes people on the boundary of the problem. Give some thought to recurrence prevention - can both the people and the processes be made more capable? Finally prepare for the next round of the cycle by identifying any necessary further improvements. And don't forget to celebrate and congratulate if gains have been achieved.

The SDCA or standardise, do, check, act emphasises stability. If variability is excessive it is difficult to distinguish between real improvement and chance variation. In this case, stabilise first before planning.

A last word: don't let the Deming PDCA cycle stand for 'Please Don't Change Anything'.

12.1.2   IDEA

IDEA is a variant on PDCA used, amongst others, by Toyota for innovation and design.

- **Investigate** – the problem, the customers, the need, the purpose, the data, previous solutions
- **Design** – the new solution, all design tools can be used (e.g. TRIZ, QFD, set based)
- **Execute** – the proposed solution and test it, check
- **Adjust** – to bring the solution closer to requirements, prepare for the next cycle.

The IDEA sequence has a great future. Not only is it easier to remember, but the stages make more sense.

12.1.3   DMAIC

The Six Sigma methodology uses a variation of PDCA known as DMAIC (or Define Measure Analyse Improve Control). This has added several useful points. See also the section on Six Sigma. You will notice that there is not a one-to-one relationship between PDCA and DMAIC. DMAIC has expanded upon the critical 'Plan' stage.

Define: Define the problem. Sub stages are identify what is important to the customer and scope the project.

Choosing the right project also means not doing an alternative project. An organisation or improvement team has limited time so should select carefully. Use Pareto. Use Cost of Quality analysis. Begin with customer priorities. Be specific on project aims; go SMART (simple, measurable, agreed-to, realistic, and time-based). Six Sigma is strong on financial returns, so a savings estimate should be made. Scoping the project is critical - where are the problem boundaries, and what will be considered outside and inside? Of course, the 'project' will be found within a process, not necessarily a department. So, 'systems thinking' is required. Typical tools: SIPOC analysis, Pareto analysis, Cost of Quality analysis, Kano model.

Measure: How are we doing? The sub stages are: determine what to measure and validate the measurement system, quantify current performance, and estimate the improvement target. Six Sigma places strong emphasis on measurement. Find a suitable measure – preferably related to the process customer or output. Six Sigma prefers to use quantitative rather than qualitative data. Think defects per million opportunities. Are current measures appropriate? Define the measure clearly, the sources of the data, the sampling plan. Think validity (is what I am measuring a good indicator - preferably a lead indicator?) and reliability (would another observer get the same result?). Think about appropriate defect classification – for instance record the total number of complaints in a hotel, by type, by location, by customer? Check the consistency in the way defects are recorded. Also, be clear on the boundary of the process. Typical tools: 7 tools of Quality

Analyse: What's wrong? The sub stages are identifying the causes of variation and defects, and providing statistical evidence that causes are real. Try to get to the root cause. Use the '7 tools' or process mapping (see separate sections). The majority of tools in this book are useful here. Creative thinking, Benchmarking, QFD, Value Analysis, Design of Experiments, are but a few of the possibilities. Six Sigma places emphasis on statistical validation of results using tests. Typical tools: 7 tools of Quality, FMEA, Design of experiments (DOE).

Improve: Fix what is wrong. Sub stages are determine the solutions including operating levels and tolerances, then install solutions and provide statistical evidence that the solutions work. Now you have to implement. 'Go to Gemba' and do it. You may use Kaizen or Kaizen Blitz. You may also have to plan by using project management tools. Typical tools: DOE, Pokayoke, Hypothesis testing.

Force Field Analysis done with the team should precede implementation.

Control: Hold the gains, and sustain. Sub stages are putting controls in place to sustain the improvements over time, and provide statistical evidence of sustainment. Verify. Measure again. And celebrate with the team. Set up SPC charts. Set new standard operating procedures. Test to see whether gains are real by going back to the old process and forward to the new. Typical tools: SPC, visual management, TPM, Standard Work.

12.1.4   8D

The 8D Cycle is another improvement cycle methodology, widely used and probably originating with Ford. The '8 D's' are eight **D**isciplines:

1. Form a team
2. Contain the symptom
3. Describe the problem
4. Find the root cause
5. Verify the root cause and select the corrective action
6. Implement permanent corrective action
7. Prevent recurrence, make the solution standard
8. Congratulate and celebrate.

12.2 'Five Whys', Root Causes and Six Honest Serving Men

12.2.1 *Root Cause Problem Solving*

The emphasis on 'root cause' problem solving is fundamental to the Lean philosophy. It means solving problems at the root rather than at the superficial or immediately obvious levels. But how do you get to the root cause? In the following sections some techniques are examined. First, we should look at the whole concept of root cause analysis.

In a thoughtful article Finlow-Bates concludes that there are no ultimate root causes. Rather, root causes are dependent upon the problem owner; there can be more than one potential root cause and the final choice of root cause cannot be made until the economics of possible solutions have been considered. He illustrates the point by the example of a delivery failure. The root cause for the customer is that the parcel is late. The root cause for the delivery company is that the problem delivery failure was due to the van not starting which can be traced to the root cause of a leaking underground tank. For the tank supplier the root cause was a failure in the solder. For the solder supplier the root cause was …. Each person along the chain is not interested in the problems of lower rungs. Each of these causes represents a failure in control or in communication. The real issue is therefore not what is the root cause, but how can the problem be (temporarily?) solved most economically and effectively to prevent recurrence.

Finlow-Bates suggests following six steps:

1. What is the unwanted effect? Finlow-Bates suggests two words; subject and deviation, for example parcels late.
2. What is the direct physical cause?
3. Follow the direct physical line of the cause. For example, parcels late, van won't start, etc. This establishes the 'staircase'.
4. Ask who owns the problem at each stage.
5. Identify where you should intervene in the staircase to effect a long-term solution.
6. Identify the most cost-effective of the solutions.

The 5 whys is a technique to try to ensure that the root causes of problems are sought out. It simply requires that the user asks 'why?' several times over. The technique is called the '5 whys' because 'why' often needs to be asked successively five times before the root cause is established.

This simple but very effective technique really amounts to a questioning attitude. Never accept the first reason given; always probe behind the answer. It goes along with the philosophy that a defect or problem is something precious; not to be wasted by merely solving it, but taking full benefit by exposing the underlying causes that have led to it in the first place. Many believe that it is this unrelenting seeking out of root causes that has given the Japanese motor industry the edge on quality, reliability and productivity.

An example: A door does not appear to close as well as it should. Why? Because the alignment is not perfect. Why? Because the hinges are not always located in exactly the right place. Why? Because, although the robot that locates the hinge has high consistency, the frame onto which it is fixed is not always resting in exactly the same place. Why? Because the overall unit containing the frame is not stiff enough. Why? Because stiffness of the unit during manufacture does not appear to have been fully accounted for. So the real solution is to look at the redesign of the unit for manufacture.

Perhaps there are even more whys. Why did this happen in the first place? Insufficient cooperation between design and manufacturing. Why so? It was a rushed priority. Why? Marketing had not given sufficient notice. Why? And so on.

Of course, it seldom works out as neatly as this. At each stage several answers are likely. It is then necessary to do a Pareto of the likely reasons before proceeding.

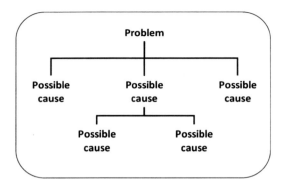

But, there are cautions! Unfortunately the 5 why procedure is widely misunderstood. The whys should not be asked in a mechanical, un-listening fashion. Rather, after each why the possibilities need to be discussed and prioritised. It should be a participative exploratory procedure, not a one-sided aggressive procedure. Even better is to internalise the procedure – ask the 'whys' to yourself.

Another major caution is that the word 'why' often comes across as a criticism, thereby eliciting defensive, concealing behaviour. So by all means think 'why' but articulate the question differently. Instead of 'Why is it late?', rephrase to 'Lets explore the reasons for lateness' or 'What might be the possible reasons for it being late?'

Even better than using the 5 Whys is to use them after the Six Honest Men discussed below.

- What is purpose? (then Why?)
- When should the work be done? (then Why?)
- Where should it be done? (then Why?)
- How should it be done? (and Why?)
- Who should do it? (and Why?)

Keep in mind Deming's '94/6' rule – 94% of problems probably lie with the system, and only 6% with the person. Thus, 'Why don't you….?' (a partly closed question) is much weaker than 'Is there another way to do that?' or 'What do you think is the cause of…?' (more open questions).

Beware: the 5 Whys can be counterproductive if you are really suggesting the solution rather than asking others to question, to think it out themselves. This is the essence of the Socratic method – leading to a far more effective, and sustained, solution because it is then their idea.

12.2.2 Six Honest Working Men

Rudyard Kipling's 'Six Honest Serving Men' remains, some 100 years after it was first written, one of the most useful problem analysis tools. The original verse is,

> 'I knew six honest serving men,
>
> they taught me all I knew;
>
> their names are what and why and when,
>
> and where and how and who'.

Such a simple little verse; so much wisdom – so often ignored!

The six men are a very useful way of defining customers, their requirements and what is really valued.

Further Reading

T Finlow-Bates, 'The Root Cause Myth', *TQM Magazine*, 10:1, 1998

Michael Marquardt, *Leading with Questions*, Jossey Bass, 2005

12.3 Organising for Improvement

There are two aspects of improvement organisation. The first is the improvement organisation itself centred on the Lean Promotion Office, champions and steering groups. The second aspect is the improvement structure itself made up of five levels from individual to supply chain project.

12.3.1 The Lean Promotion Office

When an organisation grows beyond perhaps 100 people, certainly 200, it becomes necessary to institutionalise improvement and sustainability. The Lean Promotion Office is a good name, but alternatives are a Kaizen office, or a continuous improvement (CI) office. A rule of thumb is that the Lean Promotion Office (LPO) should comprise 1 to 2% of the workforce full time during a major implementation, and 0.5 to 1% thereafter. These are the internal Lean consultants. If Six Sigma is in place, a parallel but closely linked office of a similar size is appropriate.

Many organisations, including Toyota, have found that implementing and sustaining Lean requires full time expert facilitators. They are the repository of expertise and should have general responsibility for Lean momentum. Note that the LPO cannot have authority for Lean implementation – that will always lie with line managers. So the ideal head of the LPO is a respected, Lean believer, and an influential individual who works through line managers, helping them to achieve their Lean goals. But the LPO is strictly a facilitating function – in no way should it be seen as 'the guys who do Lean around here'. Apart from the head, Lean enthusiasts or Lean disciples irrespective of age or position should of course staff the LPO.

Remember the wise words of Lao Tzu, in *The Art of War*, 'Go to the people. Live amongst them. Start with what they have. Build on what they know. And when the deed is done, the mission accomplished, of the best leaders the people will say 'We have done it ourselves''.

The LPO has specific responsibility for developing the general roadmap or Master Schedule for Lean implementation. Specific tasks that the LPO undertakes include assistance with mapping and the development of future state maps, advice on specific aspects such as number of kanbans, tailoring 5S and Lean audit assessment tools for specific value streams, preparing waste questionnaires, running short courses on specific topics such as Lean accounting, coaching on facilitation and presentation skills, and preparing newsletters and videos. Several larger organisations, for instance Ford, have established libraries and on-line information. Dell has packaged on-line training into two hour modules that can be taken in slack periods.

The LPO is a facilitating office, not a doing function. Toyota refers to 'Jishuken' (or 'fresh eyes') groups – this is what a LPO should facilitate and encourage – to look at things in a fresh way, from a different viewpoint.

In addition to the LPO, some organisations have appointed various line managers as expert internal consultants on relevant aspects such as Lean accounting, changeover, pokayoke, pull systems, and demand management. These people have the responsibility of keeping up with developments on their topic.

The question of the relationship of a Six Sigma function to the LPO is controversial. What seems to be emerging is a strong case for these two functions as separate but closely coordinated. The reason is that the Lean **Promotion** Office is a support role, not doing projects themselves. The Six Sigma office, with Master Black Belts and Black Belts is actually engaged in the more difficult improvement projects that are inappropriate for kaizen teams. The Six Sigma office, however, is also involved with training – for Green Belts and the next generation of Black Belts.

The very existence of a LPO is an indication of the organisation's commitment to Lean.

12.3.2   The Hierarchy of Improvement

Kaizen, or Lean improvement needs to be organised on five levels in most, if not all, organisations aspiring towards Lean.

Level 1: The Individual

The individual needs to be the recognised expect of her own process. This does not happen by chance and should not be built on trial-and-error. She needs to understand not only the process itself in great detail but also why the process is necessary and how it fits into the wider value stream. So, not only inserting a trunk seal in the best possible way, but knowing the necessity for keeping out damp and dust. Shingo suggested that the 'know why' or underlying philosophy is the most important stage of learning. So both improvement and sustainability begin with the individual at the workplace.

At the individual workstation level there are always opportunities for waste reduction – work piece orientation, inventory and tool location, work sequence, ergonomics, pokayoke, and on and on. Some initiatives would arise as a result of workstation-level record keeping. Toyota South Africa calls their individual program 'Eyako' – the Zulu word for 'my own'. The team leader has an important role to play here – encouraging, facilitating and recognising achievement – and bring individual improvements to the attention of others. Individual 'thank you' notes carry much weight.

Level 2: The Work Team or Mini Point Kaizen

Groups or Teams of perhaps 6, that work in a cell or on a line segment undertake improvement projects affecting their collective work area. Examples include work flows, cell layout, line re-balance, 5S, footprinting, and cell-level quality. Some activities may result from 'point kaizens' identified during wider current state mapping. These initiatives may be done 'on the fly' as a result of team meetings, or 1 or 2 day kaizen blitz activities. They may facilitated or assisted by the section leader or LPO. Many of the initiatives would arise as a result of record keeping and analysis undertaken at the cell and shown on display boards held in the team's own break area. Recognition is crucial, so the team needs to present to a wider audience. Do not make the mistake of using a level 3 blitz event when the team can comfortably do it themselves.

Scholtes recognises that there is a difference between 'teams' and 'teamwork'. 'Teams' refers to small groups of people working together towards some common purpose. Teamwork refers to an environment in the larger organisation that creates and sustains relationships of trust, support, respect, interdependence, and collaboration'. This understanding recognises that it is relatively easy to establish a team, but to establish an environment for teamworking is a lot more difficult. He quotes Petronius Arbiter in

what would be well recognised within many organisations:

'We trained hard, but it seemed that every time we were beginning to form up into teams, we would be reorganised. I was to learn later in life that we tend to meet any new situation by reorganising; and a wonderful method it can be for creating the illusion of progress while producing confusion, inefficiency and demoralisation.'

Level 3: Kaizen Blitz Group or Point Kaizen

The Blitz event is carried out in a local area, but involves both more time (typically 3 to 5 days full time) and outsiders. These events address more complex issues than the work group can handle comfortably. Examples include more substantial layout changes, the implementation of a single pacemaker-based scheduling system together with runner route, and integrating manufacturing and information flows. For many companies, blitz groups are the prime engine for improvement – to ignore is folly. Unlike the level 2 improvement teams, this type of group forms for the specific purpose of the event, and disbands thereafter. Blitz events are discussed in greater detail later in this chapter.

Level 4: Value Steam Improvements: Flow Kaizen Groups.

Flow kaizen groups work across a full internal value stream, taking weeks to 3 months for a project. They are the prime engines for creating future states. Their targets would be those set out in a future state and action plan exercise. Refer to the mapping section. Value Stream groups are normally not full time (unlike level 3 groups), although some members may work uninterrupted for a number of days. They would be led by a project manager, often assisted by the LPO, and sometimes mentored by consultants. The group would be multi-disciplinary, working along a complete process or value stream and across several areas and functions. Flow kaizen projects usually have to address process issues, system issues, and organisational issues.

Level 5: Supply Chain Kaizen Groups

Similar to Flow Kaizen Groups, these groups work in the supply chain. They will invariably comprise part-time representatives from each participating organisation. A respected project manager, typically from the OEM company would be appointed, and there is a greater role for consultants. A 'Seeing the Whole' value stream map would typically be the centrepiece.

12.4 Continuous Improvement Approaches

It has become clear that there are two elements to improvement, namely continuous improvement and breakthrough improvement. Thus Juran refers to 'breakthrough' activities, using 'project by project' improvement, to attack 'chronic' underlying quality problems as being different from more obvious problems. Davenport, in the context of business process reengineering, has referred to 'the sequence of continuous alteration' between continuous improvement and more radical breakthroughs by reengineering. And Womack and Jones discuss 'kaikaku' resulting in large, infrequent gains as being different from kaizen or continuous improvement resulting in frequent but small gains.

A traditional industrial engineering idea is that breakthrough or major event improvement activities are not continuous at all, but take place infrequently in response to a major change such as a new product introduction or in response to a 'crisis'. But during the past few years, through blitz, we have learned that effective breakthrough should be both proactive and frequent.

More senior management working across a value stream generally drives breakthrough or Flow kaizen. Incremental or Point Kaizen is led by team leaders, and sometimes by Six Sigma Black Belts working on local issues that have arisen either through value stream analysis (proactive) or from workplace suggestions (reactive).

There are therefore four types of improvement, as shown in the figure. There is, or should be, a place for all four types in every organisation. Adopting Lean manufacturing does not mean ignoring other forms of improvement to concentrate on kaizen and blitz. Passive approaches are a useful supplement and should continue. However, if all improvement is of the passive, reactive type the company may well slip behind.

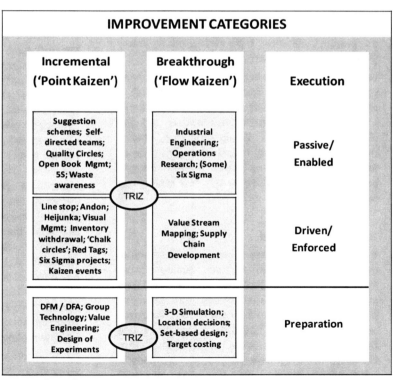

Note: An earlier version of above chart was shown in Bicheno 2002 (Cause and Effect Lean). A similar concept is used in by Hayes et al. Pursuing the Competitive Edge, Wiley, 2005.

12.4.1 A Classification of Improvement Types

Unfortunately, several British factories think they are doing kaizen, but their brand of kaizen is 'passive' or left to chance. Improvements are left to the initiative of operators or industrial engineers or managers. If they make improvements - good. If they don't - 'oh well, sometime'. Passive incremental may also be termed 'reactive'. A reaction takes place in response to a crisis. By contrast, enforced improvement is proactive. 'Crises' are actually engineered and the pressure kept on. For example, Intel brings out a new chip at regular, paced intervals and does not wait passively for technological breakthrough. 3M dictate that 30% of revenues will come from new products every year. This forces the pace.

Passive improvement has been around for many years, and it too is found in two categories - incremental and breakthrough. Classic 'passive incremental' improvement approaches are the suggestion scheme and the quality circle. Passive breakthrough is classic industrial engineering or work-study, especially where such methodologies are used for factory layout, and new technology introduction.

1. Passive Incremental

A classic team based passive incremental improvement method is the Quality Circle. Contrary to popular conception, the reward-based suggestion scheme is alive and well at many Japanese companies. At Toyota's US plant, for instance, awards are based on points and range from $10 to $10,000. Toyota has the attitude that all suggestions are valuable, so the company is prepared to make a loss on more mundane suggestions to develop the culture of improvement. At the top end of the Pareto the company reckons that the top 2.5% of suggestions pays for the entire reward programme, even though a good number of suggestions at the bottom end are loss-making, taking into account the implementation time. You are reminded of the classic statement about advertising, 'I waste half of the money I spend, but the problem is I don't know which half'.

Thomas Edison is reported to have said that the way to have great inventions to have many inventions. Toyota also insists that all suggestions are acknowledged within 24 hours and evaluated within a week. Non-acknowledgement and non-recognition have probably been the major reason for suggestions schemes producing poor results and being abandoned.

Likewise, team based Quality Circles are an integral part of the Toyota Production System. At Toyota, QC presentations to senior management occur almost every day. At Japanese companies QCs often meet in their own time. Management involvement and support are crucial elements. Edward Lawler has described a 'cycle of failure' for many Western QCs. The following sequence is typical. In early days the first circles make a big impact as pent-up ideas are released and management listens. Then the scheme is extended, usually too rapidly, to other areas. Management cannot cope with attending all these events, and is in any case often less interested. In the initial phases, the concerns of first line supervisors, who often see QCs as a threat to their authority, are not sufficiently taken care of in the rush to expand. Some supervisors may actively sabotage the scheme; others simply do not support it. Then, as time goes on, with less support from management and supervision, ideas begin to run out. The scheme fades. And it is said, 'QCs are a Japanese idea which does not work in the Western culture'. By the way, it was Deming who introduced circles to Japan, albeit that Ishikawa refined the methods.

Yuso Yasuda has described the Toyota suggestion scheme or 'Kaizen system'. The scheme is co-ordinated by a 'creative idea suggestion committee' whose chairmanship has included

Toyota chairmen (Toyoda and Saito) as well as Taiichi Ohno. Rewards for suggestions are given at Toyota based on a points system. Points are scored for tangible and intangible benefits, and for adaptability, creativity, originality, and effort. The rewards are invariably small amounts, and are not based on a percentage of savings. However operators value the token reward and the presentation ceremony itself. Note the contrast with typical Western Suggestion Schemes.

From the forgoing we learn a few important lessons. Not all improvements will pay, but creating the culture of improvement is more important. Give it time, and expand slowly. Recognition is important - management cannot always be expected to give personal support, so establish a facilitator or LPO who can. Do not underestimate potential opposition. React rapidly to suggestions. Give groups the tools and techniques, and probably the time.

2. Passive Breakthrough

Many traditional industrial engineering and work-study projects are of the passive breakthrough type, particularly when left to the initiative of the IE or work-study department. Of course, IEs also work on enforced breakthrough activities initiated by management or by crisis, but passive breakthrough activities, led by IEs, have probably been the greatest source of productivity improvement over the past 100 years. Many of Taiichi Ohno's activities could be classified as passive breakthrough. Ohno was apparently a great experimenter on his own in the dead of night. But today we recognise that many I.E. (or for that matter Six Sigma) projects done in an elitist way are unlikely to be sustained.

3. Enforced Incremental

Kaizen, as practised at Toyota, is the classic here. Waste elimination should not only be a matter of chance that relies upon operator initiative, but is driven. There are a number of ways in which this is done:

Response Analysis. At Toyota operators can signal, by switch or chord, when they encounter a problem. At some workstations, there is a range of switches covering quality, maintenance, and materials shortage. When an operator activates the switch, the overhead Andon Board lights up highlighting the workstation and type of problem. People literally come running in response. But the sting is in the tail: a clock also starts running which is only stopped when the problem is resolved. These recorded times accumulate in a computer system. They are not used to apportion blame, but for analysis. Thus at the end of an appropriate period, say a fortnight, a Pareto analysis is done which reveals the most pressing problems and workstations.

Line Stop A Toyota classic, related to the above, allows operators on the line to pull a chord if a problem is encountered. Again, the Andon Board lights up. Again, the stoppage is time recorded. But the motivation to solve the problem is intense because stopping the line stops a whole section. This means application of the 5 Whys root cause technique (see later section). Toyota in fact splits the assembly line into sections separated by small (one car?) buffers, so line stop only stops that section not the whole line.

Inventory withdrawal Many will be familiar with the classic JIT 'water and rocks' analogy, whereby dropping the water level (inventory) exposes the rocks (problems). This is done systematically at Toyota. Whenever there is stability, deliberate experimentation takes place by withdrawing inventory to see what will happen. Less well known is that this is a 'win-win' strategy: either nothing happens in which case the system runs tighter, or a 'rock' is encountered which according to Toyota philosophy is a good thing. It is not any rock, but the most urgent rock. Deliberate destabilisation creates what Robert Hall has referred to as a 'production laboratory'. However, Toyota is not averse to adding inventory where necessary. Refer to the inventory section on Ford's SMART process.

Waste Checklists Toyota makes extensive use of waste checklists in production and non-production areas alike. A waste checklist is a set of questions, distributed to all employees in a particular area, and drawn up by the LPO, asking them simple questions: 'Do you bend to pick up a tool?', 'Do you walk more than 2 yards to fetch material?', and so on. Where there is a positive response, there is waste. The result is that individuals and teams never run out of ideas for areas requiring improvement.

The 'Stage 1, Stage 2' cycle At Toyota there is a culture that drives improvement. This culture or belief stems from the widely held attitude that each completed improvement project necessarily opens up opportunity for yet another improvement activity. For want of a better phrase, Bicheno has termed this 'stage 1, stage 2' (see *Fishbone Flow*) after a list of Lean 'stage 1' activities that lead to 'stage 2' opportunities which in turn lead to stage 1 opportunities, and so on. The list of possible chains is very large, but an example will suffice. Setup reduction (stage 1) may lead to reduced buffers (stage 2), which may lead to improved layout (stage 1), leading to Improved visibility (stage 2), leading to improved quality (stage 1), leading to improved scheduling (stage 2), and so on and on.

4. Enforced Breakthrough Improvement

Active value stream current and future state mapping drive this category of improvement. They generally target a complete value stream. This type must be subject to regular action review cycles and an action plan or master schedule. Refer to the Mapping section. If a value stream map simply hangs on the wall without an accompanying master schedule it would be classified as passive breakthrough, if at all. Supply chain ('Seeing the Whole') projects would also be classified as enforced breakthrough.

Blitz or Kaizen events are a special case of enforced breakthrough and are the subject of a separate section in this book. It is breakthrough because typical blitz events achieve between 25% and 70% improvements within either a week or within a month at most. On the other hand blitz events are typically related to a small area, so are

frequently more 'point kaizen' than 'flow kaizen'. It is enforced because the expectations and opportunities are all in place. 'No' and 'it can't be done' are simply not acceptable. Concentrated resources are applied.

6. Incremental Preparation Improvement

These incremental steps are taken by designers and managers at the design or concept stage. Previous experience needs to be built upon using, for example, checklists. Deliberate steps need to be taken to learn.

7. Breakthrough Preparation Improvement

Here, dramatic breakthroughs are planned by, for example, target costing and value engineering. See the later sections on these.

Further Reading

John Bicheno, *Fishbone Flow*, PICSIE Books, 2006

Yuasa Yasuda, *40 Years, 20 Million Ideas*, Productivity Press, Cambridge MA, 1991

12.5 Kaizen

Kaizen is the Japanese name for continuous improvement. As such it is central to Lean operations. It brings together several of the tools and techniques described in this book plus a few besides. The word originates from Maasaki Imai who wrote a book of the same name and made Kaizen popular in the West. Although a registered name of the Kaizen Institute, the word is now widely used and understood and has appeared in the English dictionary. According to Imai, Kaizen comprises several elements. Kaizen is both a philosophy and a set of tools.

The Philosophy of Kaizen

Quality begins with the customer. But customers' views are continuously changing and standards are rising, so continuous improvement is required. Kaizen is dedicated to continuous improvement, in small increments, at all levels, forever. Everyone has a role, from top management to shop floor employees.

Imai believes that without active attention, the gains made will simply deteriorate, like the engineers' concept of entropy. But Imai goes further. Unlike Juran who emphasises 'holding the gains', Kaizen involves building on the gains by continuing experimentation and innovation.

According to Imai there are several guiding principles. These include:

- Questioning the rules - standards are necessary but work rules are there to be broken and must be broken with time)

- Developing resourcefulness - it is a management priority to develop the resourcefulness and participation of everyone

- Try to get to the Root Cause - try not to solve problems superficially

- Eliminate the whole task - question whether a task is necessary; in this respect Kaizen is similar to BP-

- Reduce or change activities - be aware of opportunities to combine tasks.

12.5.1   The Kaizen Flag

The Kaizen Flag is a famous diagram developed by Imai and widely copied and adapted. The flag portrays the three types of activity that everyone in a Kaizen organisation should be involved with. These three are 'Innovation, 'Kaizen', and 'Standardisation' against level in the organisation. An adapted version is discussed below. In the original, senior management spends more time on 'innovation' (to do with tomorrow's products and processes), a definite proportion on 'kaizen' (to do with improving today's products and processes). Senior managers also spend a small proportion of time on 'standardisation', that is, following the established best way of doing tasks such as, in top management's case, policy deployment and budgeting. A standard method is the current best and safest known way to do a task, until a better way is found through kaizen.

Middle managers spend less time than top managers on innovation, about the same time on kaizen and more time on standardisation. Operators spend a small, but definite, proportion of time on innovation, more time on kaizen, and the majority of time on standardisation.

Kate Mackle, former head of the British Kaizen Institute and now Principal in the consultancy Thinkflow, explains that innovation is concerned with preventing waste from entering tomorrow's processes, kaizen is concerned with getting waste out of today's processes, and standardisation is concerned with keeping waste out.

The version of the flag presented below has been developed based on Imai's original, but taking into account both experience and the ideas of the decision process developed by Ilbury and Sunter.

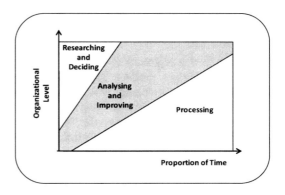

Here 'processing' is following the current standard best and safest known way.

Further Reading

Maasaki Imai, *Kaizen: The Key to Japan's Competitive Success*, McGraw Hill, New York, 1986

Maasaki Imai, *Gemba Kaizen*, McGraw Hill, New York, 1997

12.5.2 *Kaizen Events (or Improvement Events*

Kaizen events fill the gap between individual, very local improvement initiatives and bigger initiatives such as value steam improvement. They are the essential means to get cross-functional and multi-level teams involved in a Lean transformation. In that respect, kaizen events have a dual role – to make improvements but also to teach and communicate. An important and stimulating aspect of kaizen events is that they are done over a very short period of time, allowing manager involvement that might otherwise not be possible.

What comes to mind when you say 'Kaizen Event' is the 5 day variety. However, 'Mini Kaizens' taking from half a day to two days are a useful variant. What is described below is the full 5 day variety. The one or two day variety is essentially the same process, but streamlined and downsized with regard to participation.

Most kaizen events are focused on internal processes. But there is another, growing opportunity: the customer-focused kaizen event. Here the focus shifts to solving (or dissolving!) the customer's problems or improving the customer's effectiveness. This means putting yourself in the shoes of the customer. It means redefining the system boundary to include the customer.

Beware! *It is very tempting to rush into a kaizen event.* It may even produce good results and make everyone feel good. But what happens if the process is the wrong process? Then the event is simply rearranging deckchairs on the *Titanic*. So it is essential to have done an overall systems evaluation first. Such an evaluation will often come from a value stream mapping exercise, or from a capacity analysis that has highlighted a particular constraint, or from persistent failure to meet customer requirements.

Given that the right theme has been identified, and the preparation done, within one week a company and its customers could be benefiting from a leap in productivity in one area of the plant or office. Kaizen events are about 'going for it', about a preference for getting 80% of the benefit now rather than 100% of the benefit but much later, about learning by doing, by trial and error. It

is also about involvement. It is about real empowerment to 'just do it' without asking for permission to make every little change. A well-planned and followed through event has a good chance of sustaining its improvements over an extended period. Poorly planned and executed events, however, have frequently slipped back to the original state – and given such events a poor name in some organisations.

Today, 'Kaizen Blitz' (the name used by the US Association for Manufacturing Excellence) is well proven in both service and manufacturing companies. In the UK, Industry Forum (IF) has adopted a standard approach for kaizen – referred to as the Master Class process. The IF methodology has spread from automotive to, amongst others, aerospace, metals industry, and construction. Some consulting groups such as TBM and Simpler have developed their own versions.

The IF methodology was aimed at rolling out the kaizen event process. Each event would also have the aim of training facilitators to be able to run more events and introducing the concept to supervisors and others from areas intended to be targeted in future.

An appropriate quote to introduce kaizen events is:

> 'Whether you believe you can,
> or whether you believe you can't,
> you're absolutely right.'
>
> (Henry Ford)

Today we recognise that successful kaizen events require a great deal of both preparation and follow up to be successful. The following steps use the basic IF structure but based on the authors own experiences. These comprise:

A 1 day pre-diagnostic to select the area, discuss expectations, and to review measures and the measurement system that is in place. The workshop should be facilitated by an experienced practitioner, who needs to be chosen at this stage.

Then follows an initial preparation period during which measures and basic data are collected – quality and demand information. This is the scoping stage. During the next two weeks any necessary background information is collected, and how the event is to fit in with other Lean initiatives is clarified. This leads into a 3 day Diagnostic event. The aim here is to establish and clarify what the specific aims of the event itself are to be. The team is chosen. Mapping is typical here. If necessary basic education on 7 wastes and 7 tools of quality is given. The 7 DTI measures are set up. See the section on measures. The chosen team makes a presentation to management and all concerned on what they aim to do during the event. Objectives are agreed, and all necessary authorisations made - during the event, you do not want to have to seek permission to make changes. Assurances must be established on possible reduction in manning levels. Sometimes a Lean game is played by the team. Operators from the area may participate in the game and in education.

- A further period of preparation takes place over the next few weeks. The measures are firmed up. Final preparation takes place – this may include, for instance building ahead of schedule to ensure continuity of service during the event, and warning support staff such as maintenance and electricians to be on hand for the event. Any foreseen resources such as tools, tables, boards, racks, and post-its must be acquired. Arrange rooms and catering.

- A check day takes place during this period for any final arrangements

- The event itself is a five-day workshop. The idea is go around the PDCA cycle at least once but preferably a few times, ending with some tested, standardised changes. Measures are taken each day. On the last day a presentation is made. Follow up actions are given to specific people.

- After the 5 day workshop, three one-day follow up sessions are held at monthly intervals. These are to ensure that changes that were not put in place during the workshop are implemented. Examples are moving a machine embedded in concrete or

targeting a quality issue that was not 'cracked' during the workshop.

- Finally, the facilitator will stay in contact with the area for an extended period to check sustainability.

Kaizen events can be held in an area more than once, by targeting different aspects. Often, layout and 5S come first. Safety may be chosen in an environment with difficult union issues. Later, follow up events can address lead time and manning.

Here we present a proven methodology.

12.5.3   The Kaizen Event Process

Whatever the format, the following activities have to be worked through:

Some weeks ahead of the event:

- Select the area – probably from mapping and end-to-end value stream or from an accumulation of problems. But certainly taking the overall 'systems view' of the process, including the customer, to avoid working on a process that should not be there in the first place.

- In service events, give specific consideration to whether the system boundary or problem area should extend to include customer systems.

- In all service cases it is important to see the process from the customer's perspective. This aspect must be covered before the event begins.

- Select an appropriate time for the event – this is more important in an office than a factory because variation is usually larger.

- The group needs to be warned about the event, and participants from the group sought.

- Measures: Decide on relevant measures for the area, and take the measures.

- Team selection: one or more facilitators, front line managers from the area, the event owner, participants from the area, subject matter experts, people from the next most likely area for an event, outsiders. Around 12 is a good number – bigger for bigger areas, smaller for smaller areas.

Give consideration to 'tagging' in office areas where there are longer cycle activity durations. Tagging could involve placing 'travellers' on documents – or could be done electronically - and requesting people to enter arrival times and departure times on the traveller. Of course, this needs to be explained to the group.

Draw up an 'Event Charter'. This sets out

- the focus or principle concern of the event
- the aims of the event – what is hoped to be achieved
- current issues in the area to be addressed during the event
- the event boundary
- demand data – how much work does the office or area handle per day, and how does this vary across the day, week, month
- the dates of the event and duration
- who the participants are to be – including the facilitator
- what extra rooms will be used
- catering arrangements
- health and safety considerations – if electrical points or heavy items may be moved
- any approvals that may be needed should be signed off
- the extent to which work will continue in the area (if at all) during the event. If not, what to do
- any training that may be required.

The Event Itself

The following is typical for the blitz type.

- Day 1: Introductions, aims and scope, background – why is the event important, event methodology, basic Lean training including mapping, waste awareness, tools such as fishbone diagram, and if relevant practice on observation timing.
- Day 2: Go to the area and observe, map the routings, time durations, discuss the process with the people. Possibly the customers in some types of service process. Many offices have longer cycle operations so it may not be possible to observe or time all activities, so tagging (see above) or sample events may be used. If possible observe several cycles. Begin to generate ideas.
- Day 3: Idea generation, discussion around the maps; formulate plans to implement. Start the implementation.
- Day 4: The main day for implementation. Try out and adjust. Discuss with office workers and other shifts. Begin to prepare flipcharts for the presentation on Day 5. Check or estimate the measures.
- Day 5: A final check and adjust. Document the new process. List follow up items. Prepare an A3 summary sheet. Finish flipcharts for presentation. Present to area managers and senior directors. Agree next steps. Enjoy the free buffet.

After the event it is necessary to:

- Close off any outstanding points. On the last day of the event the persons responsible for doing or coordinating these mopping-up mini-projects must be identified. The event champion or a line manager MUST follow these up. At at least one service organisation, kaizen events lost credibility due to the ever-increasing list of outstanding topics that were never closed down.
- Have a review session every (say) month for a period of (say) 6 months. These may be very short meetings. But they are to look at the continuing performance of the area, and very importantly to record lessons learned. In other words they are 'after action reviews'.

An Alternative Methodology

Some organisations, like Ducati, prefer to spread an event over several weeks. The methodology is not 'blitz' type, but rather a fast version of a conventional implementation project. This allows time for adequate preparation between stages, and for disruption to be reduced. The format is as follows:

- Week 1: Gather and analyse data.
- Week 2: Generate ideas and select those for implementation during the kaizen week. Work through the implementation plan.
- Week 3: The Kaizen week. Actual implementation.
- Week 4: Check, refine.

None of the weeks is full time. During week 3, actual implementation goes on alongside regular work, but may also take place overnight and at the weekend.

12.5.4 Some lessons learned about kaizen events over recent years

- The workshop itself is the easy part. The harder and longer part is preparation and follow-up. An approximate time split is 40% preparation, 30% workshop itself, and 30% follow-up.
- The participation of managers in events is essential. Without this they will be lukewarm or even critical. Participation also helps overcome the problem of seeing authorisation.
- The participation of the supervisor or team leader from the area is essential.
- Moreover, the supervisor should attend one or two events in other areas before it is the turn of his or her area. Think how you would feel if a team descended on your area for a week and produced a 40% productivity improvement. Not too good – you may be motivated to show that what has been done was not all that good in retrospect. Ideally, when it is the turn of the supervisor to have a kaizen event in her area, she will already be the most enthusiastic participant – having experienced it elsewhere.

- Kaizen events should be co-ordinated through a Lean Promotion Office (or similar) in relation to a wider Lean implementation programme, via value stream mapping.
- All participants should have a clear understanding how the particular event contributes to the overall Lean vision or objectives.
- A good facilitator is invaluable. The ability to spot waste and opportunity builds slowly. Take the opportunity to transfer some of these skills.
- Blitz events work better in supportive companies. See the quotation from Henry Ford above. Do not set expectations too high. Under promise and over deliver.
- Sustainability remains the big issue.

12.5.5 Recording the Lessons – 'Knowledge Management'

Many useful points, both big and small come out of kaizen events. They need to be recorded so they will be available to other pats of the organisation. It is a 'drag' to have to write up the lessons but if it is not done the experience will be lost. Several organisations have set up data bases that can be quizzed on an intranet. Remember that 'knowledge management' should ideally include both 'explicit' (factual, hard) and 'tacit' (experiences and soft) information. At a minimum, set up a data base with:

- Name of Event
- Key words
- Participants.

To be able to speak to someone who has been involved in a similar project is the most valuable aspect. Thereafter, the data base could include, short notes, digital photos, value steam maps, sketches, and even voice notes.

Finally, we have identified a number of good and not so good practices:

Good:

- Having a systems view of the context into which the study area fits
- Evidence of senior management leadership and direction setting, availability to staff and recognition of successes
- Link the event with some benefit to the employees – such as space saving used for a coffee area, or improved furniture
- Have a short follow up review session every month for a few months
- Customer and stakeholder focus
- Employee involvement
- Training and development and use of the Investors in People Standard
- High energy and participation from everyone involved.
- When management is engaged road blocks are removed quickly
- When everyone involved in events is prepared to 'roll up their sleeves' – this has strong impact on the people on the office floor
- External help is invaluable in events until they become well established
- Have events across the board – at all stages, end-to-end: customer facing, clerical, administrative, operations, distribution, maintenance
- Awareness in Lean generates pull for training in some areas of the organisation.
- Learn from failures. Some events will fail – have a careful look back and seek answers
- Measurement and feedback on the measures is important

'Not so goods' and cautions:

- Lack of identification of critical success factors
- Lack of understanding of the concepts of quality and continuous improvement by some managers and employees

- Insufficient integration of continuous improvement activities
- Existence of a 'blame culture' when mistakes occur which may inhibit innovation
- Benefits not showing on financial radar, benefits don't reach the bottom line
- Need to target kaizen events on key business metrics not just in areas willing to participate
- Poor follow through – failure to close out on actions
- Lack of visibility for non participants – use visual displays/ storyboards and on the floor during and after the event – keep everyone informed of progress
- Build on what has been learned and leverage to other areas without it necessitating another kaizen event
- Need to develop internal competencies and not depend on external consultants/trainers
- Ensure all actions chosen for completion during kaizen event are directly related to achieving the charter
- Reliance on 'quick fixes' and fire fighting.

Further Reading

Nicola Bateman, *Sustainability…. A Guide to Process Improvement*, Lean Enterprise Research Centre, Cardiff University and Industry Forum, 2001

Sid Joynson and Andrew Forrester, *Sid's Heroes: 30% Improvement in Productivity in 2 Days*, BBC, London, 1996

Anthony C Laraia, Patricia Moody, and Robert Hall, *The Kaizen Blitz: Accelerating Breakthroughs in Productivity and Performance,* John Wiley and Sons, New York, 1999

Robert Hall, 'Ducati: The Lean Racing Machine', *Target*, Fourth Issue 2007

Siobhan Geary, MSc dissertation on Kaizen Blitz, LERC, Cardiff, 2006

Thanks also to Andy Brophy of Hewlett Packard.

Thanks to Bjarne Olsen of SAS who really brought home the importance of involving the customer.

The best magazine / journal on 'kaizen blitz' is *Target: The Periodical of the Association for Manufacturing Excellence.*

12.6 Mess Management

Russell Ackoff, Emeritus Professor of Systems at Wharton School, University of Pennsylvania has a useful classification for problem solving, of high relevance to continuous improvement. He says there are three levels

1. **'Resolving' problems:** This approach relies on past experience. Call a meeting and discuss the issues based on qualitative opinion. Ackoff says this is by far the most common approach, and although appropriate for truly messy problems is often ineffective. Better is:

2. **'Solving' problems:** This approach is based on the scientific approach. It uses quantitative methods. Gemba. PDCA. Six Sigma. It is far preferable where it can be used. Often, split a problem into those parts that are amenable to a scientific approach, leaving the remainder to the 'resolving' approach. A good way is to use the Ilbury and Sunter procedure. (a) Establish the 'rules of the game' over which you have no control and where there is little uncertainty. (b) Collect and analyse the facts and develop scenarios for the variables over which you have control but where there is uncertainty. (c) Develop the options. (d) Make appropriate decisions. This cycle can take a few seconds (as when driving) or months (as when making a major strategic decision). But, says Ackoff, better still is:

3. **'Dissolving' problems:** Change the nature of the problem. Take a 'Systems' view. Instead of developing complex scheduling to cope with changing and uncertain demand, influence demand variation in the first place.

So, pause to think. Are you resolving, solving, or dissolving?

12.7 A3 Problem Solving and Reports

The A3 method has grown hugely in popularity amongst Lean organisations in recent years, and with good reason.

A3 refers to the standard sheet of paper – two A4 portrait sheets, side by side. The story goes that it was the largest size of paper that it was convenient to fax.

A3 is:

- A standardised problem solving methodology incorporating the PDCA cycle.

- A standard report format. Rather than a multi-page report that may come in various formats depending on the whim of the writer, A3 forces the writer to be concise. Recall the quote from George Bernard Shaw 'I am sorry to send you this long letter – I did not have time to write a short one.' For the reader, another advantage is that he or she knows exactly where to look for the salient points. So, 'don't give me a report, give me an A3'.

- A standard documentation method, and easy filing method.

A3 is actually a family of report formats – used for planning, budgeting, communication, and problem solving. Here we consider only the generic problem solving type.

The general format of A3 is current state and analysis on the left hand side and future state and implementation plan on the right hand side. Often, along the bottom is space for 'sign off' – for people who have seen or agreed to the analysis.

An example is shown below.

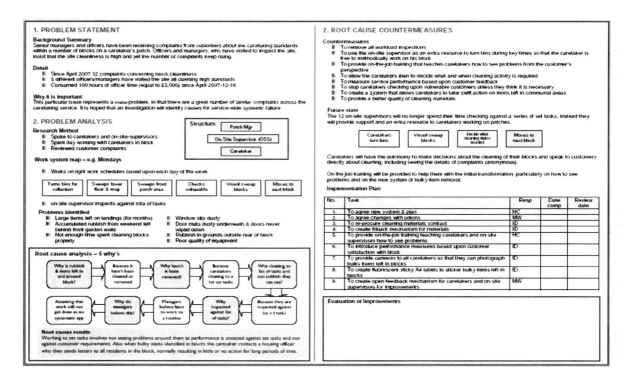

The standard layout is to have on the left hand side Check and Plan, and the right hand side is Do, Act and again Check. In the following section the various headings in the figure will be explained. Note that writing on the A3 should be clearly readable – not microscopic. About the same size font as this book or bigger.

Top right corner: The author or team, the date and the version number.

The Problem or Issue: A statement of the problem. One sentence.

Background: How did the problem arise? This is often a clue to the solution. Also, set the situation in the context of the wider value stream. A short statement of history. Two sentences maximum.

Current State: Can take various forms. A process map, possibly with kaizen bursts. A cartoon. A sketch of a product part. A data table - but beware – try to make the current state lively and interesting. Invent icons. Draw happy and sad customers.

Analysis: A standard approach is to use either the '5 whys' methodology or a fishbone diagram. When using 5 whys there may be several answers to each why. So prioritise by circling the major likely answers. Likewise on a fishbone diagram, circle the likely major causes. Sometimes a run diagram is useful. In fact, any of the classic 7 tools of quality can be used. These should be addressed to the first two of the classic 3C approach (concern, cause, countermeasure). Countermeasure is dealt with on the right hand side.

Future State: same comments as for the current state.

Countermeasures and Kaizens: These should be short sentences. It is good to have short and long term (permanent) solutions in mind. The kaizens may be short, focused actions – 'point kaizens'. Include verb plus noun to indicate the actions. 'Erect sign to direct customers'. In technical problems, a small sketch could be added.

Implementation Plan: Simply a list of actions, together with whom and by when. A more complex case may include a simple network diagram to indicate the sequence.

Cost Benefits: Not a great ROI or discounted cash flow analysis, but a simple outlay of actual cash and expected money benefits in the immediate period ahead. (State it as: £1,000 within 12 months.) Not all improvements generate cash – so include the other benefits – quality of worklife, happier customers.

Tests and verification: After the countermeasures have been implemented, the new solution should be expected to produce various results. State these, so they can be verified later. These should be related to the cost benefits.

Follow up: Wider issues might have been raised. A sentence or two would be appropriate.

Sign off: Along the bottom various line managers concerned with the issue should see and sign off the A3. Of course, they could annotate and discuss with the team.

An A3 can be used at different levels. They are used:

- For routine problem solving – point kaizens.
- As a supplement to mapping.
- As a tool and record for kaizen events. One company places kaizen event A3s on the wall near the area for a standard period of 12 months.
- As a supplement to policy deployment.
- As a test to evaluate a new employee.

Reference

Cindy Jimmerson, Presentation to MSc Lean Operations group, Cardiff Business School, April 2007

John Shook, *Managing to Learn: Using the A3 Management Process to Solve Problems, Gain Agreement, Mentor and Lead*, Lean Enterprise Institute, 2008

12.8 Communications Board

In virtually all Lean transformation situations, a Communications Board is a major device for improvement. It links directly with Lean philosophy, as it is a means of

- Communicating purpose, and ensuring consistency
- Problem surfacing and resolution
- Waste reduction
- Team-working.

The Communications Board becomes the focal point for communication, review and problem solving. Every morning there is a meeting around the Communications Board. Communication is two way – from the group leader to the group and from the group to the group leader. A meeting should take less than 15 minutes – frequently 10 or less.

Charts that are shown on the board may include:

- **The Policy Deployment matrix**. This shows the aims, projects, measures and results for the area. It will also show how the area's activities relate to the wider organisational aims and projects,
- **Concern, Cause, Countermeasure (3C) chart** – any concerns should be raised and entered, and all outstanding concerns discussed at the daily meeting. Each stage of each concern should be dated. A standard methodology such as A3 could be used with the Communications Board showing the overall status. A3s relating to concerns could be in an area next to the board.
- **Performance charts** – linked to productivity measures such as throughput. It is important that these are a source of problem identification and NOT a means of blame or competition.
- **Quality charts** – tracking problem areas, complaints, errors. Also as a means for improvement not blame.
- **Management audits** – where different levels of management are expected to visit the area at appropriate frequencies – for instance CEO

once per year, appropriate director quarterly, departmental managers monthly, section managers weekly, team leaders daily. Each turns over a magnetic counter from red to green when the visit is complete. The re-set dates (green to red) are stated. The set of activities that each is supposed to cover is kept on clipboard next to the board.

- **Total Productive Maintenance and OEE charts.** See the separate section on this topic.

- **Progress charts** – typically these will be related to the stages of the regular work cycle.

- **A skills matrix** – showing stages of development, learner, can do under instruction, can do working alone, instructor - against each skill category. These are often shown in a PDCA format cycle with the quadrants coloured in as the skill level is attained. A skills matrix is not just a 'factory thing' – it has many uses in professional environments, from lecturer to engineer, to accountant.

- **A task allocation chart** – who is working on what projects.

- **A leave roster.**

- **General company notices.**

Each chart would have a designated person responsible for chart maintenance and updating. This needs to be done by a designated time, before the daily meeting. But the communications board is to be used by the entire team.

Don't forget simple things like having a marker attached to a string, prominent location!

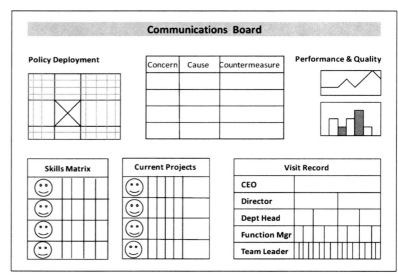

13 Managing Change

13.1 People and Change in Lean

It was Deming who said that most problems lie with the process not the person (so don't blame the person first), that there needs to be a consistent message, that there is a need to 'drive out fear', that barriers that prevent improvement and prevent 'pride in work' need to be removed. The challenge remains to this day: how do you successfully implement change in a process that consists of both people and machines? It is often easy to change the layout, move machines, and redesign material flows. Changing the people that operate this process is far from easy. Manufacturing and service operations alike are socio-technical systems, where human beings and physical equipment need to work in harmony to create the desired outcome. It is aligning this 'social system' that is the challenge when it comes to implementing change. In this section we will look into models how to achieve this.

13.2 What is the 'Social System'?

First and foremost, you need to accept that manufacturing and service operations are socio-technical systems consisting of an interplay of machines, technology and people. Addressing a subset of these only will invariably mean that change efforts will fail. Why? Because any change to the physical process is likely to affect the people in some form, and people who do not cooperate with the new way can become bottlenecks in the same way machines can. Implementing change means making changes to the social system, in a variety of ways, and all of these need to be managed.

The key features that make up the 'social system' are:

Work organisation: team structures, shift patterns, hierarchies etc. In other words, what is the structure that organises the current work? How are people grouped together, who is subordinate to whom?

Responsibilities: line of reporting, scope for making changes. Here, the main question is the extent to which the responsibility for the process is devolved down to the team level. Giving the actual team members more responsibility to improve the process is good practice, but remember that this also means that you are effectively taking responsibilities away from the team leader or supervisor, who might see this as a demotion!

Performance measurement: how are people rewarded, what incentives are given, what is the basis for promotion, etc. This is a critical point: 'what you get is what you measure!'. People will always try to look good on the performance measures that are given to them in order to make the bonus or promotion, so make sure that the measures you propose support your overall strategy! Use policy deployment or the balanced scorecard to devolve measures down into the hierarchies of the organisation.

It is very important to realise that making changes to any of the above means making changes to someone's working space and procedures, and not managing these changes will mean that the individual is likely to oppose, and in some cases even sabotage the proposed changes.

Remember, the effectiveness of change (E) is the product of the quality of change (Q), times the acceptance of change (A): $E = Q \times A$. Excelling in either quality or acceptance is not all it takes; both factors complement each other!

Crucially, remember that extrinsic motivators such as pay lose their effect with time. Intrinsic motivators, self drive, are more sustaining but require to be nurtured in the right environment.

13.2.1 Peter Senge's Systems Laws

To really know about a system, you have to know not only the entities or objects but their context – like the notes in a piece of music. Peter Senge, an influential systems thinker from MIT, has proposed 10 systems 'laws' that not only help us to understand systems better, but also are a excellent aid to avoiding implementation pitfalls. The laws are:

- **'Today's problems come from yesterday's 'solutions'**. This could be a re-statement of the 'push down, pop up' principle. Attack one problem, stemming from past actions, and another pops up. This is the fundamental problem of reductionist rather than holistic thinking. In Lean, using a new target to solve the problem often leads to unexpected behaviour.
- **'The harder you push, the harder the system pushes back'**. Or, 'systems bite back'. Most systems are in a state of natural balance. When a factor is altered others compensate. Hence the rapid growth of wildlife when predators are removed, which then stabilises due to food shortage. This happens in organisations also. Senge calls it 'compensating feedback'.
- **'Behaviour grows better before it grows worse'**. The story in the last point is an illustration. Management is often deluded (and rewarded!) by short-term results. Why? Because the whole system is not understood.
- **'The easy way out usually leads back in'**. There are many quick and easy solutions to problems in organisations – and they are all wrong! Juhani's Law states that 'the compromise will always be more expensive than either of the suggestions it is compromising'.
- **'The cure can be worse than the disease'**. Help may induce dependency – ask Africa!
- **'Faster is slower'**. Perhaps the supreme implementation law! Take time to achieve buy-in. The essence of what policy deployment should be about.
- **'Cause and effect are not closely related in time and space'**. If there is a problem in the office, the solution lies in the office.... Very likely not so.
- **'Small changes can produce big results – but the areas of highest leverage are often the least obvious'**. This is about leverage.

Malcolm Gladwell in The Tipping Point talks of 'mavens' in an organisation who have great influence despite their apparent lowly status. Find them! Likewise, timing is critical. Goldratt's 'conflict resolution diagram' may be useful to help find these influential points.

- **'You can have your cake and eat it too – but not at once'**. The essential message of Lean – you can have short leadtime and high quality and low cost – but it takes time to achieve. TRIZ (the Russian originated theory of inventive problem solving – see separate section) believes that finding 'contradictions' is the starting point for innovation. It seeks 'AND' not 'OR' solutions.
- **'Dividing an elephant in half does not produce two small elephants'**. Again, a warning on reductionism.
- **'There is no blame'**. Senge's point here is similar to Deming's preference to start with the process rather than the person. And Covey says 'win, win – or walk away' – seek ways in which both sides will win.

Further Reading

Peter Senge, *The Fifth Discipline* (revised edition), Randon House, 2006

William Dettner, *Goldratt's Theory of Constraints*, ASQ Press, 1997

Malcolm Gladwell, *The Tipping Point*, Abacus, 2000

13.3 Models for Change Management

13.3.1 The Basics

There are several models for how to implement change successfully. At the most basic level, there is the 'unfreeze, movement, re-freeze' model. The key point here is that each change process has three stages:

1. **Unfreeze**, whereby a situation is created that allows for changes to be proposed and discussed, and a stage that clearly communicates the objectives of the forthcoming change to the organisation.

2. **Movement**, where the actual changes are implemented. Again, clear communication about what is happening, and about the status of the project are crucial.

3. **Re-freeze**, whereby the status quo is 'frozen' and established as the new way of doing things. It is important to have a period of stability after every change, so that stability is regained. In the Deming cycle, this is referred to as 'holding the gains'. It is vital to create a stable process at the end of the change project, and to establish the new procedure as 'the way things are done' before embarking on a new change project. But, in today's dynamic environment, perhaps a better phrase is 'Remain slushy'.

Consider the difference between *attitude* and *behaviour*: you can change behaviour in 10 seconds by setting the right incentives -- but changing attitude takes much longer. Thus it is key to install the 'new way' firmly. Remember the experience of Chrysler in the 1990s, when they implemented Japanese manufacturing techniques in their plants, and amongst others, installed Andon cords so workers could stop the line if a problem occurred. After four months the line had not been stopped a single time – not because there had not been any problems, but simply because workers were afraid of stopping the line, as this was a punishable offense prior to the 'change'!

Also make sure to measure the before and after process performance, in order to provide the empirical evidence that the new way is indeed the better way to go. On many occasions there will be a great temptation to go back to the old ways, and having facts to show can help prevent that happening.

There are 4 C's that underlie all successful change management programmes:

1. **Commitment:** empathy and support from the top level signals that the change effort is serious and long-term.

2. **Communication:** change is often perceived as a threat, so clear and frequent communication is key to dissipate as much uncertainty as possible

3. **Co-production:** involvement of people concerned. Those affected by the change need to feel ownership of the new process, otherwise there is a great temptation to revert back to the old ways.

4. **Consistency:** or 'sticking at it'. People need to understand that this is not just a fad that will pass, but that you are serious.

These must be present at all times – and though they might sound basic, if you manage to stick to these four simple concepts then you are half way there!

Other changes models that might be useful include, for example, Ted Hutchin's, which usefully talks about five distinct stages (he refers to them as the Constraint Management Wheel of Change) – gaining consensus (on the need), gaining consensus on the direction (of change), gaining consensus on the benefits of the solution, overcoming all reservations, making it happen.

Kurt Lewin's Force Field Analysis, from the 1930's, is another simple but powerful tool for change. Draw a vertical line on a board. List and explain the forces for change. Open a discussion on the forces working against change. Do this level by level. Listen genuinely. It is a method that incorporates humility and respect.

13.3.2   The Change Iceberg

Several authors, Peter Scholtes (1998), Keki Bhote (2003), Bob Emiliani (2007), and Peter Hines et al (2008) use the analogy of the iceberg to explain Lean (and Six Sigma) change. Above the surface are the visible tools, layout, and processes. The official roles, responsibilities, plans, and standards. Below the surface lie the hard-to-see behaviours, leadership styles, and strategies. Scholtes makes the point that this informal organisation, having its own styles, values and communication links (that are often a residual of history), is what largely determines the individual worker's experience or 'culture'.

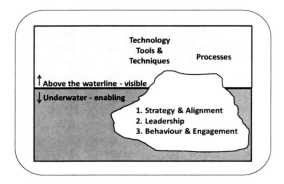

Hines maintains that 'below the waterline' a vital part is alignment of policy and measures. This is done via Policy Deployment (see separate section). The Policy Deployment procedure of buy-in and consultation certainly helps with communication and alignment. Of course, Policy Deployment relies on an appropriate policy actually being deployed.

13.3.3   Value Stream Mapping as Change Catalyst

In *Learning to See*, Rother and Shook discuss Current State, Future State and Action Plans as the trilogy in mapping and transformation. These three are also the basics of a change management programme.

Current State. The need for change must be recognised. The 'why'. This involves benchmarking current performance and identifying gaps. Future developments are even more important. Womack and Jones say that if necessary you should 'create a crisis' – certainly if you look around sufficiently well you will identify real threats. Then the need to change must be communicated. Not just communicated, but explained and discussed in detail.

Future State. Where we are going must be explained. The vision. What must be changed. 'Without a vision the people perish' says the Bible. Great visions get everyone on board 'Getting a man to the moon and returning him safely to Earth before the end of the decade', said Kennedy

(leading to even the toilet cleaner saying that his job was to help get someone to the moon). On the other hand, said Alice, 'Would you tell me, please, which way I ought to go from here?' 'That depends a good deal on where you want to get to', said the Cat. 'I don't much care where' said Alice. 'Then it doesn't matter which way you go', said the Cat.

Action Plan The plan to get there must be agreed. The 'how' and 'when'. Another of Stephen Covey's habits is essential here: seek 'Win Win'. You must and can find a way for everyone to win – reject TINA (there is no alternative), and embrace TEMBA (there exist many better alternatives). There may be no alternative to the need to change, but win-win must be sought for the how. So the change process has many similarities with the Hoshin process.

13.3.4   *Weltanschauung or Paradigms*

Peter Checkland reminds us of the power of Weltanschauung (or 'world view') on the change process. Joel Barker speaks of Paradigms. These are the views of people about what works and what does not. They are built up over a lifetime as a result of background and experience, and are fundamentally held views. They form the lens through which we interpret the world. We all have world views. Some think George Bush was great, others disagree. Some think Lean is great, others disagree. Why do they hold such strong views? Because of their background. These views take a long time to change. So embedded attitude must be expected when any new manager comes into an organisation. Don't assume they will know about Lean or believe in it. So doing can kill years of good Lean transformation work in a very short period. Witness the story of famous Lean company Wiremold, reported with great flourish in Lean Thinking, but, according to Bob Emiliani who wrote a book on the original success story, now reverted to 'batch and queue'. But remember Rosenzweig's book *The Halo Effect*. He asks 'does culture drive performance?', or does performance drive culture? We often assume the former, incorrectly. And when performance dips so does 'culture'.

13.3.5   Learning and Lean

The conventional learning curve is certainly applicable to simple tasks and to the addition of knowledge to our knowledge base. But Scholtes has observed a 'false learning curve'. The normal S-shaped learning curve of slow start but then steady progress levelling off, is assumed. But Scholtes says that true learning really begins in a second phase following the realisation that (Lean) is much more complex than first thought. This may go some way to explaining why so many change programmes fail, since to get to the point of realising 'we don't know much' (i.e. humility) can be between one and five years. This longer-than-expected timescale has been also observed by Koenigsaeker. The white flag of surrender is hung out by managers too early, when in fact good but unobserved progress is being made. This is made worse by manager mobility and short-term results orientation. In this sense a major reason for lack of sustainability lies with managers simply giving up.

Several authors have suggested that learning needs to progress through four stages:

1. **Unconscious incompetence:** you neither understand or know how to do something, nor recognise the deficit.
2. **Conscious incompetence:** though you do not understand or know how to do something, you do recognise the deficit, without yet addressing it.
3. **Conscious competence:** you understand or know how to do something, although demonstrating the skill or knowledge requires a great deal of effort.
4. **Unconscious competence:** you have had so much practice that it becomes 'second nature' and can be performed and taught to others easily.

Each stage has its problems. In the first you don't know what you don't know. Ignorance is bliss. In fact, in many organisations there is conscious or unconscious effort to keep people in this category by for example just not letting people know about Lean or saying that it is not applicable here. In the second, with realisation, may come despair and abandonment. In the third arrogance could be a problem. In the last, either one may assume, incorrectly, that most have been through the same long journey and now share the insights, or that looking back you forget the difficulties and assume that it is all so easy for others to pick up quickly.

13.3.6   'Nudging' for Lean

Thaler and Sunstein created huge interest with their book *Nudge* in 2008. This is about 'behavioural economics' and ways in which people can be positively, but gently, influenced to adopt behaviours that benefit themselves and society. But Nudge applies to Lean implementation also, even though it is not mentioned in the book. As Thaler and Sunstein point out, many good or bad practices in life result from habit or default rather than from deliberate choice. Thus 'culture' results from the 'ways things are done around here' which in turn often derive from habit or default rather than deliberate choice. Thaler and Sunstein make the point that the assumption that we are all 'economic man', always making rational decisions on the basis of weighing the evidence and considering alternatives, is highly flawed. In fact, most of us, including those with quantitative backgrounds, very often take the most convenient short cut. There is often no time, never mind the skill, to seek out and weigh all the alternatives. Herbert Simon said the same thing years ago when he said most people are 'satisficers' (good enough), not 'optimisers'. This could be an important reason why, for example, MRP does not work too well - because the defaults are chosen and the computer rather than the scheduler runs the company. In Lean, there are many counterintuitives, and many do the 'obvious', but wrong, thing. For example, larger batches and more inventory is good. In Theory of Constraints and to a lesser extent in Lean scheduling there are a lot of fairly complex calculations which need to be made and re-done when conditions change. So people will go through them and make the best decision. Right? Wrong! They will frequently make the easiest reasonable decision. So, try to 'nudge' them to make the easiest better decision.

Ways to Nudge include the following:

- The boss is regularly seen on the shop floor, always picks up any rubbish without comment, always visits the bottleneck process, always looks at A3's, spends time as a call centre operator each month, etc.

- Any out of place tools are always placed in the manager's office. No comment is made when they are collected.

- No reserved parking. Common refreshment areas.

- Restricted line side storage space, and warehouse space. No space available.

- Schedulers and designers have to walk through the plant to get to their offices.

- Setting up default values in computer scheduling systems that encourage small batches. A default may be a maximum batch size allowed without system overrides.

- Designers are given the default to select from standard components or fasteners.

- Small batch deliveries have wider delivery windows than large batch deliveries.

- Make daily improvement practices the expectation or norm.

- Sales are given the default to select from discounts for multiple regular orders rather than quantity discounts. Orders over a certain size (a month's production?) cannot be entered into the system.

- Flow lengths are displayed on the shop floor.

- Production not for delivery today is stored in an area labelled 'Racks of Shame'. No comment is made.

- Operator jobs are rotated, and appropriate accompanying multi- skill training given, as a HR default.

- Building the expectation of regularity in, for example, material handling or tugger routes, repeater product schedules, morning reviews, operators showing visitors around.

- Forklift trucks are not used. No big containers are available.

- Multipurpose machines are vetoed.

- One company uses the phrase 'make it ugly' for over-purchasing, and excessive inventory is stored in the Purchasing department.

13.3.7 Common Problems with Change Programmes

The errors that can occur and severely hamper any major change programme are plentiful. While not exhaustive, John Kotter provides a very good list of common errors that you should be aware of:

1. **Not establishing a great enough sense of urgency** – Poor financial results get people's attention, but don't be paralysed by the wealth of options available on how to proceed. Getting a transformation programme started requires the cooperation of many individuals. Without motivation people won't help and the effort is likely to fail

2. **Not creating a powerful enough guiding coalition** – Major change programmes need top-level support, but that in itself is not sufficient. In successful transformation, the chairman or CEO come together with a handful of divisional managers, plus middle management who will lead change in the respective departments, to develop a shared commitment to improvement through change.

3. **Lacking a vision** – The guiding coalition needs to develop a shared vision that is easy to communicate and appeals to all stakeholders, customers, shareholders and employees. Think about the next five years, and go beyond the numbers!

4. **Undercommunicating the vision by a factor of ten** – Transformation is impossible unless the hundreds or even thousands of employees affected are willing to help,

possibly even making short-term sacrifices. However, no one will make sacrifices unless they believe that useful change is possible. Make sure your communication is credible and regular, and remember: you cannot overcommunicate!

5. **Not removing obstacles to the new vision** – Transformations frequently hit large obstacles that middle management or employees are unable to move out of the way. Make sure that there is a communication link upwards, and that these 'elephants' are moved out of the way.

6. **Not systematically planning for, and creating short-term wins** – large transformations take time, so in order not to loose momentum make sure you put in place short-term goals, and celebrate their achievement! People need to see within 12 months compelling evidence of change that the new journey is producing results, otherwise they will lose confidence.

7. **Declaring victory too soon** – After the first battles have been won, the results will come in, and there will be a temptation to declare the war over. But remember, it takes seconds to change behaviour, but years to change attitude. The changes need to sink in deeply in the company's culture in order to be sustained!

8. **Not anchoring changes in the corporations' culture** – Change sticks when it becomes 'the way we do things around here'. New behaviours need to be rooted in social norms and shared values. The two ways to achieve that are to make a conscious attempt to illustrate how the new way of doing things has improved performance, and secondly to make sure that the new generation of senior management embodies the new vision. Promote selectively based on criteria that support the new approach!

And specific to Lean transformations, Mike Rother, co-author of the *Learning to See* mapping book gives five pitfalls of implementing Lean:

1. **Confusing techniques with objectives** – This book has also made the point that Lean is not tools.

2. **Expecting training to make Lean happen** – Mike says this is 'pure bunk'. You need to change the system. Recall here Deming's 94/6 rule whereby 94% of problems stem from the system – that only management can fix, and only 6% from operators. Change cannot come purely from the bottom.

3. **Leading from the office, via plans, maps and charts** – Mike makes the point that Lean can only be achieved through Gemba (he does not use the word but means it).

4. **Relying solely on Blitz workshops** – Issues here are the lack of the big picture, sub-optimisation and sustainability. Refer to the Kaizen Events section.

5. **Quitting after failures or too early** – As Zorba the Greek said, 'Nothing works the first time!'

To these five we can add a few of our own:

6. **Management commitment** – an old cliché perhaps, but it is difficult to think of a successful Lean transformation which has not had real commitment and involvement from the top. Sending out a clear signal helps – like asking the top team to apply for their own jobs.

7. **Cherrypicking** – pursuing a fairly random selection of tools in a fairly random selection of locations. Changeover reduction here, kanban there, mapping everywhere but with little follow through. Often these follow from the latest conference, book or meeting.

8. **The 'We are different' attitude** – so we need to re-invent our own system.

9. **The 'We can do it ourselves' notion** – Toyota did it themselves, but it took them two decades with some exceptional people. Ohno made almost no progress for a decade! Can you wait? Are you confident that you have the people, and that they will be around in a decade? Ultimately you can only do it yourself, but you will need guidance. A problem is that a fair proportion

of Lean consultants have limited exposure outside a small area. Even ex-Toyota people have crashed in environments they know little about.

10. **Not thinking that '80% is greater than 100%'**, meaning that a set of decisions 80% correct but bought into by all will be better than an optimal 100% correct solution imposed by the 'experts'.

11. **Lack of full time facilitation** – Most companies will need a Lean promotion office to keep the momentum going.

Further Reading

Peter Scholtes, *The Leaders Handbook,* McGraw Hill, 1998

Peter Checkland, *Systems Thinking, Systems Practice*, Wiley, 1973

Mike Rother, Crossroads: Which Way Will You Turn on the Road to Lean?, Chapter 14, of Jeff Liker (ed), *Becoming Lean*, Productivity, 1998

John Kotter, Leading Change. *Harvard Business Review*, 2007, Vol. 85 Issue 1, p. 96-10 (a reprinted version of: John Kotter, Leading Change: Why Transformation Efforts Fail. *Harvard Business Review*, 1995, Vol. 73 Issue 2, p59-67)

Peter Hines, Pauline Found, Gary Griffiths, Richard Harrison, *Staying Lean*, Lean Enterprise Research Centre, 2008

Bob Emiliani, *Real Lean*, Volumes 1 and 2, Center for Lean Business Management, 2007

Richard Thaler and Cas Sunstein, *Nudge*, Yale University Press, 2008

13.4 Creating the Lean Culture

The word 'culture' is a great word, and also one that is greatly misused! Scholtes describes organisational culture as the day-to-day experience of the ordinary worker. 'Culture' has become the great fallback word for why Lean is not working as it should. 'It is the culture.' There is now also a book on 'Toyota Culture'. First, let us say that we support Scholtes' scepticism about the word, and go along with his suggestion that

you substitute the phrase 'current behaviour' instead of 'culture'. But let us make a few points about Culture or Current Behaviour:

- Culture is something that you learn day by day, not by going on an outward bound course or by reading a book about moving cheese. 'Act into a new way of thinking', not 'Think into a new way of acting'.

- Do you ever hear about successful sports teams having 'cultural' problems? Seldom. Why? Because the team as a whole and all the members individually know exactly what the aim is, exactly what they need to do to get there, in detail. It is the coach who has to make these clear.

- Don't allow the 'culture' excuse. Instead, use the 5 Whys. If there is an issue and you give 'culture' as the reason it is a dead end. This is not to deny that there are cultural issues, but what is causing it?

- Psychologist Frank Devine talks about the non-negotiables, the 'limit testers' and the 'watchers'. Management must first make clear the non-negotiables – whether safety, attendance, quality, punctuality, development or whatever. There will be the 'inner committed' about whom you don't have to worry. But there are also the limit-testers who push the boundary to see if management really means what they say. The watchers watch the reaction of managers to the limit testers. If the managers give way, just a little, the downward spiral begins.

These points can all be summarised by a word – management. Not necessarily top management but also first line, face-to-face, every day management. Remember the statement 'the shop floor is a reflection of the management'. That applies to cultural issues as well. If you accuse others of being 'concrete heads', it may be a reflection on yourself.

In the excellent book, *The Halo Effect*, Phil Rosenzweig says that it is often assumed that culture contributes to success and performance. But the research does not support this. The reverse is true. Performance drives culture. If there is success, people tend to report enthusiasm for change, great support, great teamwork, great management, a futuristic organisation and a 'hero at the helm'. But very quickly the same people in the same company with the same managers report negatively when performance, due perhaps to a change in the market, declines. The same managers are now said to lack vision, be complacent, and arrogant Likewise there is a tendency to attribute success or failure to one thing – like culture or leadership - when there are almost invariably many forces at work.

So, Rosenszweig takes issue with some of the major texts such as *Good to Great*, *Built to Last*, and *In Search of Excellence*, warning that the 'lessons' may be a delusion.

Other delusions that Rosenzweig warns about include 'Connecting the Winning Dots' – where characteristics of winners are identified, but how many losers also share those characteristics? 'Timeless principles' often don't stand the test of time. And success stories are generally very poor at predicting *future* performance. So, in the authors' opinion, we need to study failures more than successes. Always ask managers from successful Lean companies about their failures....

Having said that, what might be some true characteristics?

Although Lean often involves revolutionary change, culture change is evolutionary. Day by day. And because, inevitably, managers leave, coaching future managers on their attitudes to, and interactions with, subordinates needs to be continually done. Arguably, there is no more important a task.

So, what makes a 'Lean' culture? Basically all people, from CEO to junior, share two related characteristics, both related to Learning: humility and respect.

Humility – The more you knows about Lean, the more you realise how little you really know. Dan Jones speaks about 'peeling the onion' to uncover waste – the same is true learning about Lean. A sure sign of impending failure is a manager who claims to 'know it all' or 'we tried that in 1990'.

Visitors to Dell in Ireland, which has very impressive inventory turns, are invariably struck by the humility of their managers ('we have lots to learn') and their willingness to learn from others, no matter whom. Stephen Covey says that genuine listening is the most important of his 7 habits of highly effective people.

Respect – The expert is the person nearest the actual job. But respect goes far beyond this. It is about regarding the workforce as a family. A good family downplays hierarchy, listens attentively, is genuinely interested. Parents try to develop and encourage their kids, to bring out the best in them. But it also expects contribution. Parents know that kids have new skills that they may never possess. In organisations, non-genuine respect is quickly detected.

Respect is also to do with avoiding wasting time. If you keep employees or customers waiting you are saying to them 'your time is not as important as mine'. If you allow a worker to use a machine that results in defects, you are in effect saying 'your work does matter that much'.

A psychologist once told one of the authors that, as a rule of thumb, everyone has five unrecognised relevant skills where that person is more competent than their boss. The skills may be in communicating, in humour, in drawing, in creativity, in concentration, in persistence, in accuracy, in rhythm, or whatever. It is a valuable thought.

Peter Wickens, former HR director of Nissan UK explained in *The Road to Nissan* about the necessity of reducing organisational levels, working towards single status, eliminating benefits based on position, and opening information flow.

Developing your people shows respect for them. Hansen and von Oetinger talk about 'T-shaped' managers. Toyota chairman Watanabe uses this phrase also. T-shaped because they have a broad range of skills, but at least one in-depth skill. But why only managers? Dell and Unipart, amongst others, encourage their people to improve their skills in non-strictly relevant areas such as history and cookery.

How to achieve Humility and Respect? It begins at senior levels and percolates down. Consistent demonstration over a long period of time is required. So is Gemba style management – you cannot genuinely listen if you aren't there. Both take a long time to build up, and unfortunately both are quickly destroyed. Richard Kunst tells stories about Lean managers picking up cigarette butts from the smoking area to prevent complaints thereby keeping it open, of managers waiting to greet early arrivals at work thereby encouraging on-time starts. It's all in the detail.

Further Reading

Peter Scholtes, *The Leaders Handbook,* McGraw Hill, 1998

Moreton Hansen and Bolko von Oetinger, 'Introducing T-Shaped Managers: Knowledge Management's Next Generation', *Harvard Business Review*, March 2002, pp106-117

Phil Rosenzweig, *The Halo Effect*, Free Press, 2007

13.5 Training within Industry (TWI)

Perhaps the 'granddaddy' of all improvement cycles are the Training Within Industry (TWI) steps developed during World War II, but which in turn were developed from even earlier approaches. TWI is arguably the most effective and influential training programme ever developed. TWI methods had massive impact on Japanese industry, including Toyota. The thought was, and is, that it is the front line supervisor who can have the greatest impact on day to day productivity by instructing well on how to do a job, by improving the work, and by dealing effectively with any operator problem or motivation issue. Today, many Toyota team leaders still carry quick reminder cards based on TWI. TWI methods cover what was considered to the three essential areas for a supervisor – job instruction (JI), job methods (JM), and job relations (JR). Each has a four step standardised procedure, summarised below. Those three skills are, still today, considered the essential tasks for any team leader at Toyota. The package of skills is what makes team working

effective – all three are necessary. This is why, when TWI methods arrived at Toyota in 1950, Ohno's methods began to take off. They were languishing before then.

Today, Toyota uses Job Instruction almost unchanged from the original TWI concept. TWI style Job Relations was terminated in 2000 – but is still a prime responsibility of team leaders and supervisors. TWI style Job Methods have been added to by including waste, flow, and an emphasis on System.

Huntzinger also makes the point that TWI is in fact the origin of Kaizen.

13.5.1   Supervisor Skills and TWI

As Lean develops and extends, the vital role of the first line team leader and supervisor is gradually being appreciated. In a Lean system the team leader and particularly the supervisor, is not an expeditor or a job scheduler. Instead, they are a coach, a mentor, a confidant, a problem facilitator, a manager, as well as having prime responsibility for the quality and for meeting the schedule. In the Toyota system a team leader has a span of control of perhaps 5-8 operators, with the group leader (supervisor) having a span of control of 3-4 team leaders. This may seem to be non-Lean over-staffing. In fact, they are onerous jobs requiring considerable effort. Ohno said, 'management begins at the workplace', and it is not an exaggeration to say that much of the success of the Toyota system is due to attention to detail at the frontline man-operation interface. In fact, as Art Smalley has observed, if as much time was given to supervisor training as to value stream maps, the results would be impressive.

For people familiar with TPS, the similarities with TWI are striking. Also noteworthy is the similarity between Deming's PDCA and the three methods, and the use of the Kipling's 5 Honest serving men, and waste elimination, in the job methods program. In job methods, there are also similarities with the Six Sigma DMAIC steps. Industrial Engineers will find the steps very familiar. You wonder which came first in all this?

Job Instruction	Job Methods	Job Relations
Prepare	Breakdown	Get the facts
Present	Question	Weigh and decide
Try Out	Develop	Take action
Follow up	Apply	Check results

Job Instruction takes place in practical settings, and is aimed at training new employees. It has to be done by a supervisor thoroughly familiar with the task. The first stage is to analyse the job into the important steps and within each step the key points. This is called the Job Breakdown Sheet. Then patient explanation is given. 'If the worker hasn't learned, the instructor hasn't taught'. Please refer to the section on Standard Work for greater detail on Job Breakdown.

Job Relations is aimed at giving supervisors basic behaviour in organisations, motivation, and communication skills. Get the facts, listen, don't judge, take action. In fact, use Deming 'drive out fear', with Covey, ' seek first to understand, then to be understood'.

Job Methods is aimed at giving supervisors basic and systematic problem identification and improvement skills. The industrial engineering skills of eliminate, combine, rearrange, simplify – together with 5 whys and the 6 honest serving men.

Further Reading

Donald Dinero, *Training Within Industry*, Productivity, 2005. The history, how it links with Lean, and detail on the programs. Includes a CD with the original 1940's training bulletins – still very relevant!

Patrick Graup and Robert Wrona, *The TWI Workbook*, Productivity, 2006. TWI for the 2000s

Jim Huntzinger, *The Roots of Lean: The Origin of Japanese Management and Kaizen*. Report.

There is an annual TWI conference. Papers are frequently available on the web.

13.5.2 Skills for Managers and Leaders

Leadership remains a very hot topic with hundreds of books available. In the West Leadership has assumed mythical significance – get the right top person and your problems are over. Yes, leaders are important, but be cautious of the hype surrounding leaders – at least in a Lean environment.

- You may know who the current Toyota president is (Ken Watanabe, 74 in 2007) but he is hardly a figure of the legendary status of Jack Welsh, Steve Jobs, or Larry Bossidy. Lawler and Worley point out the wealth of research studies that highlight the importance of continuity of both leadership and management. To rely on one leader is risky as numerous cases have shown.

- The Harvard case study, Jack Smith: Becoming a Toyota Manager, comes as a shock to many. Smith is recruited into Toyota with an impeccable background both academic (MSc and MBA) and high level management experience. Yet he spends his first three months on the shop floor. He learns that the team leaders are generally better than him at problem solving, he gains credibility with operators, he learns what the scientific method really means, and his eyes are opened to the hundreds of small scale improvements that are possible even in one of the best plants in the world. He learns humility. And Toyota learns about him – before 'turning him loose' as a senior manager.

- Rosenzweig's book, *The Halo Effect*, is also sobering on leadership. Many high profile, legendary, leaders fail in due course when the situation changes. Rosenzweig points out that you can always find good things to say about leaders of successful companies, and bad things to say about leaders of unsuccessful companies. But sometimes it is the same person.

So an aspiring Lean company needs to pay a lot of attention to developing its managers. Lawler and Worley, whilst not taking specifically about Lean organisations, but rather about organisations that need to be continually adaptive to change, make a number of suggestions:

- Whilst job descriptions are out, because they are wedded to the past, *person descriptions* are in. A person description lists the knowledge and skills that each manager already has but also needs to develop. The needs are based on how the organisation sees its future. The person descriptions should be searchable within the organisation on a data base.

- But, the organisation should not direct managers to learn required skills. 'People need to be responsible for their own careers and employability'. This is not to say that the organisation does not sponsor development. There is nothing as deadly as people forced to attend Lean 'education'.

- Whilst training should not be directed by the organisation, it should be available on a just-in-time basis, and rewarded appropriately.

- Organisations need to think about how jobs can be made more stable, whilst giving suitable rewards. So many Lean transformations have failed to realise their potential because of manager turnover. This may mean, for example, flatter organisations, more fluid job titles, a reward structure not based on position, and many parallel career paths.

This fits well with Lean. Working for a Lean company should be seen as a huge career boost. No organisation can guarantee jobs, although they should commit clearly to nobody losing their job as a result of improvement.

Programs	Needs	Methods
Job Instruction (JI)	Knowledge of the work	Prepare the worker
Job Relations (JR)	Knowledge of Responsibilities	Present the operation
Job Methods (JM)	Skill in instructing	Try out performance
Program Development (PD)	Skill in improving methods	Follow up
	Skill in Leading	

13.5.3 Leader Standard Work

In his excellent book, *Creating a Lean Culture*, David Mann says that there are four principal elements of Lean Management. These are Leader standard work, Visual controls, Daily accountability process, and Leadership discipline. Of these, Leader standard work has the 'highest leverage' and helps to consolidate the other three.

The concept of Leaders having standard work was spoken about by Imai in his 'Kaizen Flag' concept. (see the Improvement section). The concept turns out to be one of the most effective means for Lean transformation. It is simple to describe, but requires sustained determination for success.

Leader standard work involves drawing up a timetable of activities that each leader in the organisation, from team leader to Vice President needs to follow. Times are set aside and rigorously adhered to. A typical framework is shown in the table.

Level	Frequency	Typical activities
Team Leader	Several times per day at Gemba	Review problems, Set tasks
	Daily, with team, at Gemba or at bottleneck	Morning meeting: yesterday, today. Review charts, Problems, tasks, training, briefing, improvement, target inventory
Supervisor	Daily, with team leaders	Review with Team Leaders. Production tracking, Staff reallocation, Kaizen and training activities.
Section manager	Daily, with supervisors	Review KPI's, problems, point kaizens, progress
Value Stream Manager	Weekly with section heads and others e.g. quality, accntg	Review value stream KPI's, kaizen event plans, VSMs
Plant manager	Weekly with Section Managers and VS mgrs	Review value stream KPI's, progress, problems
Vice President	Monthly with Plant mgrs and	Review plant and value stream KPI's and progress.

At each level, there is two-way communication with those present. Issues are raised for upwards communication. Briefings allow downward communication. Escalation is incorporated – each level decides what to bring to the next level.

General agendas are fixed, but other items can be raised. Meetings are of fixed duration – typically 10 minutes, always start on-time, and may be stand-up. They take place around a board where daily performance and issues are shown. A standard problem solving method, like PDCA, is shown and used. Non-production visuals, like training and news is shown. Meetings are not missed for 'crises'.

The best examples include a visual display of the meeting timetables, and visual signals to indicate attendance satisfactory outcome of the meeting by each level of manager. Policy deployment matrices for each area appear on the board.

The procedure is followed in all areas – from canteen to design.

What leaders at all levels are expected to do, regularly, in a Lean transformation is a vital but often neglected aspect. Drawing up the level-by-level agendas and procedures, and training the leaders on how to do the task well, is a powerful unifying force.

Further Reading

Phil Rosenzweig, *The Halo Effect*, Free Press, 2007

Steven Spear, *Jack Smith (A), (B), (C),* Harvard Business School Case Study, 2004, 9-604-060

Edward Lawler and Christopher Worley, *Built to Change*, Jossey Bass Wiley, 2006

David Mann, *Creating a Lean Culture*, Productivity, 2005

13.6 The Adoption Curve and Key People

Getting the right people is a perennial theme in Lean. Jim Collins in *Good to Great*, studied various successful transformations and believes that their leaders did not do so by setting a new vision and a new strategy but 'first got the right people on the bus, the wrong people off the bus, and the right people in the right seats – and then figured out where to drive it'. The book *The Halo Effect* gives a devastating criticism of *Good to Great*, and you realise that such statements are overly simplistic. Nevertheless, there are certainly problem people and people who have great influence.

13.6.1 The Pareto in Change

Malcolm Gladwell talks about *Mavens* – people who accumulate knowledge and get into the detail. But not only do they know, they want to tell – not in a know-all kind of way but in honest assessment; they like to help. They are 'students and teachers'. So mavens spread the good or bad news. If Lean is working or not working they will know, and will say so. People listen. Then there are *Connectors*, who have lots of contacts and put people in touch. And then there are *Salespeople*, who often unconsciously sell ideas. Generally these people are known, and they carry huge influence. Change will be easier if these people are made use of. Mavens, connectors and salespeople constitute a very small but highly significant number of people. The Tipping Point is reached quite suddenly when a critical mass is persuaded. Hence the importance of the adoption curve. Gladwell also talks about how change spreads – like a disease. Not specifically about Lean change, but applicable nevertheless. There are three factors: content, carriers and context. Content is the value of the message itself. The ability to get things done. The value of the change will influence the support it receives. The message has to be powerful and relevant. Carriers are infected and move the message. They comprise the three types described earlier. Context is the environment – is it hostile or conducive to the virus. There are external and internal factors. Externally it is about the current climate that the organisation finds itself in, and internally the support that the message receives from senior management.

Andrea Shapiro has built on the ideas of Gladwell to form a change management theory which has been used in several major Lean transformations in the UK. She uses the categories of 'apathetics', 'incubators', 'advocates' and 'resistors' – a little like the Adoption Curve below. Thus advocates influence the apathetics, some of whom become incubators. Some incubators turn into advocates, others become resistors. Managing these flows is critical to achieving change. But it is an ongoing system, so the numbers in each group may grow or decline over time.

13.6.2   More on The Tipping Point: Just Do It or Analysis Paralysis

Delay can be a self fulfilling prophecy. As Johann Goethe so elegantly wrote:

'Until one is committed there is always hesitancy, the chance to draw back, always ineffectiveness. Concerning all acts of initiative (and creation) there is one elementary truth, the ignorance of which kills countless ideas and splendid plans: that the moment one definitely commits oneself, then

providence moves too. All sorts of things occur to help one that would not otherwise have occurred. A whole stream of events issue from the decision, raising in one's favour all manner of unforeseen incidents and meetings and material assistance which no man could have dreamt would come his way.' (Thanks to Lee Flinders for bringing this excellent quote to our attention.)

13.6.3   The Adoption Curve

Various writers have discussed an adoption curve covering the range of employees from early adopters to 'anchor draggers' This has merit in thinking about people aspects of Lean implementation. Here an adaptation is given, derived from the Rogers Curve, but based on the authors' experience and background. The figure shows a notional distribution of a workforce. Areas represent approximate proportions.

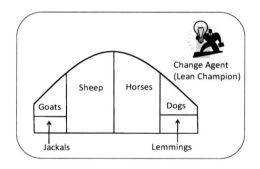

The Lean Champion is a farmer not a hunter. *Farmers* take the long view, and win in the long term. *Hunters* take the short view, get early gains but ultimately die out. Farmers are shepherds. Early adopters are found on the right hand side of the figure. These people are 'gung ho' for change. They require very little convincing. But experience shows that there are two sub-groups here. *Dogs* are faithful, but are also intelligent. This valuable group will be the core of the change initiative. By contrast, *Lemmings* are easily up for change, any change, and, in a sense, are not the people you want. ('If he thinks it is a good thing, then it must be a bad idea.') They leap in just too quickly, without thought. *Horses* are the key group. They need guidance from a Lean champion. They require to be trained, to be broken in. Horses are also intelligent. Most horses work well in teams. The strategy to be adopted with horses depends on the situation. In normal circumstances the rider is in control, and the horse will take instruction except in emergencies. When there are fences and jumping is required, horse and rider act synergistically — the trick is the right balance between guidance by the rider and initiative and judgement by the horse. On a mountain trek, however, the best strategy for the rider is to let the horse take most of the control, relying on it to pick out the safest path. *Sheep* can be led by riders with horses and dogs.

Generally they cannot be relied upon to get there without considerable guidance. Shepherding is required. Sheep are multi-function providing wool and mutton. They are adaptive to a wide range of climates. Sheep can also be led to an extent by goats, either into the abattoir or into a lush field. (Sheep is not a derogatory term — they are the backbone of much farming.) *Goats* are much more cautious. They have good reason to doubt, and some of those doubts are valuable insights. They climb trees and look around. But they can be made into valuable assistants. When they are convinced they are more useful than sheep. Goats lead sheep. Finally *Jackals* cannot be trained. They eat goats and sheep, and may scare horses. They are the true anchor draggers. Note that in this analogy, groups traditionally regarded as anchor draggers and early adopters both have sub groups. These sub groups need to be distinguished. Beware of lemmings. Listen to the goats — they may have good, thoughtful reasons for reluctance.

As an aside on what are here called Goats, Kegan and Lahey contend that a major reason why some people (and groups) are reluctant to change is 'competing commitments'. For example a manager is offered promotion but is committed to spending time with an aging relative. His superior then makes a 'big assumption' that the commitments are mutually exclusive. Destructive. To uncover this, Kegan and Lahey suggest that

managers ask a series of questions. For instance, What would you like to see changed? Then, what commitments does your complaint imply? And, what are you doing that is keeping your commitment from being realised?, leading to working out a way to reconcile this big assumption with the change. Again, it's Covey — seek 'win-win, or walk away'.

Further Reading

Robert Kegan and Lisa Lahey, 'The Real Reason People Won't Change', *Harvard Business Review*, November 2001, pp. 84-92

Malcolm Gladwell, *The Tipping Point*, Abacus, 2000, Chapter: 'The Law of the Few'

Everett Rogers, *Diffusion of Innovation*, Free Press, 1995

Edward Lawler and Christopher Worley, *Built to Change*, Jossey Bass Wiley, 2007

Mager and Pipe, *Analysing Performance Problems*, Pitman

Andrea Shapiro, *Creating Contagious Commitment*, Strategy Perspective, 2003

14 Sustainability–Making Change Stick

Sustainability can mean very different things.

- Some see it in terms of long-term viability, generally with an environmental perspective.

- Some see a dynamic nature in sustainability, whereby you continually adopt to your environment. Out of the Fortune 500 from 100 years ago, only a handful have remained there until today. Why? Adaptation is needed to sustain the business.

- Some see it simply as lasting change, which is the way in which we will use it here.

We believe that, like perpetual motion, there is no such thing as 'self-sustainability'. The Second Law of Thermodynamics, sometimes called the supreme law of the universe, says that unless you put energy into a system it will run down, degrade into disorder and, eventually, death (as in no energy). The same applies to Lean sustainability. But there are greater risks – without the right sort of energy the system will degrade very rapidly. The amazing story of Wiremold is a case in point. After 12 years of hugely successful Lean transformation the company had become one of the Lean showpieces in North America. It was written up in Lean Thinking. Yet three years after being taken over, many of the measures of performance had declined to what Bob Emiliani (who wrote a case study on the success of the company) describes as 'batch and queue'.

Sustainability is one of the great issues in management, never mind Lean. Ask GEC/Marconi or ICI. In their book *Creative Destruction*, Richard Foster and Sarah Kaplan report that of the 500 top US companies in 1957, 37% survived to 1997, and of those only 6% outperformed the stock market.

One view of sustainability is simply 'Doing a Cortes' – that is burning his boats so that his men had no choice but to adapt their old-world ways to the new world. The equivalent may be an option in a time of real crisis, but is not really available to most.

In 2007, the NHS Institute for Innovation and Improvement produced an excellent model to self-assess against a number of key criteria for sustaining change, and to identify barriers to sustainability. The model relates, of course, to health care and has 10 factors grouped into three areas: Process, Organisation and Staff. The detail factors are not discussed, but the three areas will be used here.

14.1 Process (and System) Sustainability

14.1.1 Sustainability and Tools

A large number of organisations have failed to produce the desired results from the direct and prescriptive application of Lean tools. The tools themselves have been proven to work in many situations. The difference must then be in how the tools were applied, their appropriateness, but not the tools themselves. Spear and Bowen state that observers of Lean often confuse the tools and techniques with the system itself. They point to the paradox of the Toyota Production System (TPS) that relates to a very rigid framework of activities and production flows coupled with extremely flexible and adaptable operations. This paradox is partially explained by the authors' revelation that Toyota actually practices the *scientific* method in its operational activities. Plan Do Check Act is the principle mechanism for this scientific approach. Tools and techniques are thus treated as hypotheses to be tested in the particular situation at hand. If the results are unfavourable the tools must be modified. Organisations without this understanding often fail to allow for local factors influencing the successful application and sustainability of tools and techniques. Nakane and Hall also studied TPS and came to a similar conclusion. They found companies that merely implement the techniques without developing the people and culture and hence fail to fully realise the expected gains. Emiliani agrees, calling the latter 'Fake Lean'. The point is that Lean is a system of tools and people that need to work together. A single tool by itself, like 5S or mapping, is likely to fail because its benefits are likely to be marginal or even negative.

14.1.2 Sustainability and the Systems View

Systems thinking is way of viewing and interpreting the universe as a series of interconnected hierarchies and interrelated wholes. Within organisational systems there are both technical and social interactions and interrelationships that govern the output of the system. The famous Tavistock research amongst miners, illustrated that both technical and social aspects need to be considered if a new system (in their case long wall mining) was to be successful and sustained. The 'socio' side involves understanding the system of mutually supporting roles and relationships, both formal and informal, within the organisation.

A related 'systems' aspect on sustainability is discussed by Senge who talks about feedback loops. Feedback loops are beginning to be understood in natural and eco systems but exist in human organisations also. Change generates antibodies that automatically grow to fight the change. See also Shapiro's model in the Adoption Curve section. This is like Newton's Third Law – for every action there is an equal and opposite reaction. The antibodies need to be managed or the fever will take hold. Those in favour of change may need to be neutralised. So identify the antibodies as early as possible. 'Inject' them. Antibodies that continue to react need to be moved out decisively and quickly. This is the state of 'quasi stationary equilibrium' described by Kurt Lewin that led to his diagnostic technique known as force-field analysis. For both Senge and Lewin the most important lesson from this realisation is that it is preferable to reduce the restraining forces before increasing the driving forces. In other words, time spent up front on preparing for change, means less time spent later sorting out problems. Or, to quote Frank Devine, 'if people help to plan the battle, they are less likely to battle the plan'.

14.1.3 Sustainability and Feedback: 'Nobody ever gets credit…'

Repenning and Sterman have another stimulating view using feedback loops, and dealing with how improvement efforts degrade. They talk about an 'improvement paradox' that managers face – the vast and expanding number of tools available but the inability to make effective use of the tools. Their model sees 'capability' (machines, processes, people) eroding over time. Managers are, of course, concerned with this erosion and have two ways to counter it. One way is to 'work harder', the other way is to 'work smarter'. Working harder yields quick results but not very dramatic results. Working smarter (say with Lean) is more of a risk. It takes time to yield results, but the results are often more substantial, although they sometimes fail because it is a new area. The problem is that these two are not independent. Working harder can drive out time to work smarter. A related problem is that, because working harder yields short-term results, it is often the course of action that is sought whenever there is a crisis. If management is not very careful with priorities, if they do not 'reinvest' in working smarter with the longer view in mind, all working smarter efforts will fail. A feedback loop develops – the capability gap grows, leading to more pressure to work harder. On the other hand, working smarter can gradually close the capability gap, meaning less time is needed for working harder and even more for working smarter.

Many people concerned with Lean transformation will recognise the wisdom of this analysis. It is a form of Occam's razor for sustainability – the bad 'improvement' drives out the good.

14.1.4 Sustainability of Kaizen Events

Nicola Bateman has carried out probably the most thorough study on the sustainability of kaizen events. She identified 'enablers' for both class A (situations where improvements continue to build after an event), and class B (situations where improvement builds after an event but tail off to a plateau above the level attained during the event).

Top enablers for Class B include:

- There should be a formal way of documenting ideas from the shop floor. This should not be too onerous and could be a flip chart or post-it board.

- Ensure that operators can make decisions in a team about the way they work. Emphasise teamwork and consensus decision-making.

- Make sure there is time dedicated to maintaining the 5S standard every day. Include audits and daily 5S condition checks. Managers are involved with regular audits.

- Ensure there are measures to monitor the improvements made. Continue monitoring the measure that the event was intended to improve.

- The manager (cell leader and his or her manager) should stay focused on performance improvement activity. This supports the other enablers.

Additional enablers for Class A include:

- Changes to operating methods of the cell should be formally introduced to all cell members. Introduce changes to all those who were unable to take part in the event.

- There is time dedicated to housekeeping every day. With regard to the application of 5S in a manufacturing environment Kobayashi's experience is that it cannot be sustained unless it is as part of a complete system of workplace organisation. This suggests that sustainability can only happen as part of a total Lean package.

14.2 Staff Sustainability

14.2.1 Sustainability and Managers

In the earlier Change section of the book, the tendency for managers to give up too early was identified. Sustainability begins at the top. Managers send out signals about their commitment – verbal, but even more importantly, non-verbal. The 'watchers' continually watch the manager's behaviour and 'signals' of their commitment, and, as a result, will become

sceptics or converts. What happens when the chips are down: Are workers fired? Are defectives shipped? Are schedules maintained? Is Lean training abandoned? Strongly linked to this is measures, and the priority of measures tracked by managers. ('Everybody knows that tons per day is what really counts'.) It is obvious that measures need to support the ongoing Lean initiative. Unfortunately this obvious point is often either neglected or not sufficiently thought through.

Bob Emiliani has pointed out that, surprisingly, gaining executive buy-in is not something new in performance improvement. He cites a book written by Knoeppel in 1914 much of which is applicable today. This accords closely with our experience. Why are so many top managers so hesitant, so that Lean efforts start, falter and re-start? Some reasons are:

- They are focused on finance in the case of caretaker managers, or on new product development in the case of owner-managers. Operations is low priority.

- They are risk-averse to concepts they don't fully understand. Owner-managers have made their money by means other than Lean. Why risk it?

- They see Lean as a shop floor thing, not as Lean enterprise. Lean is cost reduction, but the path to growth is via marketing.

- They see Lean as unproven in their industry, even though they accept it as a great success in automotive. This is particularly the case in process industry, having for example, fixed vessel sizes.

- They have just bought a big ERP system, and any conflict with this is to be avoided. The ERP folks 'know what they are doing'.

- Consultants have to bear much of the blame for presenting a tool-based, partial view.

These are difficult and often valid points. You need to put yourself in a CEO's shoes to appreciate the issues.

14.2.2   Sustainability and Staff Turnover

Researchers into self-directed work teams have found that such teams simply do not work when staff turnover is greater than about 30% per year. They are always in the 'forming' and 'storming' stages, never progressing to 'norming' or 'performing' stages. The same thing is very likely with Lean transformation.

14.2.3   Sustainability and Motivation

The sustainability of tools is related to the degree to which people are motivated to use them. Recall some classic theories of motivation. Herzberg's motivator-hygiene theory says that hygiene factors such as pay and conditions can de-motivate but not motivate. Only motivators, such as recognition and personal satisfaction actually motivate. And Maslow talked about the Hierarchy of Needs. This suggests that a foundation of trust and support is necessary, but sustainability requires interest and involvement.

Deming spoke about the necessity to 'drive out fear' as one of his famous 14 points. Surely it is a pre-requisite for sustainability. Fear about the short term remains one of the prime concerns of employees in Western organisations. 'So, if I do all this Lean stuff, will I really retain my job?' Forget the issue of long term (company) survival so often voiced by management – it is short term personal survival, and money, that is of far more concern. Stephen Covey, as one his 7 Habits of Highly Effective People, discusses the 'Win-win or walk away' habit. Both sides must win. A way must and can be found, or else both parties should walk away. Covey believes that without this fundamental principle there can no sustainability – in business, in personal life, or in society. Reject TINA (there is no alternative); embrace TEMBA (there exist many better alternatives – just find them).

Finally, on motivation, the excellent diagnostic booklet *Analysing Performance Problems* by Mager and Pipe, suggests that several questions need to be cleared up before getting down to motivational issues. Has training been adequate?, Is what is expected clear? Has sufficient time been allowed? Have adequate resources been provided? Only after that should you ask about the motivational issues. Is performance punishing? (Will I get extra work because I do a good job?) Is non-performance rewarding? (Slackers are rewarded with more free time.) Does it make a difference? (Will anyone notice good work?)

Of course, this is very like Deming's 94/6 rule – 94% of problems are the system, about which only managers can do things. So be very cautious of labelling someone a 'concrete head'!

14.2.4   Sustainability and Discipline

For Hirano the concept of sustainability is dependent upon discipline and in the context of 5S this means 'making a habit of properly maintaining the correct procedures'. Without good discipline the 5S system will not be maintained and the workplace will revert to chaos. The need for discipline is not restricted to the 5Ss but is also essential in all aspects of business according to Hirano. In the classic Japanese definition of 5S the fifth S or pillar is *Shitsuke* or Discipline. Refer to the 5S section. Hirano firmly believes that management and supervision must teach discipline and that problems with discipline arise when management fails to correct lapses as they occur. The workplace 'faithfully reflects the attitudes and intentions of managers, from the top brass to the shop floor supervision'. The art of correcting another person is also emphasised and the need for compassion is said to be key. The person correcting another must also acknowledge his or her own failings. This particular aspect of discipline has a cultural slant that is highlighted by the description of the worker being criticised thanking the critic for their correction followed by a bow of acknowledgement. Jeffrey Liker refers to this in his 14th Toyota Way principle, calling it Hansei. It is a scenario seldom seen in the West.

14.2.5   Other Staff Aspects of Sustainability

Choi lists the pitfalls of improvement initiatives, based on his research, as

- Alienation of line leaders – involvement and improvement being seen as interfering with their performances objective.

- Seeing improvement as the same as regular problem-solving activity – firefighting but not holding the gains or PDCA.

- Identifying improvement teams as special forces – thereby building resentment.

- Seeing improvement as a management programme – bypassing workers.

- Seeing improvement as solely a worker thing - lacking communication, interest, and involvement from management

- Intermittent – stop go initiatives – 'here we go again'.

Rosabeth Moss Kanter of Harvard says there 5 factors on sustainability:

- **Success** – People feel happier and perform better when there is a feeling of success. And vice versa. And attitude drives performance. There is a feedback loop. So managers must project confidence. War leaders know this.

- **Hard Work** – It is hard work to keep it going. This is entropy. Without it, the system runs down.

- **Emphasis on the team not the individual** – In the West we love heroes, but actually teams are more fundamental for long-term survival. Teams need to be mentored ad developed.

- **Many small wins, rather than the occasional big win** – Small wins keep up enthusiasm, and certainly add up. Certainly a TPS trait. Management needs continually to recognise small wins.

- **Attitude to failure** – Everyone fails from time to time, but what is crucial is the attitude to failure: do you punish or do you treat it as part of learning? Kantor's Law is

that everything dips in the middle (referred to as the 'valley of death' in NPD.

14.2.6   Conditions for Sustainability

Military intelligence officers know that a situation is only dangerous when there is capability and intention. Both are necessary; otherwise there is no danger. The same goes for sustainability – there must be both the capability (time, resources) and the intention (determination, drive, and insistence).

Two theories from change management seem relevant. 'Cognitive Dissonance' says that people try to be consistent in attitude and behaviour. Thus if a change is out of kilter with prevailing attitudes it will fail. The 'psychological contract' says that there is an unwritten implicit set of expectations (covering, for instance, a sense of dignity and worth) which if breached will lead to disruption and implementation failure. A related theory is that of 'the norm of reciprocity' which is what is given needs to seen as equitable with what is received. Not just money, but behaviour in general. Relations, both at work and personal, cannot be sustained without this balance.

14.2.7   Single and Double Loop Learning

The theory of single and double loop learning, from Chris Argyris, of Harvard, is particularly pertinent to Lean implementation and sustainability. Use the analogy of a thermostat: the thermostat makes continual adjustments to maintain the temperature (single loop learning), but does not question whether the temperature is appropriate for prevailing conditions (double loop learning). According to Argyris, many senior managers are excellent at single loop learning, but poor at double loop learning. This is because they have been successful throughout their career, at lower level control, but when faced with wider challenges, they fail. Moreover, they then tend to blame others.

A way around this is self-reflection and constructive feedback criticism. Set up scenarios for important interventions and get constructive, honest feedback from senior colleagues about how you come across. It is not what you say, but how you say it.

To summarize this section, there is no self-sustainability; it requires ongoing efforts in the areas of processes and systems and people. And managers, through their control of the systems, and adaptation to changing circumstances, bear much of the responsibility for the failures of sustainability.

Further Reading

Fraser Wilkinson, *Sustainability of 5S*, MSc Dissertation, Lean Enterprise Research Centre, Cardiff Business School. The author is grateful to Fraser for pointing out several of the concepts discussed here.

Nicola Bateman, *Sustainability*, Lean Enterprise Research Centre, Cardiff Business School, 2001

Peter Senge, *The Dance of Change*, Nicholas Brealey, London, 1999

Schaffer, R.,H., & Thompson, H., A., *Successful Change Programmes Begin with Results*, Harvard Business Review, Jan-Feb 1992

Nakane, J. & Hall, R. W., 'Ohno's Method – Creating a Survival Work Culture', *Target* Vol 18, No 1

Thomas Choi, 'The Successes and Failures of Implementing Continuous Improvement Programs', in Jeff Liker (ed), *Becoming Lean*, Productivity Press, Portland, 1998

Charles Standard and Dale Davis, *Running Today's Factory*, Hanser Gardner, Cincinnati, 1999, Chapter 13

Nigel Slack and Michael Lewis, *Operations Strategy*, Chapter 14, FT Prentice Hall, 2002

Michael Lewis, 'Lean Production and Sustainable Competitive Advantage', *IJOPM*, Vol 20, No 8, 2000

Bernard Burns, *Managing Change*, Third edition, FT/ Pitman, Harlow, 2000, Chapter 12

Spenser Johnson, *Who Moved My Cheese?*, Vermillion, London, 1998

Three excellent little books by David Hutchens, *Shadows of the Neanderthal: Illuminating the Beliefs that Limit our Organizations (1999); The Lemming Dilemma (2000); The Tip of the Iceberg (2001),* Pegasus Communications

Bob Emiliani, *Real Lean*, Volume 1, Centre for Lean Business Management, 2007

Lynne Maher, *Sustainability: Model and Guide*, NHS Institue for Innovation and Improvement, 2007

Jeffrey Pfeffer, *What Were They Thinking?*, Harvard, 2007

Rosabeth Moss Kanter, presentation on *thetimesonline/business* podcast, 2007

Nelson Repenning and John Sterman, 'Nobody Ever Gets Credit for Fixing Problems that Never Happened: Creating and Sustaining Process Improvement', *California Management Review*, 43:4, Summer 2001

Joan V Gallos (ed), *Organization Development*, Jossey Bass, 2006. This 'blockbuster' text contains articles by virtually every significant writer in the OD field.

15 New Product Development and Introduction

To explain how Japanese firms were able to achieve not just high productivity, but also high design quality levels with less development time and less engineering effort Clark and Fujimoto identified several organisational patterns: high levels of supplier engineering, overlapping product and process engineering, strong communication mechanisms, strong in-house manufacturing capability, wide task assignments for engineers and the heavyweight project manager system. These closely mirror the set of principles briefly outlined in 'The Machine that Changed the World' by Womack and Jones who suggested that heavyweight project leaders, or 'chief engineers', small cross functional teams dedicated to success of the project, early problem solving, cross functional communication, and simultaneous development were the core practices of the Japanese development system. The overall conclusion was that Lean is also applicable to new product development. In fact, good new product management is essential in Lean operations because up to 90% of costs may be locked in after the design and process planning stage, yet these stages incur perhaps 10% of cost. The time taken to bring a new design or product to market is where much of the competitive edge is gained or lost. And, the earlier a problem is detected the less expensive it is to solve. The new product development (NPD) and new product introduction (NPI) area is where leading Lean companies are increasingly competing.

The benefits of applying Lean to product development were first documented by Clark and Fujimoto's seminal study of product development processes in the automotive industry. Much like in manufacturing, they noticed a distinctive set of practices and organisational features that enabled Japanese companies to outperform their Western counterparts. Lean manufacturers tended also to have a Lean product development organisation.

As in manufacturing, cost, speed and quality are important criteria, but the three take on a different perspective. **Cost** – 'How much does the development cost overall?' 'How much does one unit cost?' **Speed** – How quickly can we get a product into the market, or 'time-to-market'? **Quality** – How many defects need to be rectified post product launch, e.g. in the form of costly recalls? The priorities for each of these will depend on the product characteristics. Targets for the objectives are typically determined and prioritised by management, whether consciously or not, and should be oriented toward achieving specific financial or strategic goals of the firm. For innovative, fashion-driven products time will be more critical, for commodities or utilitarian products cost will be more important. Beware, however, of simply trying to copy Toyota's approach to NPD. Instead a selection of tools is needed that matches the product characteristics and strategic objectives.

While it would be desirable for firms to excel on all of these aspects of performance, there is a growing body of literature that suggests that there are trade-offs that occur between these performance dimensions (see Figure below). For example, speed may be traded-off for development cost, since it may take more resources to get the product to the market quicker, for product performance, since increasing speed may mean curtailing performance specifications, or for unit manufacturing costs, as part integration and design for manufacturing techniques add time to the development schedule. In light of this, some authors have recently suggested that the goal of management should be to bring these trade-offs out into the open so that they may be made consciously within the project to maximise profitability or other performance criteria as appropriate.

In this section we will first give a brief overview of how to balance the multiple objectives in NPD, then illustrate 'Toyota-style' product development, before introducing the tools for each respective performance criteria. Many of the tools can be, and are being, used together. They apply across Ulrich and Eppinger's five phase process for new products:

- Concept development
- System level design
- Detail design
- Testing and refinement
- Production ramp up.

Further Reading

Karl Ulrich and Steven Eppinger, *Product Design and Development*, McGraw Hill, New York, 1995

Kim Clark and Takahiro Fujimoto, *New Product Development Performance,* Harvard Business School Press, 1991

Acknowledgments

Lee Schab and Takahiro Fujimoto kindly provided materials for updating this section. Their support is gratefully acknowledged.

15.1 Four Objectives and Six Trade-offs

In their ground-breaking work on accelerated new product development, Smith and Reinertsen identified four objectives (Development Speed, Product Cost, Product Performance, and Development Programme Expense) as being central to the management of new product development.

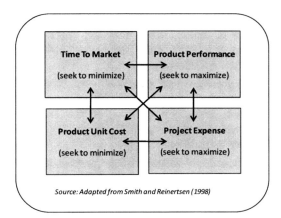

Source: Adapted from Smith and Reinertsen (1998)

The four areas interact in six ways. Smith and Reinertsen believe that it is necessary to quantify the trade-offs since every new product introduction is a compromise that needs to be understood and managed.

1. Development Speed and Product Cost – Rationalising and improving a design through part count, weight analysis, part commonality, DFM, and value engineering can save future costs. But they take time, thereby delaying the introduction of the product and possibly losing market share.

2. Development Speed and Product Performance – Improving a design can make it more attractive to customers thereby improving future sales through a larger market, a higher price, and a longer product life. These improvements take time and may sacrifice initial sales and initial market share.

3. Development Speed and Development Programme Expense – This is the traditional 'project crashing' trade-off from classic project management. Most projects have 'fixed' costs such as management and overhead that accumulate with time. On the other hand, within limits, adding extra resources decreases project duration but costs more. Is it worthwhile spending more to finish earlier? There are non-linear effects – digging a trench with six men does not take one sixth of the time it takes with one man.

4. Product Performance and Product Cost – Adding or redesigning a feature may improve performance but at what product cost? What marginal performance are customers prepared to pay for?

5. Product Cost and Development Programme Expense – By spending more on development, say through value engineering, we may be able to reduce cost. Is it worth it?

6. Development Programme Expense and Product Performance– Improving a design and improving performance may result in improved sales. But improving performance may involve additional cost. In a less complex way, Mascitelli suggests a 'least discernable difference test' which is to ask whether customers will pay a penny more for a feature that is being considered. There is a conceptual maximum number of satisfied customers that reduces on the one hand because of reduced benefits and on the other hand because of price increases.

Similarly, Smith and Reinertsen suggest that a spreadsheet model be developed to quantify these six relationships. Remember to use the time value of money, trading off immediate project costs against discounted future sales, essentially using Net Present Value (NPV). Quantifying the six relationships forces marketing, design and engineering to think carefully about additions and rationalisations. Such a sensitivity analysis can be used as a powerful management tool, as it creates a visual prioritisation of performance objectives. Based on such NPV data, you can illustrate the effects of questions such as 'how much does one month delay cost?', or 'if you miss unit cost by 3%, what will this do to profitability?', or 'what does missing a features or performance specification mean for sales?' Playing with such scenarios will focus everyone's attention on the truly important factors.

However, there are no religious rules to be followed – as Michael Cusumano and Kentaro Nobeoka show in their comparisons of Japanese approaches to car development between 1980 and 1990, the answers to these tradeoffs change with time and environment.

Further Reading

Michael Cusumano and Kentaro Nobeoka, *Thinking Beyond Lean*, The Free Press, New York, 1998

Preston Smith and Donald Reinertsen, *Developing Products in Half The Time*, Van Nostrand Reinhold, New York, 1991

Ronald Mascitelli, *Building a Project-Driven Enterprise: How to Slash Waste and Boost Profits through Lean Project Management*, Technology Perspectives, 2002

15.2 Wastes in New Product Development

Many professionals want to do their professional work, but spend inordinate amounts of time on frustrating secondary work. An estimate of waste of time, often shocking, can be obtained by activity sampling done by the professionals themselves. A priority, then, is to maximise the time that designers spend designing, or engineers spend engineering.

Waste awareness, as in manufacture, is an excellent point of departure. Two years spent in a design office showed us that wastes were endemic but that there was minimal awareness of waste. Allen Ward, in his seminal book on the Lean Product and Process Development, which is strongly based on Toyota design principles, gives three categories of 'knowledge waste'. First, Scatter Waste which disrupts flow through poor communication barriers, and inappropriate complex tools – Ward illustrates this with FMEA! Second, Hand-off waste which results from the separation of knowledge, responsibility, action, and feedback. Third, 'Wishful Thinking' waste. This includes discarded knowledge and making decisions without data or thinking – this is partly what Reinertsen describes in his tradeoff model – see above.

Here we attempt our own list, with inspiration from one of the great gurus of Lean Design, Ron Mascitelli.

- **Sorting and Searching** – Basically, 5S in design and NPD.

- **Inappropriate targets** – These lead to either cutting corners where targets are too tight, or to loafing where they are too lax. The latter is 'Parkinson's Law' – work expands to fill the time available.

- **Underload and Overload** – An important managerial task is to load level the work. If not done, the results are the same as inappropriate targets. See the section below on Critical Chain.

- **Inappropriate prioritising** – particularly from designs that are shelved due to policy changes. Note that this is not the same as 'bookshelving' – see below.

- **Interference** – From 'dropping in', e mails, socialising, noise, etc.) Rules need to be established to allow for thinking time.
- **Inappropriate trade-offs** – see Reinertsen, above.
- **Excessive part proliferation** – This results from not using a standard set of, for example, fasteners or components but 'starting from a blank sheet of paper'. Also not building on appropriate previous designs.
- **Presence** – At meetings where 90% of the time a person has no role.
- **Waiting** – for decisions, for stage gates, for tests, for data, etc. This is particularly important for critical resources, where flow needs to be maintained, and appropriate time buffers used.
- **Starting too late; stopping too early** – NPD needs to be end-to-end – from customer need to well into product life.
- **Inappropriate involvement** – Omitting or delaying the involvement of key functions such as marketing, production, manufacturing engineering, tooling, quality, packaging, distribution.

The following two, not only waste time and resources on rework, but are compounded when learning and recording is not built in.

- **Lack of feedback** – resulting in a low learning rate.
- **Not recording lessons learned**.
- **Mistakes, defects and errors**.

The following three wastes are a set. The 'Service Gaps' model is useful here (see Zeithami et al). They talk about the gap between what is actually wanted and what it's assumed is wanted, the gap between what is wanted and what is specified, the gap between specification and performance, and the gap between actual performance and what is said about performance.

- **Co-ordination**. A good design is holistic. Optimising the parts does not necessarily optimise the whole. A very important idea for the lead designer.
- **Communication** – All participants need to have the same, clear, goals.
- **Ill-defined product requirements**.

An effective, and relatively quick, way to reduce these wastes is, first, to recognise their presence, and, second, collectively to work out what each professional in the group needs to do. Many designers value their creativity and individuality – so bureaucratic rules don't work. However, voluntary participation does work, particularly if it makes their life easier.

Further Reading

Ronald Mascitelli, *The Lean Product Development Guidebook*, Technology Perspectives, 2007

Ronald Mascitelli, *The Lean Design Guidebook*, Technology Perspectives, 2004

Allen Ward, *Lean Product and Process Development*, LEI, 2007

Valerie Zeithaml, et al., *Delivering Quality Service*, Free Press, 1990

15.3 Toyota's Approach to Product Development

Toyota's approach to new product development is based on a mixture of techniques related to cost and lead-time reduction, as well as quality assurance. In their study of Toyota, Clark and Fujimoto found several 'best practices': high levels of supplier engineering, overlapping product and process engineering, strong communication mechanisms, strong in-house manufacturing capability, wide task assignments for engineers and the heavyweight project manager system. These closely mirror the Lean principles, but applied to new product development. Cusumano also published a list of Lean product development practices based on Honda's model that reinforced these findings and additionally noted Honda's skilful use of computer aided design tools. This bundle of concepts became widely accepted as 'Lean product development' and rapidly diffused beyond the automobile producers into the manufacturing industry in general. The following features draw on both Allen Ward and on Morgan and Liker.

First, let us make a general point. Putting in front-end effort on project clarification, working concurrently and meticulously, even if much more slowly, to avoid sub-system technical conflict is thoroughly worthwhile. You want to minimise the number of post launch problems. The tortoise and the hare.

Ward, and Sobek, start with a useful pneumonic, LAMDA – the product developer's version of PDCA. Look, Ask, Model, Discuss, and Act. (Notice this – much time is spent on understanding customer needs, and modelling and discussing alternatives, before starting actual design work.)

- **Chief engineers** – sometimes referred to as 'Heavyweight project managers'. They are not project managers in the conventional administrator sense but, as Ward calls them, Entrepreneurial System Designers (ESD). Moreover, they are experienced ENGINEERS, not managers. Their task is both holistic (integrating all parts) and end-to-end (from need to use). Thus if an ESD brings a product to market on time but the product itself fails in the market, he has failed. These chief engineers have only a small staff but work through functional managers using their influence, reputation and considerable experience to avoid the many delays so common in traditional product development. The chief engineer or ESD has such strong influence on the final product that it often becomes known internally as 'Mr. Ohashi's car'. ESDs also 'represent customers' – not relying only on market survey but uncovering needs that customers have not yet articulated. See, for example, notes on Ideal Final Result in TRIZ. As such they are responsible for specification, cost targets, layout, and major component choice in order to make sure that the product concept is accurately translated into the technical details of the vehicle.

- **Functional managers are responsible for developing 'towering expertise'** in their own areas – maybe engines, suspension, or controls. They bring state-of-the-art solutions into new products. Functional staff are as much researchers as designers, so they enjoy high status and professional development in their own area. In new product development they work for their functional engineer, not for the project manager. The latter needs to negotiate for their time but be concerned with integration. The ESD, however, signs every drawing. Integral with the use of functional expertise are the following:

- **Set based design** – The concept here is to keep options open as late as possible. There are 'sets' of options for the various system elements, and these are gradually narrowed down as the design clarifies. As a result, there are fewer 'stage gates', but broader 'milestones'. Between the milestones there is considerable flexibility. Concepts are gradually narrowed using 'Concept Screening' techniques (see later). Possible 'solution sets' are explored in parallel, but once a particular solution is decided upon it is frozen unless a change is absolutely necessary. This is similar to the Lean concept of postponement - delaying freezing the specifications until the last possible moment.

- **Bookshelving** – Starting with sets may sound wasteful. Not so, if unused designs, and basic research, are bookshelved for future use. This also allows rapid new product development. It is a form of modularity.

- **Trade-off curves** are developed by the functional experts. Examples may be strength against thickness, number of cycles against type of alloy, noise level against insulation thickness. Sets of choices are quickly established. Having trade-off curves helps the ESD and the engineer to make appropriate the trade-offs, as described by Reinertsen, above.

- **Check sheets** – the simple but powerful idea of recording experiences systematically as a project proceeds. What works and what does not. As a result the wheel is not re-invented. Expertise does not walk out the door when a person leaves. Next time a similar product or sub-system is stated, start with what was learned last time.

- **In-house tooling and manufacturing engineering** – alas, a skill set that so many have outsourced. Outsourcing may seem cost efficient, but remember product features are easy to copy, but how to make them, and how to make them fast, is not. Good new product development rotates staff from design to manufacturing engineering. For example, prototypes are built by regular manufacturing engineers, so that they experience what it takes to produce this product at full volume.

- **Concurrency (or cross-functional teams)** – Quite a well-accepted idea is to work concurrently on stages rather than 'over the wall' (for example, research to design to engineering to production). So, multi-discipline teams work together. The chief engineer is the facilitator. While car design is proceeding, engineering and die production also proceeds. During the early concept stages, engineering makes from 5 to 20 one-fifth scale clay models. The engineering team begins full-scale clay modelling at intermediate stages as well as at final stage, unlike other car manufacturers who only make half size models during early stages. This enables tool and die designers to begin work. They too use engineering check sheets, built up from experience, about what can and cannot be done. Difficulties are fed back immediately to the design team.

- **Project levelling** – Bring the Heijunka concept into new product development. This means careful thought on time phasing. Traditional critical path software, even using the resource levelling feature, is seldom good enough. Set based design,

bookshelving, and functional development can all be used. The ideas of Goldratt on Critical Chain (see later) may be useful.

- **Project flow** – Avoiding hold-ups by critical resources whilst waiting for other stages, is an important role for the chief designer's team. A leaf can be taken from Lean Construction where the 'Last Planner' methodology aims to do just this by developing checklists for other functions before critical activities are due to start.

- **Visibility** – The good Lean principle of visual management is even more important in product development. Toyota uses an 'Obeya' (Big Room) for each new product where all activities and progress is shown on charts. Co-ordination meetings are held in the Obeya.

- **Supplier involvement** – A critical decision, by the chief engineer is what to insource and what to outsource. See the Supply chain section. Clark and Fujimoto found that Japanese companies tended to sub-contract out much larger fractions of the engineering work to a group of suppliers with whom they had developed close relationships. This practice allowed projects to be kept compact and simplified the amount of internal project coordination required which, in turn, contributed to shorter lead times and higher development efficiency. In addition, since it was the suppliers who were eventually to manufacture the parts anyway, this practice allowed them to develop specialised knowledge and design the components themselves with their own manufacturing capabilities in mind, lowering the cost of components. This is also known as 'open spec'. A seat must fit into the space, but you design the seat.

- **Front loading** – The later in the design process a problem is fixed, the more effort is required, hence the more expensive and lengthy fixing the problem becomes. Front-loading aims to address this pulling key decisions forward, whist retaining set-based flexibility. The early identification and solving of problems can help reduce development time and cost, and frees up resources to be more innovative in the marketplace. According to Thomke and Fujimoto, front-loading can be achieved by (1) project-to-project knowledge transfer, which leverages previous projects by transferring problem and solution-specific information to new projects; and (2) rapid problem-solving that leverages CAD and other technologies to increase the overall rate at which development problems are identified and solved.

Further Reading

Kim Clark and Takahiro Fujimoto, *New Product Development Performance,* Harvard Business School Press, 1991

Durward K. Sobek II, Allen C. Ward and Jeffrey K. Liker, 'Toyota's Principles of Set-Based Concurrent Engineering', *MIT Sloan Management Review* Winter 1999, Vol. 40, No. 2, pp. 67–83

James Morgan and Jeffrey Liker, *The Toyota Product Development System,* Productivity, 2006

Allen Ward, *Lean Product and Process Development*, LEI, 2007

Allen Ward and Michael Kennedy, *Product Development for the Lean Enterprise*, Oaklea Press, 2003

Lawrence P Leach, *Lean Project Management: Eight Principles for Success*, Advanced Projects, 2006

Stefan Thomke and Takahiro Fujimoto 'The Effect of 'Front-Loading' Problem-Solving on Product Development Performance' *Journal of Product Innovation Management* Vol. 17 No. 2, p. 128-142, 2000

15.4 Cost

15.4.1 *Value Analysis (VA) - Value Engineering (VE)*

Value engineering and analysis (VE/VA) has traditionally been used for cost reduction in engineering design. But the power of its methodology means that it is an effective weapon for quality and productivity improvement in manufacturing and in services.

Today the term value management (VM) recognises this fact. The first step in any VA/VE/VM project is Orientation. This involves selecting the appropriate team, and training them in basic value concepts. The best VE/VA/ VM is done in multidisciplinary teams. 'Half of VE is done by providing the relevant information,' says Jaganathan. By this he means that clarity of communication about customer (internal or external) is half the battle, particularly if customer needs have changed without anyone taking notice.

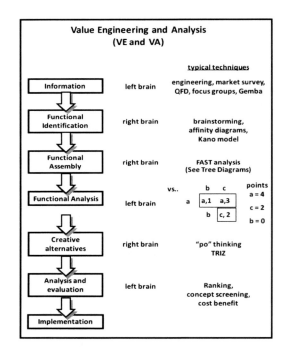

VM proper begins by systematically identifying the most important functions of a product or service. Then alternatives for the way the function can be undertaken are examined using creative thinking. A search procedure homes in on the most promising alternatives, and eventually the best alternative is implemented. You can recognise in these steps much similarity with various other quality techniques such as quality function deployment, the systematic use of the 7 tools of quality, and the Deming cycle. In fact these are all mutually reinforcing. VM brings added insight, and a powerful analytical and creative force to bear.

Value engineering was pioneered in the USA by General Electric, but has gained from value specialists such as Mudge and from the writers on creative thinking such as Edward de Bono. Today the concepts of TRIZ are most relevant. The Society of American Value Engineers (SAVE) has fostered the development.

VM usually works at the fairly detailed level of a particular component or sub-system, but has also been used in a hierarchical fashion working down level by level from an overall product or service concept to the detail. At each level the procedure described would be repeated. Like many other quality and productivity techniques, VM is a group activity. It requires a knowledgeable group of people, sharing their insights and stimulating one another's ideas, to make progress. But there is no limitation on who can participate. VM can and has been used at every level from chief executive to shop floor.

What appears to give value management particular power is the deliberate movement from 'left brain' (linear) analysis to 'right brain' (creative) thinking. (Stringer has explained this.) Effective problem solving requires both the logical step forward and the 'illogical' creative leap. Edward de Bono, of lateral thinking fame, talks about 'provolution' - faster than evolution but more controlled than revolution.

Functional analysis

Functional analysis is the first step. The basic functions or customer requirements of the product or service are listed, or brainstormed out. (Right brain thinking.) A function is best described by a verb and a noun, such as 'make sound', 'transfer pressure', 'record personal details' or 'greet customer'. The question to be answered is 'what functions does this product/service undertake?' Typically there will be a list of half a dozen or more functions.

There is a temptation to take the basic function for granted. Do not do this; working through often gives very valuable insights. For instance, for a domestic heating time controller, some possible functions are 'activate at required times', 'encourage economy', and 'supply heat when needed'. The example shows a hamburger design. Pair wise comparison or points distribution may be used to weight the functions or requirements. In pair wise comparison, each function is compared with each other function, and 1 point is given to the most important of the two, or zero points if the functions are considered equally important.

Customer Requirements	Importance	Percentage importance	components							
			beef		bun		lettuce		ketchup	
Taste good	6	46	23	.5	11.5	.25			11.5	.25
Provide nutrition	2	15	10.5	.7	4.5	.3				
Appeal visually	1	8	2.4	.3	1.6	.2	2.4	.3	1.6	.2
Value for money	4	31	15.5	.5	15.5	.5				
		100%								
Overall influence		%	51.4		33.1		2.4		13.1	
Cost (%)		%	62		30		20		8	
Value index		Influence/ cost	0.83		1.1		0.12		1.6	

Adding up the scores gives the relative weightings These relative weightings need to be converted to percentages such that the sum adds up to 100%. Now the components of the product are listed as columns in the matrix. See the figure. Then the importance of each component to each function is estimated and converted to a percentage of total cost. In the example, the function 'taste good' is estimated to be influenced 50% by the beef and 25% each by bun and ketchup. The influence is written in the bottom left hand corner of each cell. The weighted influence (i.e. the weight x the influence) is written in the top right of each cell. The overall influence of each component is the sum of the top right hand cell entries, and is written in a row below the matrix. Then the cost of each component is estimated, and written in a row. The last step is to calculate a value index, which is the influence % divided by the cost %. Ratios less than one are prime candidates for cost reduction. Ratios substantially greater than one indicate the possibility of enhancing the feature. Back to the right brain.

Creativity

Now the creative phase begins. This is concerned with developing alternative, more cost effective, ways of achieving the basic functions and reducing costs of the most important components. Here the rules of brainstorming must be allowed - no criticism, listing down all ideas, writing down as many ideas as possible however apparently ridiculous. Various 'tricks' can be used; deliberate short periods of silence, writing ideas on cards anonymously, sequencing suggestions in a 'round robin' fashion, making a sketch, role-playing out a typical event, viewing the scene from an imaginary helicopter or explaining the product to an 'extra terrestrial'. Humour can be an important part of creativity. See also TRIZ, later.

A particularly powerful tool is the use of the de Bono 'po' word. This is simply a random noun selected from a dictionary to conjure up mental images that are then used to develop new ideas. For instance 'cloud' could be used in conjunction with the design of packaging where the basic

function was 'give protection'. The word cloud conjures images of fluffiness (padding?), air (air pockets?), rain (waterproof?), silver lining (metal reinforcement?), shadow (can't see the light, leads to giving the user information), cloud is 'hard to pick up' (how is the packaging lifted?), wind (whistling leading to a warning of overload?), moisture (water/humidity resistant? water can take the shape of its container – can the packaging?), obscuring the view (a look inside panel?), and so on. Do not jump to other 'po' words; select one and let the group exhaust its possibilities.

Analysis and Evaluation

Now back to the left brain. Sometimes a really outstanding idea will emerge. Otherwise there may be several candidates. Some of these candidates may need further investigation before they can be recommended. (In the above example, is it feasible or possible to introduce some sort of metal reinforcement?). Beware of throwing out ideas too early - the best ideas are often a development from an apparently poor idea. So take time to discuss them. In some cases it may be necessary for the team to take a break while more technical feasibility is evaluated or costs determined, by specialists. There are several ways to evaluate. Pairwise comparison with multi-discipline group discussion is good possibility. Another possibility is to write all ideas on cards, then give a set of cards to pairs of group members, and ask each pair to come up with the best two ideas. Then get the full group to discuss all the leading ideas. Yet another is to draw up a cost-benefit chart (cost along one axis, benefits along another). Ask the group members to plot the locations of ideas on the chart. There is no reason why several of these methods cannot be used together. Do what makes the group happy - it is their project and their ideas.

Implementation

Implementation of the most favourable change is the last step. One of the benefits of the VM process is that group members tend to identify with the final solution, and to understand the reasons behind it. This should make implementation easier and faster.

Further Reading

Kaneo Akiyama, *Function Analysis*, Productivity Press, Cambridge MA, 1991

G. Jaganathan, *Getting More at Less Cost*, Tata McGraw Hill, New Delhi, 1996

See also, J Jerry Kaufman and Roy Woodhead, *Stimulating Innovation in Products and Services with Function Analysis and Mapping*, Wiley, 2006. This gives a detailed description of FAST (Function Analysis System Technique) modelling.

Acknowledgements

Many of the original ideas on value engineering are due to Art Mudge. These and other ideas have been further developed by Dick Stringer and Graham Bodman of the South African Value Management Foundation.

15.4.2   Design for Manufacture (DFM) – Design for Assembly (DFA)

Design for manufacture (DFM) is a key 'enabling' concept for Lean manufacture. Easy and fast assembly has an impact right through the manufacturing life of the product, so time spent up front is well spent. A wider view of DFM should be considering the cost of components, the cost and ease of assembly, and the support costs.

Cost of Components should be the starting point. Much will depend upon the envisaged production volume: for instance, machined components may be most cost effective for low volumes, pressings (requiring tooling investment) best for middle volumes, and mouldings (requiring even higher initial investment but low unit costs) best for higher volumes.

Other considerations include

- Variety as late as possible
- Design for no changeover or minimal changeovers

- Design for minimum fixturing
- Design for maximum commonality (Group Technology)
- Design to minimise the number of parts.

Complexity of Assembly

Boothroyd and Dewhurst have suggested a DFA index aimed at assessing the complexity of assembly. This is the ratio of (the theoretical minimum number of parts x 3 seconds) to the estimated total assembly time. The theoretically minimum number of parts can be calculated by having each candidate part meet at least one of the following:

- Does the part need to move relative to the rest of the assembly?
- Must the part be made of a different material?
- Does the part have to be physically separated for access, replacement, or repair?

If not theoretically necessary, then the designer should consider the physical integration with one or more other parts. And why 3 seconds? Merely because that is a good average unit assembly time. Once this is done, then Boothroyd and Dewhurst suggest further rules for maximum ease of assembly. These are:

- Insert part from the top of the assembly
- Make part self-aligning
- Avoid having to orient the part
- Arrange for one-handed assembly
- No tools are required for assembly
- Assembly takes place in a single, linear motion
- The part is secured immediately upon insertion.

Boothroyd and Dewhurst now market a software package to assist with DFM. They suggest that for both DFA and Design Complexity (see below), not only should there be measurement and monitoring of one's own products, but that the measures should be determined for competitor's products as well. This is a form of benchmarking. Targets should be set. Such measures should also be used for value engineering.

Continuous improvement is therefore driven by specific targets, measures, and benchmarks, and not left to chance. It should be possible to create design and assembly indices for each subassembly, to rank them by complexity, and Pareto fashion to tackle complexity systematically. Further, it should be possible to determine, from benchmarking competitor products, the best of each type of assembly and then to construct a theoretical overall best product even though one may not yet exist in practice. This is a form of Stuart Pugh's 'Concept Screening' method (see later section).

Assembly Support Costs should be considered at the design stage. This includes consideration of

- Inventory management and sourcing
- The necessity for new vendors
- A requirement for new tools to be used
- A requirement for new operator skills to be acquired
- The possibility of failsafing.

More recently C Martin Hinckley has estimated that assembly defects are directly proportional to assembly time. To this end he has developed so-called Quality Control of Complexity that is a straightforward way of estimating assembly time. Details are given in his book.

Assembly alternatives can now be considered relatively easily for their impact on assembly defects due to mistakes.

Complexity of Design

Boothroyd and Dewhurst have also suggested a measure for the assessment of design complexity. They use three factors: the number of parts (Np), the number of types of parts (Nt), and the number of interfaces between parts (Ni). First, the numeric value of each of these factors is determined by addition. Then, the factors are multiplied and the cube root taken. This yields the design complexity

factor. Note that reducing the number of parts usually also reduces the number of interfaces, which are points at which defects and difficulties are most common. Also, reducing the number of types of part has a direct impact upon inventory management and quality.

Other Types

'Design for' is not limited to 'assembly' (DFA) or 'manufacture' (DFM). There are other types too:

Design for Performance (DFP)

Design for Testability (DFT)

Design for Serviceability (DFS)

Design for Compliance (DFC), and, of course,

Design for Six Sigma (DFSS), see section below.

Further Reading

G. Boothroyd and P. Dewhurst, *Product Design for Assembly Handbook*, Boothroyd Dewhurst Inc., Wakefield, RI, 1987

Karl Ulrich and Steven Eppinger, *Product Design and Development*, McGraw Hill, New York, 1995, Chapter 3

Subir Chowdhury, *Design for Six Sigma*, Dearborn Press, Chicago, 2002

C Martin Hinckley, *Make No Mistake!*, Productivity Press, Portland, 2001

15.4.3   Modularity, Platforms and Component Carry-over

In order to save development cost, there are essentially three ways to cut down. First, increase the 'component carry over' by using components from the previous model. This might not always be a possibility, in particular in high-clockspeed industries where technology moves on. Second, choose to adopt a modular product architecture that allows for flexibility in sourcing and customisation. Third, you can create product platforms, which each house multiple end products. Here development cost is saved by creating larger economies of scale for the 'basic design'. We will discuss the last two in turn. Also see Cusumano and Nobeoka in their book on 'multi-project management' for more details on the benefits of modularity and component carry over.

Modularity

Modularity essentially means that a 'one-to-one mapping' between components and functions exists, and that interfaces are standardised. In computers, for example, the hard drive only has the function to store data. It is plugged into a standard slot, and the interface it connects to is also standardised. Essentially, you can use any hard drive you want. This is different in 'integral' products, where one component is connected to multiple functions: for example, the brake system in a car is there to slow the vehicle down, when applied, but it also links to the handbrake and ABS-ESP systems, that apply the brakes without the driver requesting it, in order to keep the vehicle on the road. Here, one component takes on multiple functions, and interfaces are generally specific to a given firm.

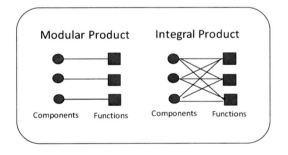

Modularity can be used as a means for outsourcing, but also for product customisation. This long-established form of customisation simply involves assembly to order from standard modules. Examples are legion: calculators or cars having different appearance but sharing the same 'platform', aeroplanes, and many restaurant meals. Pine lists six types: 'Component sharing', where variety of components is kept to a minimum by using Group Technology (see separate section), Design for Manufacture (see

separate section), 'Component swapping' (cars with different engines), 'Cut-to-fit modularity' (a classic example being made to order bicycles), 'Mix modularity', combining several of the above, and 'bus modularity', where components, such as on a hi-fi are linked together. Baldwin and Clark take this further, believing modularity to be a fundamental organising principle for the future. Thus today Johnson Controls makes the complete driver's cockpit for Mercedes, and VW runs a truck factory using not only the modules of suppliers but their operators also.

Gilmore and Pine (1997) have gone on to state that there are four approaches to module customisation (see Creating the Lean Supply Chain section). 'Collaborative customisers' work with customers in understanding or articulating their needs (a wedding catering service). 'Adaptive customisers' offer standard but self-adjusting or adapting products or services (offering hi-fi, or car seats which the customer adjusts). 'Cosmetic customisers' offer standard products but present them differently (the same product is offered but in customer specified sizes, own-labels). 'Transparent customisers' take on the customisation task themselves often without the customer knowing (providing the right blend of lubricant to match the seasons or the wear rate).

Johnson and Bröms give a description of Scania Trucks' approach to modularity. Four basic modules – generate power, transmit power, carry load, house and protect driver – have been developed as modules to meet various environmental conditions encountered around the world. A matrix has been developed for parts and modules. These remain standardised until an improvement takes place, but are used 'Lego brick' style to act as the foundation for any new model. Scandia builds to order using the matrix.

Also sometimes the German term 'Baukastensystem' is used to describe modularity, which refers to the inter-changeability of parts.

Platforms

A product platform is a design from which many derivative designs or products can be launched, often over an extended period. So, instead of designing each new product one at a time, a product platform concept is worked out which leads to a family of products sharing common design characteristics, components, modules, and manufacturing methods and technology. This in turn leads to dramatic reductions in new product introduction time, design and manufacturing staff, evaluation methods such as FMEA, as well as inventory, training, and manufacturing productivity; in short, Lean design.

There are similarities to the Modularity concept and GT (group technology), but product platforms go far wider. Product platforms are found from calculators (e.g. Casio) to Cars (VW / Audi). The Apple Macintosh uses the platform of a common operating system (MacOSX) and common microprocessors (the Intel Duo) for a variety of computers. A classic case, cited by Meyer and Lehnerd, is Black&Dekker. In the 1970s, Black and Decker's product portfolio comprised scores of different motors, armatures, materials, and tooling, most of it simply evolving piecemeal one at a time. The company began their product platform strategy by bringing together design and manufacturing engineers to work in teams. They started, Pareto fashion, with the motor, and created a common core design with a single diameter but varying length, able to be adapted to power output from 60 to 650 watts. The range of motors now shared common manufacturing facilities, reduced changeover, and dramatically better quality. After motors, armatures and then drill bits were tackled. This led to reductions of 85% in operators, and 39% in cost. The cycle time for new product introduction was slashed, eventually enabling B&D to introduce new products at the rate of one per week. The extra investment needed for process technology was repaid within months. Lower costs were passed on to customers, and new markets created, resulting in a surge of demand and the decimation of competitors. Notice the similarities with 'The Essential Paretos', described in the 'preparing for

Flow' section of this book. Meyer and Lehnerd, identify five principles. These are:

- Families should be identified, which share technology and are related to a market. Note again similarity with GT which shares manufacturing process steps, but says nothing about markets. From this basic family feature-rich derivations are developed over time.

- Product design takes place alongside production design. Product platforms are a strong argument in favour of the Lean 'small machine' principle; too often, because of the existence of an expensive big machine, product designs are undertaken which seek to make use of the machine rather than giving first consideration to the customer. Simultaneous design or concurrent design (a good Lean principle) should be used.

- Try to design for global standards, logistics, and component supply. An example is the internationally adaptable power requirement used on HP printers.

- Try to capitalise on latent demands that a product platform can create. VHS VCRs, and CDs are outstanding examples.

- Seek design elegance, not mere extension. (Java seems to be the reaction against bloated software packages with more and more features seldom used.)

The building blocks of product platforms, according to Meyer and Lehnerd are:

- Market orientation, designers close to the customer — Lexus' design team living in California, Olympus designers working periodically in camera shops. And derivative products should be aimed at different market segments using the same basic platform; this should be part of the initial concept — example VW and Audi using the same platform but aimed at different segments, and through derivatives in engine size, and features, at different groups.

- Using design building blocks - both internal and external, capitalising on the best product modules, rather than starting afresh.

- Using the most appropriate manufacturing technologies, and

- Using the whole organisation - 'total' innovation.

Like Quality, product platforms require a cross-functional total approach along the whole chain. Product platform thinking is also totally compatible with the ideas of target costing, where derivatives are rolled out over time whilst capitalising on platform economies. All the derivatives form a stream of target-costed products. And, as Robin Cooper has pointed out, this is done through Value Engineering.

Further Reading

Carliss Baldwin and Kim Clark, Managing in an Age of Modularity', *Harvard Business Review*, Sept-Oct, 1997, pp 84-93

Marc Meyer and Alvin Lehnerd, *The Power of Product Platforms*, Free Press, New York, 1997

Behnan Tabrizi and Rick Walleigh, 'Defining Next Generation Products: An Inside Look', *Harvard Business Review*, Nov-Dec 1997, pp116-124

David Robertson and Karl Ulrich, 'Planning for Product Platforms', *Sloan Management Review*, Summer 1998, pp 19-31

H Thomas Johnson and Anders Bröms, *Profit Beyond Measure*, Nicholas Brealey, 2000.

Anon, 'Why Detroit is Going to Pieces', *Business Week*, Sept 3, 2001, p 60

See the ECR Journal. www.ecr.org

15.5 Speed and Levelling: Critical Chain & Lean Project Management

Eli Goldratt's ideas on project management, as explained in his book *The Critical Chain*, are most relevant for reducing the time and variation in new product introduction. His ideas represent a big advance on traditional project management critical path analysis. Mascitelli has added usefully to ideas on Lean project management. Mascitelli

and Goldratt make a powerful combination for projects and new product development.

Goldratt explains that, in many projects, safety times are added to estimates of activity durations to allow for variation and other contingencies. In traditional project planning the critical path is the longest path by time through the network, but along this path there is 'float' as a result of a propensity to over-estimate time durations. Of course there is natural variation of time in activities – some taking longer than estimated, some taking shorter. But in traditional project management activities that take a shorter time than planned often result in the next activity having to wait for non-critical resources before proceeding. Activities that take longer than planned of course extend the entire project duration. Putting these two situations together means that most projects end up late. Using conventional project management software such as Microsoft Project does not address these issues of variation.

The Critical Chain solution is to recognise and manage constraints. The TOC rules of recognise, exploit, subordinate, elevate, and repeat (see the Theory of Constraints (TOC) section) is exactly what is done. As with TOC theory, completion of the entire project on time depends on the management of the critical resources. The project activity network logic is built up and validated by participants as in conventional project management. But from that point there are differences.

First, activity times need to be estimated as realistically as possible but without allowing for any 'float' or contingencies. Second, the critical chain is the longest path through the network by *resource usage*. In other words if there is one of a particular resource it cannot be used in two places at once, so the critical chain may be longer than the conventional critical path. A complex project may require special software to determine the critical chain, but for many projects it can be determined by inspection. Third, time and resource buffers must be located in order to protect the critical chain.

Appropriate resource buffers must be located in front of critical chain activities, for two reasons. One reason is to prevent the delay of having to wait for critical resources. They should be warned in good time to be available if needed.

Another is to make sure that if the activity finishes early, the next activity is able to be started without delay. Resource buffers are necessary whenever there is a change in the resource used along the critical chain. This is 'exploiting' the constraint and 'subordinating' non-critical activities.

Finally, buffers are grouped together at the end of the project network to allow for variation in time estimates. A project time buffer protects the whole project, and a feeding buffer protects non-critical activities. The size of each buffer is discussed as a totality by the project team, depending on the risks to the project as a whole, and not to individual activities. Try to arrange high risk activities as early as possible, so there is more time to catch up.

Having now determined the project duration, the duration is compared with the allowable or target completion time. If the project duration is longer, the constraints must be 'elevated' or broken by providing additional resources.

When project execution begins, concentrate on the critical chain and 'exploit' the critical resources.

Monitor the buffers. Keep a focus on the time remaining in the project as against the allowable time remaining. If necessary, 'elevate'. Ignore activities that have been completed. Monitor the buffers ahead, most of which are likely to have changed against the original plan. Time buffers can be monitored in a statistical process control (SPC) fashion — in other words some natural variation is to be expected and is ignored, but large variations signal special remedial actions. In fact, it is a good idea to ignore time deviations within a certain predetermined range of natural variation. This is management by exception, saving huge time and bother.

Mascitelli adds 12 points for Lean project management. These include, amongst others,

- 'testing for customer value' — is the customer prepared to pay for extras? – refer to the Four Objectives section above

- being clear on the deliverables and hand-offs of each activity, decreasing and eliminating design reviews to retain only those that are essential

- 'staged-freezed specifications' – similar to the 'Set Based' concurrent engineering described in a section above

- using visual management and standard work – two Lean basics

- risk buffering – a variation on the critical chain idea

- 'dedicated time staffing' whereby resources working on critical chain activities are allowed uninterrupted blocks of time

- and the Goldratt concept of resource managers and project managers (the latter taking over control of resources where activities become critical).

Further Reading

Eli Goldratt, *The Critical Chain*, North River, 1997

Ted Hutchin, *Constraint Management in Manufacturing*, Taylor and Francis, 2002

Ronald Mascitelli, *Building a Project-Driven Enterprise: How to Slash Waste and Boost Profits through Lean Project Management*, Technology Perspectives, 2002

Ronald Mascitelli, *The Lean Product Development Guidebook*, Technology Perspectives, 2007

John Arthur Ricketts, *Reaching the Goal: How Managers Improve a Services Business Using Goldratt's Theory of Constraints*, IBM, 2008

15.6 Quality

15.6.1 *Design for Six Sigma*

Design for Six Sigma (DFSS) is a methodology that parallels Six Sigma improvement, with the former focusing on new designs and products and the latter on improving existing processes. The experience of many is that conventional Six Sigma can achieve around 4 or 5 sigma performance, but to get to Six Sigma requires attacking the design side. An analogy is the public health engineer (DFSS) working with the doctor (Six Sigma) who ultimately relies on good sanitary conditions. It is useful to see DFSS as working together with Design for Manufacture, described later. Six Sigma uses the DMAIC steps. Refer to the Quality chapter. DFSS uses the IDDOV steps (identify, define, develop, optimise, verify). Both have a strict sequence of steps. Both aim at robust, low variation processes. Both share the same organisational hierarchy of Master black Belts, Black Belts and Green Belts and both require senior management commitment. Probably a DFSS project takes longer, but potentially has larger and longer-lasting payoff, and likely has greater sustainability.

Identify – The first step concerns clarifying the scope of the project. It involves doing a preliminary cost benefit / business case and scoping exercise. It will often include a project charter statement.

Define – This is concerned with clarifying the requirements for the product or service. Customer requirements are the starting point. Two principle tools are the Kano model and Quality Function Deployment (QFD). The Kano model is discussed in the Lean Philosophy section; QFD is discussed later in this section. In fact, QFD is a 'meta technique' which runs all the way through a DFSS exercise.

Develop – With desired customer requirements and ideas on design and processes clarified, the next step is to identify and evaluate various product design options. Three techniques are common – brainstorming, TRIZ and concept screening. The latter two are discussed later in this section. Also, Failure mode and effect analysis (FMEA) will usually be included – compulsorily so in some industries such as aerospace and automotive.

Optimise – This step is analogous to the Six Sigma steps of analyse and improve. In DFSS the Taguchi Loss Function is used, which postulates that customers suffer a loss proportional to the square of the distance from the optimal point. In DFSS, 'parameter design' involves maximising the function (energy use or efficiency, for instance), rather than reducing variation. Six Sigma uses design of experiments to identify the most sensitive variables, and to reduce the spread of the variation.

DFSS uses parameter design to find the best combination of variable factors to maximise the function and to test the robustness against various operating conditions. The next stage is Tolerance Design, which looks for the most critical tolerances – which have to be tight and which less tight – in other words, which tolerances have the greatest impact on overall variability.

Verify – The final step involves testing the design. Substeps involve looking at the capability of the manufacturing process, testing prototypes, and establishing process control plans. The latter may involve looking at statistical process control and failsafing (pokayoke) aspects – refer to the Quality section.

The last two steps can be viewed against the Hinckley framework of variation, mistakes and complexity against people (men), machines, method, measures, materials, mother nature. DFSS is an elegant way of addressing most of these cells at the design stage. For the Hinckley framework refer to the Quality section.

Further Reading

Geoff Tennant, *Design for Six Sigma*, Gower Press, 2002

Subir Chowdhury, *Design for Six Sigma*, Dearborn Press, Chicago, 2002

15.6.2   *Quality Function Deployment*

Quality Function Deployment (QFD) is a 'meta' technique that has grown in importance over the last decade and is now used in both product and service design. However, it is understood by the authors that Toyota no longer uses it. QFD is a meta technique because many other techniques described in this book can or should be used in undertaking QFD design or analysis. These other techniques include several of the 'new tools', benchmarking, market surveys, the Kano model, the performance - importance matrix, and FMEA. Customer needs are identified and systematically compared with the technical or operating characteristics of the product or service. The process brings out the relative importance of customer needs which leads, when set against the characteristics of the product, to the identification of the most important or sensitive characteristics. These are the characteristics that need development or attention. Although the word 'product' is used in the descriptions that follow, QFD is equally applicable in services. Technical characteristics then become the service characteristics.

Perhaps a chief advantage of QFD is that it uses a multidisciplinary team all concerned with the particular product. QFD acts as a forum for marketing, design, engineering, manufacturing, distribution and others to work together using a concurrent or simultaneous engineering approach. QFD is then the vehicle for these specialists to attack a problem together rather than by 'throwing the design over the wall' to the next stage. QFD is therefore not only concerned with quality but also is concerned with the simultaneous objectives of reducing overall development time, meeting customer requirements, reducing cost, and producing a product or service which fits together and works well the first time. The mechanics of QFD are not cast in stone, and can easily be adapted to local innovation.

The first QFD matrix is also referred to as the 'House of Quality'. This is because of the way the matrices in QFD fit together to form a house-shaped diagram. A full QFD exercise may deploy several matrix diagrams, forming a sequence that gradually translates customer requirements into specific manufacturing steps and detailed manufacturing process requirements. For instance, a complete new car could be considered at the top level but subsequent exercises may be concerned with the engine, body shell, doors, instrumentation, brakes, and so on. Thereafter the detail would be deployed into manufacturing

and production. But the most basic QFD exercise would use only one matrix diagram that seeks to take customer requirements and to translate them into specific technical requirements.

The 'House of Quality' Diagram

In the sections below the essential composition of the basic house of quality diagram is explained. Refer to the figure.

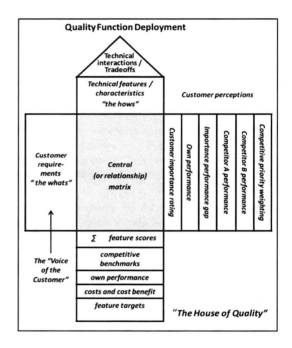

Customer Requirements

The usual starting point for QFD is the identification of customer needs and benefits. This is also referred to as 'the voice of the customer' or 'the whats'. Customers may be present or future, internal or external, primary or secondary. All the conventional tools of marketing research are relevant, as well as techniques such as complaint analysis and focus groups. Customers may include owners, users, and maintainers, all of whom have separate requirements. After collection comes the problem of how to assemble the information before entering it into the rows. In this the 'new tools' of affinity and tree diagrams have been found to be especially useful. It results in a hierarchy; on the primary level are the broad customer requirements, with the secondary requirements adding the detail.

Marketing would have responsibility for assembling much of the customer information, but the team puts it together. Marketing may begin by circulating the results of surveys and by a briefing. It is important to preserve the 'voice of the customer', but the team may group like requirements using the affinity diagram. The team must not try to 'second guess' or to assume that they know best what customers need.

Rankings or Relative Importance of Customer Requirements

When the customer requirements are assembled onto the matrix on the left of the house diagram, weightings are added on the right to indicate the importance of each requirement. Market research or focus groups establish weightings or, failing these, the team may determine rankings by a technique such as 'pairwise comparison'. (In pairwise comparison, each requirement is compared with each other. The more important of the two requirements gains a point, and all scores are added up to determine final rankings.) The Kano model (see separate section) is very often used with QFD as an aid in determining appropriate weightings.

Technical Characteristics & Associated Rankings

Customer requirements and weightings are displayed in rows. The technical characteristics (or 'hows' or 'technical responses') form the columns. These characteristics are the features that the organisation provides in the design to respond to the customer requirements. For a kettle this may include power used, strength of the materials, insulation, sealing, materials used, and noise. Once again these could be assembled into groups to form a hierarchy, using the Tree Diagram. Here the team will rely on its own internal expertise.

There are at least two ways to develop technical characteristics. One way is go via measures that respond to customer needs. For instance a customer need for a kettle may be 'quick boil'. The measure is 'minutes to boil' and the technical response is the power of the heating element. Another is to go directly to functions, based on the team's experience or on current technology.

The Planning Matrix

To the right of the central matrix is found the planning matrix. This is a series of columns that evaluate the importance, satisfaction, and goal for each customer need. See the QFD figure. The first column shows importance to the customer of each need. Here a group of customers may be asked to evaluate the importance of each need on a 1 to 5 scale (1=not important, 5=vital, of highest importance). In the next column the current performance of each product or service need, is rated by the group of customers. The difference between the columns is the gap - a negative number indicates possible overprovision, a positive number indicates a shortfall. The next few columns give the competitor's current performance on each customer need. The aim of this part of the exercise is to clearly identify the 'SWOT' (strengths, weaknesses, opportunities, threats) of competitor products as against your own. For example, the kettle manufacturer may be well known for product sturdiness, but be weak on economy. If economy is highly ranked, this will point out an opportunity and, through the central matrix, show what technical characteristics can be used to make up this deficiency. The gap (if any) between own and competitors performance can then be determined. Since the QFD team now has detail on the gap for each need and of the importance of each need, they can then decide the desired goal for each customer need – normally expressed in the same units as the performance column. Deciding the goal for each need is an important task for the QFD team. These goals are the weights to be used in the relationship matrix. Note that in some versions of QFD there are additional columns.

The Central (or Relationship) Matrix

The central matrix lies at the heart of the house of quality diagram. This is where customer needs are matched against each technical characteristic. The nature of the relationship is noted in the matrix by an appropriate symbol. The team can devise their own symbols; for instance, numbers may indicate the relative strength of the relationship or simply ticks may suffice. The strength of the relationship or impact is recorded in the matrix. These relationships may be nil, possibly linked, moderately linked or strongly linked. Corresponding weights (typically 0, 1, 3, 9) are assigned. Thereafter the scores for each technical characteristic are determined. The team, based on their experience and judgement, carries out this matching exercise. The idea is to identify clearly all means by which the 'whats' can be achieved by the 'hows'. It will also check if all 'whats' can in fact be achieved (insufficient technical characteristics?), and if some technical characteristics are not apparently doing anything (redundancy?). A blank row indicates a customer requirement not met. A blank column indicates a redundant technical feature. In practice, matrix evaluation can be a very large task. A moderate size QFD matrix of 30 x 30 has 900 cells to be evaluated. The team may split the task between them.

Technical Matrix

Immediately below the relationship matrix appears one or more rows for rankings such as cost or technical difficulty or development time. The choice of these is dependent on the product. These will enable the team to judge the efficacy of various technical solutions. The prime row uses the customer weightings and central matrix to derive the relative technical characteristic rankings. A full example is shown on the previous page.

Next, below the relationship matrix in the QFD figure, comes one or more rows for competitive evaluation. Here, where possible, 'hard' data is used to compare the actual physical or engineering characteristics of your product against those of competitors. In the kettle example these would include watts of electricity, mass, and thermal conductivity of the kettle walls. This is where benchmarking is done. By now the QFD team will know the critical technical characteristics, and these should be benchmarked against competitors. See the section on Benchmarking - especially competitive benchmarking. So to the right of the relationship matrix you can judge relative customer perceptions and below the relative technical performance.

The bottom row of the house, which is also the 'bottom line' of the QFD process, is the target technical characteristics. These are expressed in physical terms and are decided upon after team discussion of the complete house contents, as described below. The target characteristics are, for some, the final output of the exercise, but many would agree that it is the whole process of information assembly, ranking, and team discussion that goes into QFD which is the real benefit, so that the real output is improved inter-functional understanding.

The Roof of the House

The roof of the house is the technical interaction matrix. The diagonal format allows each technical characteristic to be viewed against every other one. This simply reflects any technical tradeoffs that may exist. For example with the kettle two technical characteristics may be insulation ability and water capacity. These have a negative relationship; increasing the insulation decreases the capacity. These interactions are made explicit, using the technical knowledge and experience of the team. Some cells may highlight challenging technical issues - for instance thin insulation in a kettle, which may be the subject of R&D work leading to competitive advantage. The roof is therefore useful to highlight areas in which R&D work could best be focused.

Using the House as a Decision Tool

The central matrix shows what the required technical characteristics are that will need design attention. The costs can be seen with reference to the base rows. This may have the effect of shifting priorities if costs are important. Then the technical tradeoffs are examined. Often there will be more than one technical way to impact a particular customer requirement, and this is clear from rows in the matrix. It may also be that one technical alternative has a negative influence on another customer requirement. This is found out by using the roof matrix. Eventually, through a process of team discussion, a team consensus will emerge. It may take some time, but experience shows that time and cost are repaid many times over as the actual design, engineering and manufacturing steps proceed.

The bottom line is now the target values of technical characteristics. This set can now go into the next house diagram. This time the target technical characteristics become the 'customer requirements' or 'whats', and the new vertical columns (or 'hows') are, perhaps, the technologies, the assemblies, the materials, or the layouts. And so the process 'deploys' until the team feels that sufficient detail has been considered to cover all co-ordination considerations in the process of bringing the product to market.

Note that QFD may be used in several stages in order to 'deploy' customer requirements all the way to the final manufacturing or procedural stages. Here the outcome of one QFD matrix (e.g. the technical specifications) becomes the input into the next matrix which may aim to look at process specifications to make the product.

Assembling the Team

A QFD team should have up to a dozen members with representation from all sections concerned with the development and launch of the product. Team composition may vary depending on whether new products or the improvement of existing products is under consideration. The important thing is that there is representation from all relevant sections and disciplines. There may well be a case for bringing in outsiders to stimulate the creative process and to ask the 'silly' questions. Team members must have the support of their section heads. These section heads may feel it necessary to form a steering group. QFD teams are not usually full time, but must be given sufficient time priority to avoid time clashes. The team leader may be full time for an important QFD. The essential characteristics are team leadership skills rather than a particular branch of knowledge.

Relationship with other Techniques

As mentioned, QFD is a 'meta' technique in that several other techniques can be fitted in with it. For example, value management may be used to explore some of the technical alternatives, costs and tradeoffs in greater detail. Taguchi analysis is commonly used with QFD because it is ideally suited to examining the most sensitive engineering characteristics in order to produce a robust design. Failure mode and effect analysis (FMEA) can be used to examine consequences of failure, and to throw more light on the technical interactions matrix. In the way the QFD team carries out its work, weights alternatives, generates alternatives, groups characteristics, and so on, there are many possibilities. QFD only provides the broad concept. There is much opportunity for adaptation and innovation.

Further Reading

Lou Cohen, *Quality Function Deployment: How to make QFD work for you*, Addison Wesley, Reading MA, 1995

John Terninko, *Step-by-Step QFD: Customer Driven Product Design*, Second edition, St Lucie, 1997

15.7 Additional Tools for Lean Product Development

15.7.1 TRIZ

TRIZ is a family of techniques, developed originally in Russia, for product invention and creativity. It is superb for innovative product design, and for production process problem solving. TRIZ is a Russian acronym for the theory of inventive problem solving. In 1948, the originator of TRIZ, Genrich Altshuller, suggested his ideas on improving inventive work to Stalin (a big mistake!), and was imprisoned in Siberia until 1954. His ideas once again fell into disfavour and only emerged with perestroyka. The first TRIZ ideas reached the U.S. in the mid 1980s. TRIZ is already linking up with Lean in Policy Deployment and QFD, and with Six Sigma in Design for Six Sigma. Lean TRIZ Six Sigma seems destined for a big future.

The fundamental belief of TRIZ is that invention can be taught. All (?) inventions can be reduced to a set of rules or principles and that the generic problem has almost certainly already been solved. The principles, relying on physics, engineering, and knowledge of materials can be learned. A TRIZ team uses the basic principles to generate specific solutions. Here, only a brief overview or flavour of some of the 40 principles can be attempted. TRIZ is bound to, and deserves to, become better known. We hope this will be a stimulant to you to acquire some TRIZ publications.

Darrell Mann has summarised TRIZ into five main elements.

- **Contradictions** – TRIZ believes that 'the world's best innovations have emerged from situations where the inventor has sought to avoid conventional trade-offs', for example, composite materials that are strong and light. TRIZ uses a matrix to identify which of the 40 principles are most likely to apply in any contradiction.

- **Ideality** – TRIZ encourages problem solvers to begin with the Ideal Final Result and work backwards, rather than moving forward from the current state.

- **Functionality** This is an extension of value engineering principles. (See earlier.) 'Solutions change, functions stat the same' – like holes not drills. This concept was discussed in the Value section.

- **Use of Resources** – TRIZ encourages making best use of any resource that is not being used to its maximum potential. 'Turning lemons into lemonade'.

- **Thinking in Space and Time**. Don't just think about the current state in the present, but also about the past and the future, and about wider states (supply chains?) and narrower states (sub processes?). This gives 9 boxes for understanding and projection.

A partial list of some of the 40 inventive problems includes:

- partial or overdone action (if you can't solve the whole problem, solve just a part to simplify it)
- moving to a new dimension (use multi layers, turn it on its side, move it along a plane, etc.)
- self-service (make the product service itself, make use of wasted energy)
- changing the colour (or make it transparent, use a coloured additive)
- mechanical vibration (make use of the energy of vibrations or oscillations)
- hydraulic or pneumatic assembly (replace solids with gas or liquid, join parts hydraulically)
- use porous material (make the part porous, or fill the pores in advance)
- use thermal expansion (use these properties, change to more than one material with different coefficients of expansion)
- copying (instead of using the object use a copy or a projection of it)
- thin membranes (use flexible membranes, insulate or isolate using membranes)
- regenerating parts (recycle), use a composite material.

This is a powerful list - just reading them can stimulate ideas.

Altschuller emphasises thinking in terms of the 'ideal machine' or ideal solution as a first step to problem solving. You have a hot conservatory? It should open by itself when the temperature rises! So now think of devices to achieve it: bimetallic expansion strips, expanding gas balloons, a solar powered fan.

A general methodology comprises three steps. First, determine why the problem exists. Second, 'state the contradiction'. Third, 'imagine the ideal solution', or imagine yourself as a magician who can create anything. For example, consider the problem of moving a steel beam. Why is it a problem? Because it cannot roll. The contradiction is that the shape prevents it from rolling. So, ideally, it should roll. How? By placing semi-circular inserts on each side along the beam. Finally, invention requires practice and method. Like golfer Gary Player who said the more he practised the better his luck seemed to get, Altschuller suggests starting young and keeping one's mind in shape with practice problems. Also keeping a database of ideas gleaned from a variety of publications.

Many TRIZ ideas require some technical knowledge, or at least technical aptitude. Therefore it will not work well with every group. However, it is most useful for designers, technical problem solvers, persons involved with QFD, and for implementation of Lean manufacturing (particularly the technical issues).

Further Reading

G. Altschuller, *And Suddenly the Inventor Appeared*, Technical Innovation Centre, Inc., Worcester MA, 1996

Darrell Mann, *Hands on Systemic Innovation*, CREAX, 2002

Darrell Mann, *Hands on Systemic Innovation for Business and Management,* IFR Press, 2004

Semyan Sarransky and Ellen Domb, *Simplified TRIZ*, St Lucie, 2002.

15.7.2   Pugh Analysis

Pugh Analysis is a quick and easy way to compare various alternatives and possibly to come up with a yet better alternative. The method was named after the late Stuart Pugh, Professor of Design at Strathclyde University, Scotland.

The method involves drawing up a table. The columns are the alternatives. Rows are the criteria upon which the alternatives are to be judged. The first column is the base case or existing alternative or simply the first alternative. A column of 0's is written in this first column next to each criteria. Then the second alternative is considered. For each criteria, alternative 1 is compared with the base case. No great debate is required, but the team must simply decide whether alternative 1 is better (+), same (0), or worse (-) on each criteria. Repeat for each alternative, comparing against the base case.

Simply adding up the + and − signs gives a quick judgement of the strength of each alternative. Then, go through each row or criterion in turn. For each row identify the + signs. Ask what makes this alternative superior. Make a note on the right hand side. When all rows have been worked through try to generate an alternative that incorporates all the superior features − a 'best of the best' analysis.

Pugh Analysis can also be done using weighted criteria. First the criteria must be weighted relative to one another. Then, as before identify the +, 0, - for each alternative and criteria relative to the base case. Now simply add the criterion weight together and the + or − sign. As before, judge better alternatives and try to generate a 'best of the best' additional alternative.

Further Reading

Stuart Pugh, *Creating Innovative Products using Total Design*, Addison Wesley, 1996

15.7.3 New Product Ramp Up

There appear to be distinct differences between traditional and Lean ways of doing new product introduction and ramp up. The following are a few:

- The Lean way is to separate out the variables. Take the famous catapult exercise in Six Sigma training. Pugh's Concept Screening Technique dramatically illustrates the necessity for removing as much variation as possible before attempting to maximise the length of throw. Otherwise you don't know if changes are the result of changing the variables or inherent variation in the process. The same applies in ramp up. First, concentrate on the machines and methods. When these are proven capable, move to the next stage. Second, concentrate on the people and the materials. Third, concentrate on meeting the takt time, and reducing it as real production begins. Do not try to sort it all out at once. Extending this idea, a complete new product – say a car with new body and new engine is risky. Master one major introduction at a time. Moreover, do not overlap major new product introductions. Finish one off, and learn the lessons before introducing the next major new product.

- In early days, before actual manufacture of customer-destined products begins, hold short sharp blitz-type production runs as several separate exercises using non-saleable product. Learn the maximum from each. This is PDCA – with deliberate stops in between.

- Ramp up the capacity by adjusting the takt time. Don't try to prove the maximum rate is possible and then settle down to a much slower rate. In other words, control the learning curve. In slowly decreasing the takt time, try to get quality right for each takt rate.

- Accumulate the lessons learned systematically – and use as a check list next time around.

- Have specific individual responsibility and accountability for each aspect of new product introduction. This must be completely sorted out before transferring to the next stage gate. Do not allow carry over.

- Separate out new product development from ramp up. To attempt to combine them risks many cycles of adjustment and missing the target launch time. Well before the start of saleable product production, set targets concerning design freeze, except for safety, which is the responsibility of design, and for options freeze which is the responsibility of marketing.

Acknowledgements

Emma Rigler (nee Wilson) highlighted many of these points during her MSc dissertation at Cardiff Business School.

15.7.4 The '2P' or '3P' Method

The 2P ('Process Preparation') or 3P ('Product and Process Preparation') method is a useful tool for assessing complex design choices. This can be done either early on in the development process, or later on with any major design change. Its main component is a multi-day event where a cross-functional team considers all possible design choices, assessing their impact on customer requirements, quality and cost. It also takes the impact on the production processes into account. Mock-ups and simulations are used frequently. See Mascitelli for a structure of a 2P/3P event.

Further Reading

Ronald Mascitelli, *The Lean Product Development Guidebook*, Technology Perspectives, 2007

16 Creating the Lean Supply Chain

16.1 What is supply chain management?

Supply chain management is a relatively new discipline: only in the early 1980s did firms realise that their competitiveness was not just determined by what they do, but also by what their upstream suppliers and downstream suppliers were doing. Out of that insight came the notion that it is equally as important to manage your supply chain, as it is to manage your own operation. Having a fast cycle-time in manufacturing is great, but when you have a slow distributor, the customer does not enjoy the overall benefit. In fact, for most manufacturers, their products are a function of their suppliers' processes as much as of their own processes. Vehicle manufacturers for example will buy about 60-80% of the value of their product ex-factory from suppliers. The actual assembly plant only accounts for about 12% of the cost of manufacturing a vehicle. On top of that comes the cost for distribution, retail and marketing operations, which can account for up to 30% of the list or retail price. So it is important to note that:

- Supply chain capabilities are a significant determinant of competitiveness. Just think of Dell, versus Compaq/HP, or Wal-Mart versus K-Mart.

- A final product is not the sole achievement of the OEM, but the customer experience is co-determined by the supply chain in terms of quality, cost, delivery.

- A significant proportion of the value of the final product is generally sourced from suppliers.

- The performance of one tier in the supply chain is a function of the supply and distribution functions, i.e. surrounding tiers.

Think of the case of Cisco, which had to write off $2.5 billion worth of inventory, as it was too slow to adjust its supply chain to a slowing demand in the marketplace. Or of Airbus, which had to delay the delivery of its new A380 flagship product for almost two years because it was not able to deliver correct product specifications to its suppliers – and ended up with cables that were too short. Back to the drawing board as the first aircraft was already in production!

So, as Martin Christopher says, 'Value Chains compete, not individual companies'. In traditional Operations Management, you optimise the processes with a single factory, and the assumption is that by linking these local optima, you get a global optimum at the supply chain level. That of course is wrong. In fact, some very costly dynamic dysfunctions can develop in supply chains (the Bullwhip effect is one of them), which lead to amplified orders, demand variability, poor capacity utilisation, stock-outs of some items, and overstocking on others.

The key trick in supply chain management is to consider the entire *system* of suppliers, manufacturing plants, and distribution tiers, and to aim for synergy: **2+2=5.** In other words, to analyse the system by looking at the connections or interfaces between firms, and then manage the system *as a whole*. The differential benefit of supply chain management then is the value you derive by not simply managing individual pieces, but the entire system. According to Martin Christopher, the goal is

'..to manage upstream and downstream relationships with suppliers and customers in order to create enhanced value in the final market place at less cost to the supply chain **as a whole.**'

In this section we will show the basic design of supply chains, show the root causes for the costly distortions that can occur, show how to work with supplier and logistics firms, how to deal with customer orders, the need to customise products, and finally, some general frameworks for designing Lean supply chains.

Further Reading

David Simchi-Levi, Philip Kaminsky, Edith Simchi-Levi, *Designing and Managing the Supply Chain*, Irwin McGraw Hill, Boston, (Second edition) 2003

Martin Christopher, *Logistics and Supply Chain Management*, FT Prentice Hall, 3rd edition, 2004

16.1.1 Aligning incentives

Once you have identified that the supply chain does not run optimally, how do you persuade the various companies in the system to change their behaviour? How do you motivate a firm to compromise (sub-optimise) their own operation, for the greater good of an overall more efficient supply chain? Essentially there are only two mechanisms at hand: **power**, and **shared rewards**.

In the first, the more powerful entity in the system simply dictates the changes, and punishes the supplier on non-compliance. This is common in the automotive industry, where few car makers buy from many smaller suppliers. This approach is not Lean. It does not respect the supplier, and creates bad feelings which in the long run will hurt the car maker.

A better approach is to share rewards, as is common in other sectors. In the grocery retail business for example, Coca-Cola or Unilever are equal to a Tesco or Wal-Mart in power, and shared rewards are needed to motivate partners to change the supply processes. Here, both parties collaborate and either directly share the savings from the process improvement, or the long-term lock-in (essentially the prospect of a renewed contract) persuades suppliers or retailers to comply.

Narayanan and Raman outline four steps for aligning supply chain processes.

1. **Acknowledge that an incentive misalignment exists.** Use demand amplification mapping to show the current dysfunctions, and highlight the waste.

2. **Diagnose the cause for the misalignment,** using root-cause analyses.

3. **Change incentives** (contracts, performance measures) to reward partners for acting in the supply chain's best interests.

4. **Review periodically,** and educate managers across tiers, so that they understand the implications of their decisions on the other partners in the system.

Further Reading

Narayanan, V.G., Raman, A. 2004. Aligning incentives in supply chains. *Harvard Business Review* 82 (11), p.94-103

16.1.2 Efficient versus responsive supply chains

There are several basic types of supply chain design, and here Marshall Fisher's Harvard Business Review article is already a classic - with great relevance to Lean strategy. Fisher claims that the reason why so many supply chain (and Lean?) implementations fail is that that they are wrongly configured according to demand. He claims that there are two categories of demand: Functional – typically predictable, low margin, low variety, with longer life cycles and lead times, and no need to mark down at end of season, and Innovative – typically less predictable, high margin, high variety, shorter life cycles and lead times, with end of season discounting common. Functional demand requires an efficient process or supply chain, innovative requires a responsive process. Mismatches between demand and process give problems, as do transferring successful managers from one type of supply chain to the other.

Fisher says that the loss of contribution as a result of stockout is markedly different between functional and innovative products. In the former, a typical contribution margin of 10% and stockout of 1% translates to 0.1% of sales (negligible) but in the latter, rates of 40% and 25% respectively translate to 10% of sales (very significant).

For Lean strategy the implications are that a product requiring a responsive supply chain would be inappropriate in a low cost distant location, but that may be just the right thing for stable, functional demand. This may well mean having more than one type of facility and demand chain to cope with different demand segments. Can you equate efficient with Lean and responsive with 'agile'? Certainly not!

Both need many Lean principles in place. But a responsive chain requires, in addition to fast, flexible flow strategic inventory buffers. Better still is reducing order lead times and uncertainty by faster information flows (EDI, EPOS, ECR, see later section on collaboration) including process re-engineering. Fisher gives three alternatives – reduce uncertainty (faster information or mass customisation), avoid uncertainty (reduce lead-time) or hedge against uncertainty (buffer inventory). Of course this is just like the Factory Physics concept that there are only three ways to counter uncertainty: inventory, capacity, or lead time.

Fisher says that many companies find products in the mismatch quadrant of 'innovative products' and 'efficient supply chain'. If located in this mismatch segment there are two possibilities – either make products and demand more functional (product line rationalisation, or design) or make the supply chain more responsive by the three approaches mentioned. An organisation would need to balance the options. SAB Miller is a case in point. They have massively efficient, low flexibility, high volume 'base' breweries and more flexible but less efficient flex breweries. An individual brewery therefore cannot be judged on its own. The Systems approach.

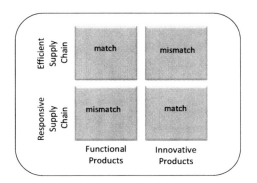

Further Reading

Marshall Fisher, 'What is the Right Supply Chain for Your Product?' *Harvard Business Review*, March/April 1997

16.2 Dynamic distortions

Think of dynamic distortions as waves on the ocean – the smoother the waves are, the less energy the ship loses going in one direction. Supply chains are the same: the more volatile the demand and delivery patterns, the more inventory, expedited shipment and under- or over-utilised capacity you can expect. In a study of the grocery retail sector, Kurt Salmon and Associates found that there was 12.5%-25% excess cost in supply chain, with a $30bn cost saving potential in the US grocery sector alone (on a $300bn turnover). Key to understanding these 'ripple effects' is to distinguish the root causes, and the system reaction that follows. Let's start with the most important root cause, uncertainty.

16.2.1 Types of Uncertainty

There are three basic types of uncertainty which can have a negative impact on any process, factory or supply chain.

1. **Demand uncertainty** – This type is related to the marketplace that is what the customer orders. This can be variable because of the weather (e.g. ice cream), seasonality (e.g. sales of lawn mowers), or follow a general trend (e.g. customers buying more and more flatscreen TVs, rather than conventional tube ones). Other factors that impact on the demand are sales promotions (which generally cause temporary upswings, but also downswings for related products), new product introductions or new technologies (see section on disruptive technologies), and competitor action (sales promotions, new products). Apart from promotions and dynamic pricing there is generally little influence over demand uncertainty. But beware of assuming that demand is uncertain, before you have explored all the root causes! See also the section on Demand Management.

2. **Conversion or throughput uncertainty** – This is any type of process uncertainty that hits throughput, such as producing defects, machine stoppages and breakdowns, long

change-overs, as well as unpredictable lead-times for a given process. These can all be eliminated, and it is of course where TPM and Six Sigma and many Lean tools are relevant.

3. **Supply uncertainty**– any type of uncertainty related to the delivery of materials and components. This could be in the form of variable quality, poor on-time delivery performance, and variable lead-times. Supplier co-operation is relevant.

In addition to these basic types it is important to distinguish between **actual** uncertainty (i.e. caused by the end customer) and **self-created** uncertainty (i.e. created by poor coordination in the supply chain). For example, when a promotion increases sales for a certain product, generally consumers will pull purchases forward to make good use of the low price. Thus, the high demand during a promotion is generally followed by an artificial and self-created slump in demand. Wal-Mart uses this argument not to have any promotions at all, and instead runs a stable supply chain using 'Everyday low prices' (EDLP). They successfully avert the negative consequences of promotions, and instead lower all prices equally based on the savings from running a stable supply chain.

Further Reading

Davis, T. 'Effective supply chain management', *Sloan Management Review* 34 (4), 1993, p35-46

16.2.2   The 'Bullwhip Effect'

The Bullwhip effect is a supply chain phenomenon in which fluctuations in orders amplify as they move along a supply chain (see also section on Demand Amplification Mapping). The 'bullwhip' is an increase in order variance as the signal is transmitted from one to the next tier in the supply chain.

There is a vertical dimension to do with instability and growth in magnitude and a horizontal dimension to do with fluctuations over time. The Bullwhip effect can seriously damage the performance of a supply chain, however Lean an individual player in the chain may be. The effects are the need to keep overcapacity, fluctuations between low and high demand (even when there is little fluctuation at the customer end!), and poor customer service.

How does it happen? The basic problem applies to any system that has a delay in responding to a change in the input signal. As the system takes some time to respond and adjust the output upwards or downwards, there is a slight over- or undershoot that is eventually passed on to the next tier. Here, the reaction has to be even greater. Have you ever wondered why traffic on the motorway sometimes comes to a complete stop, for no apparent reason? The answer is the bullwhip: as the cars are driving too close to one another, the reaction time for drivers to reduce speed is insufficient, so eventually the 'reaction' of the system amplifies up to the point where traffic comes to a halt. The same effect happens in the supply chain, which we are showing here as a set of two linked water tanks: one representing the retailer (downstream), the other one the supplier (upstream). The amount of water represents the inventory in the system, the valves that control the water flow into and out of the tanks are the ordering decisions.

As there is a delay in the information flow, the preceding tier (the supplier in this case) has to react even more strongly to a change in demand: overall, the variance of the order increases (see figure).

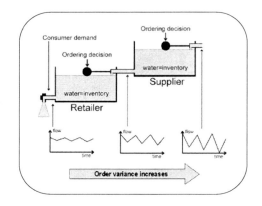

Thus, the demand signal is perceived to be more variable, the higher up in the supply chain you go. Research has shown that on average, the order variance increases by a factor of 2 for each tier in the supply chain!

It is important to remember here that 'structure drives behaviour' (as shown very nicely by Jay Forrester and John Sterman at MIT in the Beer Game). That is, many of the elements that cause demand amplification or the bullwhip effect are built into the system, and are not driven by the end customer!

Why does it happen? Lee et al. (1997) identify five inter-related factors as causes, building on the earlier work by Forrester and Burbidge:

1. **Demand forecasting and signal processing** – This is one element of the so-called Forrester or amplification effect. Forecasters at each stage of the supply chain try to hold and adjust safety stocks to buffer against variation. A chain reaction takes place as a result of a minor disturbance leading to greater variation and hence more safety stock all along the chain. Signal processing amplification also results from the way orders are interpreted, so is linked to batching discussed below. Improving the forecasts and reducing uncertainty by sharing information is an effective counter.

2. **Lead times** – The other element of the so-called Forrester effect, results directly from the fact that safety stocks and order quantities are calculated from lead times and variability. Reducing the lead-time improves performance.

3. **Batching** – Also known as the Burbidge effect, this results from orders being placed in batches – a large batch followed by no orders, in a repeating cycle. Batches may be ordered for transportation or order cost reasons. The EPE and milkround concept can help here.

4. **Price fluctuations and promotions** – Supply chain players may try to anticipate price increases or take advantage of quantity discounts. Information sharing and coordination of response to increases can help. Elimination of inappropriate quantity discounts in favour of 'every-day low pricing (EDLP)' is effective.

5. **Rationing and Inflated orders** – Also known as the Houlihan or 'Flywheel' effect, this results from supply chain partners trying to anticipate shortages or distributors rationing supplies in the interests or fairness. Over ordering can lead to a vicious cycle where the increase is interpreted as an increase in ultimate customer demand rather than a safety stock policy change. Orders lead to shortages leading to higher orders and so on until extra capacity is installed, leading to a collapse in orders. Again, sharing of information is a way to go. Note that these bullwhip factors are usually thought of in a supply chain context but may also occur internally within a plant.

In simple terms, there are three enemies of a stable (and Lean) supply chain:

1. **Inventory & delays** that worsen any 'swing' of amplification. The longer the response time to a change, the worse the swing upstream. This is a particular problem for global supply lines. Also, decision delays require stock, as the longer the forecasting horizon the worse the forecast becomes, and the more buffer has to be held. Finally, safety stock decisions can send false signals, so beware of this trap: inform your supply chain partners if you adjust stock levels

2. **Unreliability or uncertainty** – Any kind of uncertainty needs to be covered with inventory. Here, unreliable processes cause unreliable delivery, and ultimately uncertainty at the receiving end. The starting point needs to be a reliable and capable process.

3. **Hand-offs or decision points** – Every hand-off or tier in the system bears the danger of distorting the demand signal, as planners tend to have 'misperceptions' of the actual demand. So the more people who interfere with the demand flow, the worse the swings

generally become. In particular be aware of 'double-guessing', by creating a forecast on someone's forecast.

Always remember Michael Hammer's quote: 'Inventory is a substitute for information' – where you have perfect information, you do not need any buffer stock. The less reliable information you have, the more inventory you need to hold.

Further Reading

David Simchi-Levi, Philip Kaminsky, Edith Simchi-Levi, *Designing and Managing the Supply Chain*,(Second edition), McGraw Hill, 2003

Steve Disney and Denis Towill, 'Vendor-managed inventory and bullwhip reduction in a two-level supply chain', *International Journal of Operations and Production Management*, 23: 5/6, 2003

Sterman, J.D. *Business dynamics: systems thinking and modeling for a complex world*, Irwin McGraw-Hill, Boston, 2000

16.3 Managing supplier relations

16.3.1 Supplier Selection

For Lean supply to work there must of necessity be few or even single suppliers per part. The idea is to work with a few good, trusted suppliers who supply a wide range of parts. During the last decade drastic reductions in many a company's supplier base have taken place. An objective is to remove the long tail of the supplier Pareto curve whereby perhaps 10% of parts are supplied by 80% of the suppliers.

Generally, collaborative long-term supplier partnerships make sense for 'A' and possibly 'B' parts; less so for commodity items, where commodity purchasing via internet auctions may be developing. Part criticality and risk also influence the rationalisation decision; you would not risk partnership with a company having poor industrial relations, or weak finances, or poor quality assurance. This means that a team approach is necessary in supplier selection. The Purchasing Officer may co-ordinate, but throughout the partnership Design would talk to

their opposite number in Design, Quality to Quality, Production control to Production control, and so on.

There should be little risk of 'being taken for a ride' (also referred to as 'opportunistic behaviour') because the supplier has too much to lose. But there are ways around this too: having one supplier exclusively supplying a part to one plant, but another supplier exclusively supplying the same part to another plant. This spreads the risk whilst still achieving single supplier advantages.

Alternatively there is the Japanese practice of cultivating several suppliers simultaneously but then awarding an exclusive contract to one supplier for a part for the life of the product, and selecting another supplier for a similar part going into another end product.

Four Models for Supplier Strategy

There are at least four models for thinking about supplier selection and sourcing that should be considered by Lean supply chain managers. Several logistics managers make use of more than one of the following to help structure their selection and rationalisation process.

1. The **Runner Repeater Stranger / ABC inventory model** is discussed under inventory considerations in the Future State section of the book. Clearly a difference in sourcing policy between A category runner parts (close partnership?) and A category stranger parts (loose partnership?), should be considered. Also a difference between A category runners and C category runner parts (arms-length?). And so on.

2. **Part complexity and logistics supply chain complexity** – Consider a two by two matrix with part or process complexity along one axis and logistics supply chain complexity along the other axis. Logistics complexity may also refer to the needs for flexibility and responsive lead times. Long lead time in itself does not indicate logistics complexity. High-High with both process and logistics complexity: here close partnerships are a possibility. The 'low-low' segment may suggest

global low cost purchasing via perhaps E Bay. Complex parts and processes with low logistics difficulties may suggest partnership sourcing on a worldwide basis. Finally, high logistics difficulties but with low part or process complexity may suggest local sourcing but from arms-length suppliers.

3. **Jeffrey Dyer suggests three categories – internally manufactured, partner suppliers, and arms-length independent suppliers**. His analysis shows the huge advantages gained by Toyota in sourcing approximately half its component costs from partner suppliers. But the other two categories each make up approximately one quarter of total component costs. Dyer refers to this mix as the governance profile. But the ideal profile differs by industry favouring a higher proportion of partner suppliers in high-tech adaptive industry.

Note however that Toyota does not seem to have joined the web-based purchasing revolution, preferring to deal with a limited number of suppliers in a traditional way. One of the reasons may be the way in which the Internet-based purchasing platforms have evolved. The much-praised COVISINT platform launched by several OEMs in 2000 never delivered on its grand promise of saving $1,000 per car, and was soon abused by vehicle manufacturers for auctions that were entirely decided on by the lowest price, and where rules were frequently bent to the advantages of the OEMs. Not surprisingly, suppliers were reluctant to join, and soon came up with their own purchasing portal, SupplyOn, which assures fair rules and is used successfully by the automotive industry.

4. Clayton Christensen has a concept based on his **'disruptive technologies'** thesis (see the Strategy section). This is not a supplier selection concept, but more to do with sourcing strategy. The performance of a product type (such as a PC) or major subassembly type (such as hard disk) improves with time. The needs of customers also grow, but generally at a slower rate. In the early days of the life cycle, when the needs of customers are above the product performance curve, Christensen calls this 'not good enough'. But as performance grows, the product or subassembly outstrips the needs of customers, even demanding ones. PCs are now in this category for most customers. Christensen calls this 'good enough'. When a product is strongly good enough it is vulnerable to disruptive technologies (see the Strategy section). Christensen believes that a sea change occurs as products or assemblies move from the not good enough to the good enough category. In the former case, integration is critical to success, (say the early days of Ford), because R&D, design, and manufacture have to be tightly integrated. The integrators make the money. But in the good enough category, a company must compete on new dimensions of speed and flexibility. Modules and interfaces are more clearly specifiable. In this case disintegration is required - being able to source the current best components from the appropriate suppliers. Here, power shifts to those who are able to supply the needed modules with the required flexibility. But the OEMs have power also – forcing suppliers to develop more innovative components; in fact saying to them that they are 'not good enough'.

Further Reading

Jeffrey Dyer, *Collaborative Advantage*, Oxford University Press, 2000

Clayton Christensen et al, 'Skate to Where the Money Will Be', *Harvard Business Review*, November 2001, pp72-81

16.3.2 *Modes of Supplier Relations*

There are essentially two basic and opposing models of how to relate to suppliers: the cost-driven adversarial model, and the long-term collaborative model. The former is the traditional Western model where you aim to negotiate hard, get the best unit cost. And if next year another supplier offers a better price, you switch. The Japanese model is very different. Here the relationship is built on trust, and long-term commitment. In Japan it was further cemented by cross-ownership (the *keiretsu* in Japan, or *chaebol* in South Korea). However it would be a mistake to assume that there is no competition in a Japanese-style relationship. In fact the reason the Japanese supplier relationship model is so much more successful (for A- and B-parts), is because it merges the benefits of long-term collaboration and trust, with a persistent element of market-pressure.

The Partnership philosophy is that, through co-operation rather than confrontation, both parties benefit. It is a longer-term view, emphasising total cost rather than product price. Cost includes not only today's price of the part or product, but also its quality (defect / ppm rate), delivery reliability, the simplicity with which the transaction is processed, and the future potential for price reductions.

But partnership goes further: long-term, stable relationships are sought rather than short term, adversarial, quick advantage. The analogy of a marriage is often used. It may have its ups and downs, but commitment remains. In a partnership, contracts will be longer term to give the supplier confidence and the motivation to invest and improve. Both parties recognise that the game where low prices are bid and then argued up on contingencies once the contract is awarded is wasteful and counterproductive. Instead, it may be possible for both parties to co-operate on price reduction, sharing the benefits between them. Such co-operation may be achieved through the temporary secondment of staff. See the next section on Supplier Associations.

The features of a Japanese-style supplier partnership are:

1. **Long-term collaborative relationships**, where trust and commitment, as well as respect of the right of mutual existence are the prime directive. There is no opportunistic behaviour ('screwing up the supplier') for a short-term advantage. The focus is long-term.

2. **Dual sourcing**: each component will have few, but at least two sources. The proportion of the volume is adjusted every year according to supplier performance. So there is a long-term commitment and security, but also an element of market pressure in the relationship.

3. **Joint Improvement activities**: there is a strong collaboration with suppliers on operational improvement. For example, Toyota has a dedicated Supplier Support Center (TSSC) in Kentucky to educate suppliers in Lean. Also, while Toyota demand annual cost reductions, these are realised by collaboration, not in isolation.

4. **Operations and logistics:** Level production schedules are used to avoid spikes in the supply chain. Also, milk-round delivery systems that can handle mixed-loads, and small-lot deliveries needed for Just-in-Time or Just-in-Sequence supply. The disciplined system of JIT delivery windows at the plant means that suppliers deliver only what is needed, even if this compromises load efficiency in transport.

The Supplier-Partnering Hierarchy

The collaborative supplier relationship model is essential for supporting a Lean supply chain, and can be applied in the Western world as much as it has been used in Japan. The argument that the Eastern keiretsu or Chaebol structures are essential to support long-term relationships has long been disproved. Japanese vehicle manufacturers are working efficiently with Western suppliers, where they do not own parts of the company.

Liker and Choi illustrate how the Japanese manufacturers have built equally strong supply chains to support their US plants. They illustrate a set of principles for building 'deep supplier relations'.

Conduct joint improvement activities.

1. Exchange best practices with *suppliers*.

2. Initiate kaizen projects at *suppliers'* facilities.

3. Set up *supplier* study groups.

Share information intensively but selectively.

4. Set specific times, places, and agendas for meetings.
5. Use rigid formats for sharing information.
6. Insist on accurate data collection.
7. Share information in a structured fashion.

Develop *suppliers'* technical capabilities.

8. Build *suppliers'* problem-solving skills.
9. Develop a common lexicon.
10. Hone core *suppliers'* innovation capabilities.

Supervise your *suppliers*.

11. Send monthly report cards to core *suppliers*.
12. Provide immediate and constant feedback.
13. Get senior managers involved in solving problems.
14. Turn *supplier* rivalry into opportunity.
15. Source each component from two or three vendors.

Create compatible production philosophies and systems.

16. Set up joint ventures with existing *suppliers* to transfer knowledge and maintain control.
17. Understand how your *suppliers* work.
18. Learn about *suppliers'* businesses.
19. Go see how *suppliers* work.
20. Respect *suppliers'* capabilities.
21. Commit to co-prosperity.

Trust, Partnership and Dedicated Assets

Another view on the collaborative supplier relation model was given by Jeffrey Dyer in his seminal book in 2000. He identified three key characteristics which make the Toyota US supply chain so effective. Although much has changed since Dyer's analysis - for instance Ford has hived off Visteon, GM has hived of Delphi, and Mercedes has acquired Chrysler, the points made remain valid and important in many industries. The three interrelated characteristics are:

The critical role of Trust where the trustworthy partner builds confidence in its promises and commitments and does not exploit the vulnerabilities of its partners. Building trust takes time (for example in selecting and favouring suppliers) but then allows fast, flexible flow in new product introduction and in the supply chain. Bureaucracy and waste in the form of transactions can be dramatically cut. Dyer gives impressive evidence of the extent of transaction costs, and the cost of mistrust. He points out that trust also encourages investment, innovation, and stable employment. Dyer shows Toyota in the USA well ahead in trustworthiness.

Investment in dedicated assets Building on trust allows investment in dedicated assets. Dyer shows that during the 1990s Ford and GM internally manufactured about twice as much as Toyota, had approximately the same proportion of arms-length suppliers, but had approximately one fifth the proportion of supplier partners of Toyota. Things have changed, since. Dedicated assets are possible with partners, and in turn allow better productivity, quality, design, and speed.

Incidentally, Dyer points out that the advantages of dedicated assets and partnership are much more important in complex industries. This is where Japanese industries are much more efficient. But they are far less important in simple product industries where arms-length relationships may be beneficial.

The development and transfer of knowledge throughout the network Again, made possible with trust and dedicated assets, knowledge transfer of both explicit and tacit knowledge is a key factor in improvement in productivity and quality. In our experience Toyota cells in the UK are often far more productive than cells run for other manufacturers within the same supplier site. This is because they enjoy more assistance, get more stable schedules, have more confidence in the future, have simpler procedures, often get better terms, enjoy better coaching, and are less fearful about visits from Toyota improvement experts and engineers than most other customers.

In Japan, and increasingly in the rest of the world, supplier partnership is now expanding down from relationships with first tier suppliers, to second and even third tier. Larger firms in the car industry have been leaders, but other industries and smaller firms are following. In common with TQM, the thought is that quality is only as good as the weakest link.

Further Reading

Jeffrey Dyer, *Collaborative Advantage*, Oxford University Press, 2000

Liker, J.K., Choi, T.Y., 'Building Deep Supplier Relationships', *Harvard Business Review* December 2004, p 104-113

16.3.3   Supplier Associations

The supplier association concept is an extension of the supplier partnership concept. Supplier associations are 'clubs' of suppliers who form together for mutual help and learning. Members may all supply one company, or are all from one region serving different customers. The associations seek to learn best practices from other members or to gain competitive advantage and/or productivity through co-operation. In Japan, Supplier Associations are known as kyoryoku kai.

There are three types of association: for operations (to gain cost, quality, delivery improvements), for purchasing (to gain from economies of scale), and for marketing (to gain from synergistic practices or by pooling expertise). Peter Hines defines the first type as 'a mutually benefiting group of a company's most important subcontractors brought together on a regular basis for the purpose of co-ordination and co-operation as well as (to) assist all the members (by benefiting) from the type of development associated with large Japanese assemblers: such as kaizen, just in time, kanban, U-cell production, and the achievement of zero defects.'

The aims (following Hines) are:

- to improve skills in JIT, TQM, SPC, VE/VA, CAD/CAM, Flexibility, Cost
- to produce a uniform supply system
- to facilitate the flow of information
- to increase trust
- to keep suppliers in touch with market developments
- to enhance the reputation of the customer as a good business partner
- to help smaller suppliers lacking specialist trainers and facilities
- to increase the length of relations
- to share developmental benefits
- to provide an example to subcontractors of to how they should develop their own suppliers.

The company-sponsored variety may benefit from the parent company's expertise and resources, often given free. The regional variety simply shares resources such as training seminar costs and training materials, but also will share expertise by lending key staff experts to other member companies for short periods. The regional type may be partially funded from government, and may have a full-time facilitator. In Japan it is considered an honour to be asked to join a prestigious supplier association, as run by a major corporation. Joint projects, assistance in areas of expertise, development of common standards, training, courses, an interchange or secondment of staff for short periods, benchmarking, hiring of consultants or trainers, factory visits within the association, joint visits to outside companies or other associations, are all common.

The type of supplier who may join an association is not necessarily dependent on size - in fact, larger suppliers with their own corporate resources may benefit less. Also, suppliers of common or catalogue parts may not be invited. Suppliers that are usually targeted are those dependent upon a parent for a significant (perhaps 25% or more) proportion of their business. The purchasing department of the parent company often plays a key role, but some supplier associations have been set up on the initiative of lower tier suppliers or academic groups such as the Cardiff's Lean Enterprise Unit. Often, a supplier association will hold an annual or biannual assembly to look at performance figures. Ranking of suppliers by different measures is presented. This is often sufficient motivation for lower ranking members to ask for help or to take action on their own.

A supplier association usually will have its own set of rules and regulations and be run by (perhaps) a retired senior engineer from the parent company or increasingly by a full- or part-time co-coordinator from one of the companies. Support staff are seconded for short periods, depending on projects and needs. Often member companies pay a subscription fee. At the top level, the association will have a steering group at MD level, which meets perhaps annually. Some functional directors may meet quarterly. Engineers and front line staff may meet more frequently or may form temporary full-time task groups to address particular problems. Some associations consider social events to be important icebreakers. Within the association there may be a functional split by product category, or by area of concern cost, quality, delivery, production planning, etc.

Purchasing Associations

A variation is an association that bands together for mutual purchasing advantage, gaining from improved quantity discounts and greater 'clout' than a single company can bring to bear. A database of required materials and goods is usually maintained, sometimes by a third party. These have been successful in Australia, often on the initiative of a purchasing consultant. A purchasing association does not necessarily go in for all the activities of an operations association, and may be confined to purchasing staff. A type that has become fairly common in JIT plants is where a contractor takes on the responsibility for the inventory management and supply of numerous small items. This is a form of 'vendor

managed inventory'. Because such contractors operate in different regions they may be able to gain quantity discounts some of which are passed on. Typically such a contractor supplies one large plant, but there are variations where a contractor supplies numerous small firms in a region. This is almost like having a co-operative shop, except that the contractor is a professional inventory manager and re-stocker.

A **Marketing Association** may have characteristics similar to 'Agile Manufacturers'. That is, they pool resources for synergistic gain or to win large contracts. Such groupings, often known as consortia, have been common in defence, computing, and construction.

Further Reading

James Womack, Daniel Jones, Daniel Roos, *The Machine that Changed the World*, Rawson Associates, 1990

Jeffrey Dyer, *Collaborative Advantage*, Oxford University Press, 2000

Richard Schonberger and Edward Knod, *Operations Management* (Sixth Edition), Irwin, Illinois, 1997, Chapter 9.

Peter Hines, *Creating World Class Suppliers: Unlocking mutual competitive advantage*, Pitman, 1994

Richard Lamming, *Beyond Partnership*, Prentice Hall, 1993

Donald Fites, 'Make your Dealers your Partners', *Harvard Business Review*, March/April 1996

For a case study on the establishment of a supplier association in Wales see Dan Dimancescu, Peter Hines, Nick Rich, *The Lean Enterprise*, AmaCom, New York, 1997

16.4 Supply Chain Collaboration

16.4.1 Vendor Managed Inventory (VMI)

A centralised information system, with actual demand forecasts provided by the first stage to all players in the chain is an effective method of significantly reducing the bullwhip effect. Not quite as good is where each player determines target inventory levels determined from moving averages from the next stage downstream, and uses this target as the basis for orders to the next stage upstream. Disney and Towill suggest that the appropriate use of VMI (vendor manager inventory) may be a solution (see also section below). Here the customer passes inventory information to the supplier instead of orders. The actual inventory at the customer is compared with a pre-agreed reorder point (ROP), set to cover adequate availability. Both parties also agree an order-up-to point (OUP). When actual inventory is at or below the ROP the supplier delivers the difference up to the OUP. This system can work well between each tier in a supply chain, and is made more effective using milkrounds.

Applying this logic to the water-tank model from above, what happens in VMI is that the supplier now takes over the ordering decision from the retailer. This is beneficial, as it provides the supplier with direct visibility of 'what is going on' at the retailer in terms of stock levels, and most importantly, it also eliminates one decision-tier from the supply chain. As we have seen above, the bullwhip effect is driven by lead-time, uncertainty, and hand-offs or decision points. VMI is a powerful tool in reducing the bullwhip effect. It reduces uncertainty by allowing additional visibility of consumption at the retail tier, it cuts lead-times as the supplier does not have to wait for a formal order, and it eliminates a decision point.

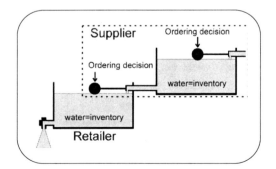

16.4.2 Information Sharing

Information sharing can happen in two ways: the retailer or manufacturer can share its actual sales data ('EPOS', or electronic point of sale data, that retailers generally share with their suppliers), or they can share and align their forecasts with their suppliers (collaborative forecasting). These two types serve very different purposes. EPOS data can be very useful to plan short-term execution and to drive the replenishment signal (where it works like a kanban: sell one, replenish one). Shared forecasts have no value in the short term, but are essential to align capacities and avoid bottlenecks and overproduction in the future. Also, sales promotions need to be communicated well in advance, so that the entire supply chain is aware of the likely short-term increases, but does not overreact when the spikes go through the system.

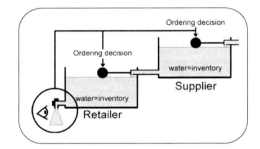

16.4.3 Collaborative Planning, Forecasting and Replenishment (CPFR)

The collaborative planning, forecasting and replenishment approach (CPFR) was piloted by the grocery retail sector (see vics.org), and effectively merges the VMI and collaborative planning elements, to form a model of close supply chain collaboration. Shown here in our water tank model, CPFR uses tools to increase the demand visibility (collaborative forecasting), EPOS data to drive the replenishment (continuous replenishment), as well as reduced decision tiers for inventory and order management (VMI). Thus, it is a powerful tool to manage high-volume supply chains in the fast-moving consumer goods arena. The CPFR model since has also been used in many other sectors, but remember these points:

- There is a cost for setting these systems up, so do use a Pareto analysis to determine whether it is worth it
- The system only works where you can include close to 100% of the demand. When some suppliers or customers do not collaborate, the value you derive from CPFR will be considerably less
- Make sure to use the additional information gained not just for sales planning, but also communicate this to production. Link the production schedule to the customer forecasts. A common mistake is to have the information, but not to use it!

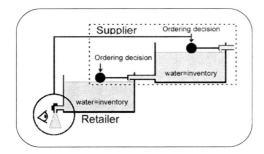

Further Reading

Matthias Holweg et al., 2005, Supply Chain Collaboration: Making Sense of the Strategy Continuum, *European Management Journal,* Vol. 23, No.2, p. 170-181

16.5 Lean Logistics

16.5.1 Milk-round collection and delivery

The long-established Milkround concept is widely applied, across many industry sectors. The idea is that a vehicle travels frequently around a set route starting and ending at the plant, and visiting several suppliers en route. At each supplier a small (daily?) batch of (several?) parts is collected in a particular window slot – typically a half hour. A milkround may also be found in distribution. The milkround concept is similar to the waterspider or

runner concept within a plant. Runners have proved a hugely effective concept within plant. Similarly, milkrounds are proving a hugely affective way of reducing amplification and encouraging steady flow between supply chain members. The runner is the internal drumbeat; the milkround is the external drumbeat. The greater the degree of mixed model, or the lower the 'EPE', the better it will all work. Milkrounds are also an aid to problem surfacing and improvement. See the separate section on runners and waterspider.

Milkrounds can reduce the waste of transport, improve fast, flexible flow and reduce lead-times. They encourages confidence, and as a result reduce buffer inventories and encourage synchronised scheduling. Perhaps a small batch of several parts is collected every day rather than a large batch of one part number every week. Moreover, an efficient routing calling at several suppliers can reduce total distance. If the company is really clever it can deliver finished products, return totes or even move parts from one supplier to another. Some milkrounds include cross docking, whereby parts are picked up on a milkround from more distant suppliers, perhaps using smaller vehicles consolidated into a larger vehicle.

Synchronisation is needed to minimise the length of time inventory spends on the cross dock. The marginal cost of joining a milkround circuit may be small. This idea should be sold to supplier meetings. The more suppliers or distributors who join, the less the cost to everyone. Today, milkrounds are 'owned' by either OEMs or first tier suppliers, although the vehicles may be owned by a third party contactor.

16.5.2 Risk Management, Risk Pooling

Wherever there is an investment, there is risk. So a key ability in management is to assess risk, and to devise strategies for mitigating or hedging against it. There are several stages to risk management: **prevention,** which aims at lowering the odds of the risk occurring, **control,** which reduces the damage if it does occur, **transfer via**

insurance, where you pay someone else to take on your risk, **diversification,** whereby you aim not to put all your money on one card, and finally, **hedging,** where you contract for a future price, which prevents a loss, but also prevents any extraordinary gain. In addition to these, there are also operational means how to reduce risk, called **risk pooling.**

With regards to supply chain management, the risk pooling idea is to redesign the supply chain, the production process, or the product, either to reduce the uncertainty the firm faces or to hedge uncertainty so that the firm is in a better position to mitigate the consequence of uncertainty. The basic ways this can be done are three:

1. **Location pooling,** whereby the inventory from multiple territories and locations is combined into a central or regional facility, which minimises the risks of stock-outs or overstocking.

2. **Product pooling or postponement,** whereby product configuration is delayed using a modular design, which can serve overall demand with fewer products variants. HP implemented this approach very successfully within its printer division to counter demand uncertainty across markets.

3. **Capacity pooling,** whereby each production facility produces several models, in order to counter any peaks or troughs for individual models. Volvo is using so-called 'swing models', which are produced in both of its plants to counter any demand fluctuations over the life cycles of its other models.

Overall Risk Pooling follows the two rules of forecasting: postponement (and aggregation) increases quality of the demand signal, and reduction of lead-time (or forecasting horizon) increases the quality of the signal. Risk-Pooling strategies are most effective where demands are negatively correlated (i.e. as demand for one product goes up, the demand for another one goes down), as then the uncertainty with total demand is then much less than the uncertainty with any individual item.

Further Reading

David Simchi Levi et al., *Designing and Managing the Supply Chain*, MacGraw-Hill, (2nd ed), 2003

16.6 Order Fulfilment and Product Customisation

16.6.1 Responsiveness, Flexibility or Agility?

There has been a great deal of buzz around the concept of 'agility' and 'agile manufacturing'. Although it is generally perceived to be an opponent to Lean ('you can either be Lean, or agile..'), the interesting fact is that agile actually was invented by Lee Iacocca at Chrysler in the early 1990s as a synonym for Lean; for political reasons you did not want to be associated with 'Japanese' manufacturing techniques. Since, the term has been used in a different context by Martin Christopher and his group at Cranfield University, where it has become synonymous with 'responsiveness', i.e. how to create supply chains that are able to adapt to changes in demand, product or technology in a short timeframe. It is important to understand the distinction between flexibility, responsiveness, and agility, as very often these terms are used in a confusing manner.

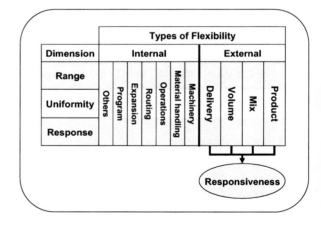

So, flexibility is the basic ability to react in a given dimension. This can be internally, where you have flexible machines that allow for small batches, or externally, where a company may be able to adjust its volume of output quickly (volume flexibility). The chart above shows the main types of flexibility, and their dimensions.

Responsiveness only applies to the market or external side, and is the ability of a company to be flexible in terms of its delivery lead-time, its volume or output, its product mix, and ability to introduce new products into the market.

Agility is a concept that aims to help firms become responsive, for example by suggesting additional inventory buffers, spare capacity, or by postponing the product customisation. Agile concepts work best in uncertain or fast-moving sectors (such as fashion), and can be easily merged with Lean techniques. Christopher and Towill suggest using Lean for the stable or base demand, and using agile techniques for the strangers or volatile products. Often consultants and academics will try to highlight conflicts between Lean and agile – but this is nonsense! Lean is driven by customer value; if the customer values a short response lead-time which is enabled by holding component inventory, then this inventory is not waste. Think about Dell, which runs a very Lean supply chain – enabled by 2 weeks worth of component inventory on site near its factories. This inventory might appear wasteful if you only consider the factory level, but in fact it adds considerable value when you look at it from the customer's point of view.

Further Reading

Andreas Reichhart and Matthias Holweg, 'Creating the Customer-responsive Supply Chain: A Reconciliation of Concepts', *International Journal of Operations and Production Management,* Vol. Vol.27 No.11, 2007, p.1144-1172

Ben Naylor et al., 'Leagility: integrating the Lean and agile manufacturing paradigms in the total supply chain' *International Journal of Production Economics* Vol, 62 No. 1-2, 1999, p.107-118.

Nigel Slack, 'The flexibility of manufacturing systems' *International Journal of Operations & Production Management, Vol.*7, No. 4, 1987, p.35-45.

Martin Christopher and Denis Towill, 'An integrated model for the design of agile supply chains.' *International Journal of Physical Distribution and Logistics Management* Vol.31, No.4, 2001, p.235-246

Kidd, P., *Agile Manufacturing - Forging new Frontiers.* Wokingham, Addison Wesley, 1994

16.6.2 Order Fulfilment Strategies

Order fulfilment is essentially the concept a company uses in responding to a customer order. Depending on the customer willingness to wait, and the cost of providing customised products, there is a range of strategies from pure make-to-forecast, to building all products to order.

The most basic strategy here is to make products to forecast (make-to-forecast, MTF), and to sell all products from stock. This is most common in the retail sector, where items are sold from the shelf in the high-street store. There is no customisation. This approach has the advantage of creating a stable production schedule, but bears considerable risks in stock obsolescence and stock-outs.

Another approach is the Assemble-to-Order (ATO) model, which uses component inventory to assemble products to customer order. Dell is the classic case. Computers are not built to stock to be put into stock at the retail stores (although Dell is also thinking about opening stores), and instead are only built when a firm order is received. This works well for modular products built from only a few components, but does not work well for products that use customised components.

Here, the build-to-order (BTO) or make-to-order (MTO) model works best. In a BTO model, components are only ordered when a customer order is received, and then the product is assembled from these customised components. This is the typical approach for luxury vehicle manufacturers, where products can be built from a pool of billions of possible specifications. The main problem for BTO manufacturers is to manage demand so that the factory is well utilised. As a result many firms will show a mix of

MTF and BTO production in the auto industry, which balances the cost of under-utilisation and stock holding.

Finally, where not only the production, but also the designs are customer-specific, we speak of Engineer-to-Order (ETO) strategies. Here not only are the assembly and components customer-specific, but the design is also customised. Typical examples are oil rigs or Formula 1 cars, which are designed to fit a specific purpose. ETO is common in the construction and machine tool sectors.

These are the basic strategies; there are several ways in which the approach can be refined, for example by using web-based search tools to locate inventory (called 'Locate-to-Order'), or by mixing some MTF for Runners with BTO for Repeaters and Strangers (called 'Hybrid BTO'). Each of these approaches has certain advantages and disadvantages, as shown in the table below.

Decoupling Point or 'Push-Pull' Boundary

Depending on the order fulfilment strategy used, some parts of the supply chain are driven by customer orders, other parts by forecasts. This boundary between Push and Pull is called a 'decoupling point'.

Decoupling points are always inventory locations, which are needed to counter any forecast error. In the grocery supply chain, the decoupling point is the inventory on the shelves in the supermarket. In Dell's case, it is at the component inventory level. In a true BTO system, the decoupling point generally lies in the 2nd or 3rd tier of the supply chain, as both components and products are made in response to a customer order. There can be several decoupling points in one supply chain.

PUSH				PULL
	Make-to-Forecast (MTF)	**Locate-to-Order (LTO)**	**Amend-to-Order**	**Hybrid Build-to-Order**	**True Build-to-Order (BTO)**

	Make-to-Forecast (MTF)	Locate-to-Order (LTO)	Amend-to-Order	Hybrid Build-to-Order	True Build-to-Order (BTO)
Goals	Decoupled production and relative stock management to allow for efficient production	Use of stock visibility to widen customers' choice at the expense of extra transportation cost	Sophisticated push system with limited flexibility and high risk of pushing when demand drops	Compromise between stable production and cost of inventory in the market	Customer-driven value chain using active demand and revenue management
Benefits	• Efficient production • Local optimisation of factory operations	Higher chance of finding right vehicle in stock	Higher degree of custom-built vehicles in production	• Stable base production • Relatively short OTD times on average • Less discounting needed	• No stock apart from showroom and demonstrator. • No discounts, but active revenue management to maximise profit
Weaknesses	• High stocks in market, discount-based selling for ageing stock and alternative specification • Customer orders compete with forecast for capacity • Complete de-coupling from customer	• Still high stock levels needed to source • Extra transportation cost for transfer between dealers	• Customer orders built only when they fit • Unsold orders are built anyway, so same problems as MTF • High temptation to revert to MTF if demand drops	• Still stock in market • Still requires discounting of ageing stock to cope with forecast error • Danger of reverting to complete push	System is sensitive to short-term demand fluctuations, hence will not work without proactive demand management

Further Reading

Matthias Holweg and Frits Pil, 2001, 'Successful build-to-order strategies start with the customer.' *MIT Sloan Management Review* Vol.43, No.1, p.74-83.

Frits Pil and Matthias Holweg, 2004, 'Linking Product Variety to Order Fulfilment Strategies.' *Interfaces* Vol.34, No.5, p.394-403

16.6.3   Mass Customisation

The basic notion of 'mass customisation' was raised by Stan Davis in 1987, when he put the conceptually opposing terms 'mass production' and 'individual customisation' together, by suggesting that for manufacturers to survive they need to develop the capability to deliver customised products – at the same price as current mass produced items! Jospeh Pine took up this idea, and proposes four methods to implement mass customisation:

1. **Customise services around standard products or services.** Although standard products are used, customisation takes place at the delivery stage. For example, airline passengers may be offered different meals or in-flight entertainment, and pizza customers are offered substantial choice. On the Internet it is possible to receive a customised news service. Standard hotel rooms may nevertheless be offered in non-smoking, secretarial support, quiet-during-day, or close-to-entertainment or pool varieties. Customisation through service may be the way, for instance through individual support of a standard computer or through knowledge of the specific requirements of individual customers using a dry cleaner. The key to this method is good information on customer, especially repeat customer, needs.

2. **Create customisable products and services** Here, customisation is designed into standard products which customers tailor for themselves. Examples are adjustable office chairs, automatically adjustable seats, steering wheels or even gearing on some cars, or the flexible razor, which automatically adjusts to the user's face. The key here is often technology, but technology which follows customer need. The automatic teller machine offering a variety of services is a prime example.

3. **Point of Delivery Customisation.** Here variety is built in just prior to delivery, even later than the first type of customisation. For instance, software specific to a customer's requirements may be added. In-store point of delivery customisation - spectacles, photo developing, quick-fit tyres - are now commonplace. This type of customisation often requires 'raw material' or semi-processes inventory to be held at the point of delivery, but the advantage is zero finished goods inventory and improved speed of response.

4. **Quick Response** usually involves integration along much of the supply chain. A classic example is Benetton's 'jerseys in grey' which are kept un-dyed until actual demand is communicated, often via electronic data interchange (EDI), and supplied on a quick delivery service. Inventory is only kept in a partly processed state at a central factory, none in the distribution chain, and minimal is kept in shops. (By the way, Quick Response is also known as Efficient Consumer Response (or ECR), the former associated with apparel, and the latter with groceries.)

Further Reading

Joseph Pine, Mass Customisation, *Harvard Business School Press*, Boston, MA, 1993

James H. Gilmore and B. Joseph Pine, 'The Four Faces of Mass Customisation', *Harvard Business Review*, Jan/Feb 1997, pp 91-101

16.6.4   Demand and Revenue Management

Understanding and Managing Demand

The nature of demand needs to be understood. This relates to:

- Kipling's 'Six Honest Serving Men', What, Why, When, Where, How, Who. It amounts to detail on market segmentation. Take a hotel example: who are the customers (businessmen and tourists), what do they need (support and leisure), when (daytime and evening), why (working away from home, having a break), where (customised by location), how (focused hotels? separate blocks? secretarial support for business, video for tourists, etc).

- The Runners, Repeaters and Strangers concept (see the Essential Paretos). The Expected response time of different customer groups.

- The Possibility of influencing demand patterns by discounts and promotions.

- Trend, Seasonality and Variation – the last four taken together have an important influence on location of supply chain nodes, and on supply chain scheduling. It may be possible to fill-in troughs in demand with customer groups having longer response time and price expectations. For example, scheduling car assembly for hire cars in non-peak months, or pricing according to lead time like airlines.

Revenue or Yield Management

Another approach commonly used in service industries is revenue management (often also referred to as yield management or dynamic pricing). The basic idea is to manage capacity by altering the price. This is in particular critical for service providers such as airlines and hotels, where the capacity is lost if not used. An airline seat not filled cannot be 'put into the warehouse', and passengers cannot be transported 'just in case'. So service firms rely on altering demand via price to manage their capacity. Many manufacturing firms have not woken up to this idea yet, and instead prefer to rely on regular stock-clearing promotions to rid themselves of their overproduction.

16.7 Creating High-performance supply chains

In this final section we will introduce some basic frameworks that sum up the concepts we have introduced in this chapter. The first one is Richard Wilding **'3 T's** of highly effective supply chains', which are **Time**, as any lead-time results in bullwhip and inventory, **Transparency**, as a lack of forward visibility creates uncertainty and ultimately stock or unused capacity, and **Trust,** as long-term collaborative supplier relationships by far outperform short-term adversarial ones.

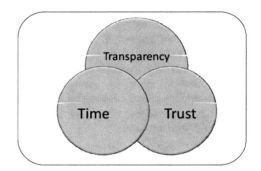

Another framework that is powerful and easy to remember is Hau Lee's **Triple-A** supply chain, which features **Agility** to respond to changing customer needs, **Adaptability** to long-term changes in markets and technology, and **Alignment** of incentives to enable cooperation and coordination across tiers in the chain. Together, these frameworks cover the basic elements of what it takes to design a Lean supply chain. See references below for more detail.

Further Reading

Hau Lee, Creating the Triple-A supply chain, *Harvard Business Review,* October 2004, p.105.

Richard Wilding, The 3Ts of Highly Effective Supply Chains, *Supply Chain Practice,* 2003, Vol. 5 Issue 3, p.30-39

17 Accounting and Measurement

17.1 Lean Accounting

First of all, you need to distinguish between 'Lean accounting', and 'accounting for Lean'. Lean accounting tries to minimise the number of transactions and the efficiency of the process; accounting for Lean tries to improve decision making to enable Lean operations. We will cover both.

David Cochrane and Thomas Johnson, amongst others, have made the point that many companies are now managed the wrong way around. They start with measures or targets, then work out the physical solutions (the 'hows' and the 'whats'). By contrast, becoming Lean should start with the purpose, derive 'hows' and 'whats', and then choose to reinforce achievement.

There are a few points to remember before reading this section about Lean accounting. A strong reason accounting for Lean gained prominence is because traditional accounting systems were essentially backwards looking (i.e. only reporting on past performance, but giving few (if any) real pointers how to improve in the future), and are not able to reflect accurately the improvements made through Lean. The point to remember is that Accounting for Lean is undoubtedly a major improvement over the status quo, but must still not be used to control the financial performance of the firm. Accounting information is always descriptive, not prescriptive! Even though Accounting for Lean will provide you with much better information, it is important to understand that the financial performance is an emergent outcome of the relationships among the organisations' parts. As Thomas Johnson points out, managers who strive to improve financial results by encouraging their staff to chase financial targets, will invariably achieve worse results than those who help improve the system that generates these results! As Deming pointed out, managers should not use financial targets to control financial results, instead, manage the relationships that produce these results.

Nonetheless, accounting is a vital instrument to control the organisation, not least because of the legal requirements and shareholder accountability requirements. But it is against this backdrop that you should read this chapter.

Further Reading

Thomas Johnson, Management by Financial Targets isn't Lean, *Manufacturing Engineering,* December 2007, p 1-5

Johnson, T., Kaplan, R.S. 1987. *Relevance Lost - The Rise and Fall of Management Accounting,* Harvard Business School Press, Boston

David Cochrane, 'The Need for a Systems Approach to Enhance and Sustain Lean', in Joe Stenzel (ed), *Lean Accounting: Best Practices for Sustainable Integration*, Wiley, 2007

17.1.1 Warnings and Dilemmas

While the benefits of Lean are generally obvious to the Operations people, they are far from obvious to the traditional Accounting world. Hence a few warnings and dilemmas that you need to be aware of.

Much of the 'conflict' between traditional and Lean thinking on accounting and measures derives from a fundamental difference in assumption about the system. Traditionalists believe that parts are separate and that improving the parts will lead to improving the whole. Lean thinkers, by contrast, take an end-to-end, or value stream view. Thus a traditionalist might automate a warehouse, or favour speeding up a machine, but a Lean thinker would ask about eliminating the warehouse or slowing the machine. This fundamental difference in viewpoint leads to other differences:

- Inventory is seen an asset in our current accounting systems – so its reduction can appear unfavourable on the Balance Sheet.

- Stopping work because there is too much inventory or it is not needed for a while will mean that budgeted activities will not take place. Activities 'absorb' overhead, so that not carrying them out means that the overhead will not be absorbed and there will be an unfavourable variance. There will be under-recovery. Unfavourable variances, in turn, show up on the profit and loss (or income) statement.

- A reduction in lead time may result in customers cutting orders for a while. This is a short term negative effect, but must be placed against longer term competitive advantage.

- Overproduction, at least in the short term, may generate positive variances and increase the book profit. However, your cash flow is likely to improve. If profit is seen as more important than cash, this is an issue.

- Two products are made – one with high labour content and high contribution, the other with high automation, low labour and low contribution. As overhead is allocated, the high labour, high contribution product may turn into a loss maker. It may then become a candidate for outsourcing. If this happens, the overhead will now have to be allocated to the low contribution product, eventually also driving it out of the business.

- Saving operators as a result of Lean activity may be a dilemma. If people's jobs are threatened, they are unlikely to participate in improvement. Even though an assurance has been given that 'no-one will lose their job as a result of improvement', no saving is made until a person actually leaves. This can be managed in a situation of growth or where there is labour turnover. However, there will be a delay before the saving reaches the 'bottom line'. If there is no growth or where there is small labour turnover the dilemma is worse. Moving them to an 'indirect' category will create an unfavourable variance in that category. Thus, many claimed savings are 'fake'.

- In a low growth scenario it may be dishonest to say that no-one will lose their job due to improvement.

- If supplies can be acquired at a discount this will generate a positive variance. But what if that means that delivered batches will be bigger and more inventory will need to be stored?

- When labour is reduced skills and training are lost – but this goes uncosted. But Lean should be about growth. With growth, those skills are more valuable – and much less expensive than if acquired from scratch.

- A cell is implemented. The previous system had process layout. The cell is more labour-intensive but lead time is slashed. So a huge reduction in lead time is a little more expensive. Competitiveness and delivery performance are much improved, but the financials are unfavourable – at least in the short term.

- Through pokayoke and self inspection it is possible to slash the number of inspectors. Inspectors are 'indirect' because they work in several departments. Overhead goes down but the standard cost in the area increases.

- 'Kaizen results don't show up on the bottom line!' Of course they don't! Actual savings are only made when people actually leave or less material is purchased. A statement like the opening one reflects the view that Lean is about cost reduction. In fact, it is about growth and competitiveness. However, it is also true that many Kaizens will have no impact on growth or competitiveness, and are waste.

- There is invariably a lag between actual performance and the financials.

Accounting for Lean is a developing field. After several centuries of little change the basic assumptions of accounting are at last being questioned in the light of Lean and Theory of Constraints. We need to distinguish between Financial Accounting which is required for tax and shareholder purposes and is subject to GAAP, and Management Accounting which is used for decision-making. Accounting for Lean falls into the second category.

In some ways, what has happened in Lean manufacturing is beginning to happen in accounting in Lean environments. Lean manufacturing lifted the focus from the activity to the value stream. End-to-end performance became much more important than the efficiency of an individual operator or machine. It is the non value adding steps between operations that get the focus. Economies of scale were important pre-Lean. But with Lean, economies of flow and time are more relevant. Similarly with accounting for Lean: it is not the person, machine, or department but the end-to-end value stream that matters. More was better, and producing more, irrespective of demand, was rewarded with positive variances and greater apparent profit. Lean (and TOC) has begun to show the fallacy of this non-systems view.

17.1.2 What should Accounting for Lean give us?

- More relevant information for decision-making. More relevant means the ability to identify factors that are becoming uncompetitive, and where there are potential opportunities for improvement.

- Positive support and evidence for doing the right things – fast, flexible, flow. For reducing inventories and lead times, for improving quality, and for improving delivery performance.

- Financial numbers that are transparent to non-accountants without having to go through several days of education. 'Plain English' profit and loss statements that exclude variances, show actual operations profit, but maybe have an additional line to show changes in overhead, labour and inventory.

- A simplified system that cuts waste and unnecessary transactions. A Lean accounting system needs to be a minimalist system – tracking only the absolute minimum transactions with the lowest frequency possible.

- A system that highlights when to take action, as importantly as when not to.

- Guidance on medium term product costing and target costing.

Although accountants, planners and managers probably may not like it, recall that Ohno said that an aim of Lean / TPS should be to make the system so simple and visible that there would be little need for complex controls. Real Lean must cut overhead! Ohno also said, *'Excess information must be suppressed!'*

17.1.3 What should Accounting for Lean NOT give us?

- Evidence that implementing Lean is exactly the wrong thing to do. The warnings at the beginning of this section are illustrative.

- Product costing on a month-by-month basis. There is a Western obsession with detailed product costing brought about by the belief that costs can be controlled by the financials. They cannot. Only productivity improvement can make a difference. Variances often encourage game playing managers who spend an inordinate amount of time manipulating figures rather than focusing on improvement. Plant and machines are sunk costs. These costs cannot be changed in the short term, only manipulated. In fact, there is no such thing as the true product cost, at least in the short term.

- Detailed variance analysis. Variances are tracked in detail against standards, resulting in for example labour efficiency variances, volume variances, material usage variances, and purchase price variances. Variance analysis is almost pure waste. Worse, it can generate non-Lean behaviour. Many non-accountants do not understand where 'unfavourable' negative variances come from. (In fact, they are based on assumptions and forecasts about future operation levels.) But they learn that overhead is absorbed as labour and machine hours accumulate, and they do not want to be caught with unfavourable variances. This encourages overproduction. The point is, what can a manager do about an unfavourable variance in the short term? Answer: almost nothing favourable for Lean. Senior managers need to ponder that one.

- Being 'precise' but late (and worse, expensive) is far worse from a Lean perspective from being approximate but fast.

Maskell makes the point that Lean should not be regarded as a short-term cost cutting strategy, but rather as a long term competitive strategy. Cutting waste creates opportunity for growth, and accountants have an important role to help identify what is to be done with freed up capacity. Use marginal costing? Certainly. But standard costing? Often inappropriate.

17.1.4 An Accounting for Lean and Lean Accounting System

Some pointers for a Lean accounting system follow.

- Work towards direct costs. Rather than trying to 'solve' the overhead allocation problem by some 'elegant' procedure such as Activity Based Costing (ABC), set an objective to decentralise overhead functions so that they can be directly associated with cells or product lines. So for example have schedulers, quality, maintenance, purchasing, and training associated with particular products.

- Allocate the shared overhead in a way that supports Lean – a good way is allocation based on lead-time, not on labour hours or activities.

- Have a general overhead pool for all overhead that is not directly associable with a product or service.

- As far as possible, record costs by end-to-end value streams. Avoid transfer costs between sections of the value stream.

- Eliminate variance reporting.

- Eliminate detailed product cost reporting. Instead, do a periodic estimation exercise together with line managers. Look ahead rather than back. Direct cost association helps considerably. ABC (cost driver) methodology can be used to identify cost drivers.

- Do not do regular or continuous ABC. Report product contributions rather than product costs.

- Use accountants, together with designers and marketers to look at costing alternative materials and features.

- Reduce the number of transactions. Use the runners, repeaters and strangers concept to take advantage of repeating operations where transactions like purchasing can be simplified. Backflush transactions. Use blanket transactions for purchasing.

- Reduce reporting time. As with changeover reduction, do as much prior work as possible prior to period end. Then make adjustments only. Do a Pareto analysis on transaction size – do not delay reporting on many small items that can be carried over to the next period with minimal consequence. Be fast and approximate rather than slow and 'precise'.

- Encourage accountants to think about variation of costs rather than cost variances. Cost variation means looking into the distribution of product costs. What is the spread from worst case to best case. Then, why are the worst cases occurring? This is much like the Six Sigma methodology. Tackle the worst cases. This concept is developed by Johnson and Bröms who maintain that many a product line has been abandoned due to unfavourable average costs that could have been saved by appropriate analysis and pruning.

- Report by exception. Get accountants to think common cause and special cause (SPC). Only report special cause events.

- Reduce the frequency of reporting intervals. Ask what are the benefits and costs of various reporting intervals.

- Clarify the presentation of accounts so all can read them. This means that the word 'variance' should not appear on a Profit and Loss statement. Specifically report actual increases and decreases in inventory.

- Record inventory valuations in terms of raw material value only. Do not accrue value. Do not show 'deferred labour and overhead' as costs that have been accrued into inventory.

- Go through the implications of Lean implementation with senior managers beforehand. They need to know and expect the consequences of inventory reduction, and labour and machines changes on Profit and Loss and Balance Sheet. But also incorporate lead times, defect rates, and customer satisfaction alongside the financials.

- Highlight changes in cash flows. It is money going into and out of the company that is of prime importance. Know the implications on cash of Lean implementations.

- If the company has constraint resources (and most do), focus costs around these constraints. Calculate contribution per constraint minute. Know what the opportunity cost of an hour lost or gained at a constraint will be, and get the accountants to cost it.

- Get accountants to participate in the assembly and evaluation of Future State maps. There should be parallel value stream maps that examine the financing periods (time to pay suppliers, time to finance operations, time to get in cash). In other words, end-to-end cash flows or cash turns – not just inventory turns.

Further Reading

Adrian Gordon, *The Lean Control Book*, MSc Lean Ops dissertation, Cardiff Business School, 1999

Richard Schonberger, *Let's Fix It!*, Free Press, 2001, Chapters 4 to 7

Orest Fiume, 'Lean Accounting and Finance', *Target*, Fourth quarter, 2002

Jean Cunningham and Orest Fiume, *Real Numbers*, Managing Times Press, 2003

Jim Huntzinger and Robert Hall, Measurement Conundrums, *Target*, v23 n4, Fourth Issue, 2007

H Thomas Johnson and Anders Bröms, *Profit Beyond Measure*, Nicholas Brealey, 2000

John Darlington, *Notes on Costing*, MSc Lean Operations, Cardiff Business School.

Brian Maskell and Bruce Baggalay, *Practical Lean Accounting*, Productivity, 2004

Joe Stenzel (ed), *Lean Accounting: Best Practices for Sustainable Integration*, Wiley, 2007

17.2 Performance Measures

17.2.1 Measurement Basics

Begin any discussion on Lean measurement by recognising that measurement is waste. It should be limited and minimised. 'You cannot fatten the calf by weighing it'. At the same time, recognise that an effective measurement system is one of the most powerful tools for change, and for Lean transformation, that exists. Measurement should

- Provide short-term indicators of problems – and no problems.

- Be part of a feedback loop of surfacing and resolving problems.

- Relate to learning or capability of the process or people.

- Focus on improving performance.

- Be capable of being acted upon.

'A science is as mature as its measurement tools', said Louis Pasteur (as quoted in Dean Spitzer). Is it not time for Lean to develop more maturity?

Two basics are:

- **Measures not Targets** – Measures help you to decide what to do. But Targets are often associated with rewards, punishments and motivation. Targets thereby almost invariably encourage deviant behaviour. The many examples from the British Health Service illustrate – from ignoring patients who have passed the target wait-deadline, to removing wheels from hospital trolleys so they don't count as patients waiting on trolleys. Moreover, when targets are associated with rewards, often ever bigger rewards have to be given. Targets were a Deming pet-hate. Motivational measures (or targets) frequently result in cheating, but informational measures can assist improvement.

- **The Process, not the Person** – Deming spoke about the 94/6 rule – 94% of problems can be traced to the process, but only 6% to the person. But often it is the person who is measured, not the process. Start with the assumption that it is the process that is broken and most times you will be right. Almost everyone has experienced negative measurement – errors, cost overruns, lateness – and almost everyone has responded by negative emotions – blame, threats, defensiveness. Most of this can be avoided if you start with the process not the person – so the manager, not the subordinate, needs to correct the process.

Informational measures come in two types: Objective – based on facts that can be observed and verified, and Subjective – based on opinions or judgement, which therefore have the opportunity for distortion, prejudice, and even revenge. Subjective measures may be hard to avoid, but the user or interpreter needs to be aware of their shortcomings. The philosopher Wittgenstein talked about a ruler: if you use a ruler to measure a table you may also be using the table to measure the ruler. The less you trust the ruler's reliability, the more information you are getting about the ruler and the less about the table. So in asking for a rating, are you measuring the thing or the measurer?

'What gets rewarded gets done' says Michael LeBoeuf as his GMP 'greatest management principle in the world', but better is Spitzer's statement, **'You get what you measure'**. Think about it, and beware!

Michael Hammer's '7 Deadly Sins of Performance Measurement' are a salutary list. Briefly, with Lean transformation examples, they are: Vanity (measures that are aimed at making the manager look good – a partial lead-time improvement, not end-to-end); Provincialism (measuring within department boundary not the value stream); Narcissism (measuring from your point of view, not the customer's – delivery performance against promised date not customer's request); Laziness (assuming one knows what is important to measure – cost when delivery performance is more important to the customer – not 'going to gemba'); Pettiness (measuring only a small part – delivery on time, but not in full); Inanity (measuring without thought of the consequences – prioritising OEE – OEE improves but schedule attainment decreases and batch sizes increase); Frivolity (not being serious – 'we can't stop the line to look at problems').

A Good Measurement System

Dean Spitzer says there are four keys to the success of a measurement system.

1. **Context** – Effective measurement can only occur in a positive, supportive context. This is the culture that surrounds the measurement – informational or punishment, process not person. The attitude of the boss. An unfavourable measure is an opportunity not a threat. We want to surface issues, not suppress them.

2. **Focus** – Measure the right thing. Be aware of measuring too much. Pareto. Derive many of the measures from participative policy deployment, not sucked out of the air. As Nassim Taleb says, 'It is important to be aware that the following is fallacy: The more information you have, the more you are confident about the outcome.'

3. **Integration** – There must be an integrated system for measurement. Maybe the Balanced Scorecard, although better in a

Lean context would be the Create Flow, Maintain Flow, Organise for Flow framework or Policy Deployment. In any case measures need to be aligned, balanced, and adaptive.

4. **Interactivity** – Measures need to be acted on in real time. Two-way interaction. Actually setting up the measures is only a small part – how they are used and reviewed is at least as important. Perhaps a daily meeting around the Communications Board. Spitzer says this is a social process, not a technical process.

Further Reading

Dean Spitzer, *Transforming Performance Measurement*, AmaCom, 2007

Nassim Taleb, *Fooled by Randomness*, Random House, 2005

Deming's and Shewhart's Counsel

A quotation from W Edwards Deming's famous book, *Out of the Crisis*, serves as a salutary warning on measures. 'Rates for production are often set to accommodate the average worker. Naturally, half of them are above average and half below. What happens is that peer pressure holds the upper half to the rate, no more. The people below the average cannot make the rate. The result is loss, chaos, dissatisfaction, and turnover.'

Deming illustrated his frustration with managers and measures with his famous red bead game. Six volunteers draw 50 beads at a time from a container having red and white beads, using a paddle. The reds are defects. The participants are urged to produce fewer defects. Of course there is variation between the participants, but it is out of their control. The 'good' performers are praised, the 'bad' ones given a warning. Some improve ('warnings work!'), some don't and are fired. A better way is to set up a control chart. Then you realise that all variation is 'common cause'.

Shewhart's Insight

Shewhart, Deming's teacher, said that measures should be seen as continuing and self-correcting, following the PDCA cycle. For Shewhart, measurement was part of a prediction cycle. Shewart saw measurement having three elements, the data, the human observer, and the conditions. Note that all three are subject to variation. The past is used to interpret the present in order to predict the future. Everyone develops a theory or model (or a filter) as to the relationship between data and effect. We all implicitly use models, good or bad – and they are uncertain. Since we are dealing with uncertainties in data, observation, and interpretation we should use control charts to assist in understanding the variation – whether special cause or common cause. And we should try to improve on the model and understand the system via Plan Do Check Act.

Virtually all measures should be tracked on an SPC-type chart, in order to distinguish common cause from special cause variation. Beware of misleading measures. 'Drowning in a river of average depth 3 feet', and 'The next person to walk through the door will have more than the average number of legs.'

Further Reading

W Edwards Deming, *Out of the Crisis*, Cambridge, 1986

Walter Shewhart, *Statistical Method from the viewpoint of Quality Control*, Dover, 1986

17.2.2   Relevant and Distant Measures

Richard Schonberger warns that bombarding employees with measures about which employees can do nothing and which the customer does not care about, can have adverse affects. Low productivity may not be the employees' fault – a point made by Deming. Schonberger argues that measures should not be primarily for 'control' purposes (negative connotations, often lagging the events, but sometimes required), but for encouragement and innovation. So 'controls' should be minimised. As with control charts, do not take action unless the process is out of control. On the other hand, measures that encourage improvement should be emphasised – like a receiving dock operator discussing delivery performance with drivers on a daily basis – rather than simply viewing delivery schedule achievement ratio, or plotting inventory time profiles at constraints – rather than viewing stock turns, or measuring minor stoppages and their reasons – rather than viewing last week's OEE data.

Likewise, Schonberger says that measures should be time appropriate. Category 1 measures can be influenced in the short term and a short-term expectation is appropriate. Examples are scrap, flow distance, WIP turns, changeover time, unsafe acts, capability index, skills mastered. Category 2 take longer to respond. Examples are throughput, raw material and finished goods turns, delivery performance, labour productivity, OEE, and employee satisfaction. Measuring and expecting short-term results could be counterproductive – improvement may be possible in the short term, but at the expense of medium term deterioration – like running inventories down, now but ignoring next quarter. Measurement trends should be interpreted over the medium term and are the responsibility of middle managers. Category 3 measures require longer-term interpretation, not short term reward or punishment. Examples are market share, customer retention, share price, new products launched. These are the responsibility of senior management.

Further Reading

Richard Schonberger, *Let's Fix It!*, Chapters 6 & 7, Free Press, 2001

Richard Schonberger, 'Performance Measures for a World Class Workforce', *Target*, Vol 15 No 4 1999

17.3 The Basic Lean Measures

Arguably, there are four basic or prime measures for Lean. Each of them encourages 'all the right moves'. Each can be implemented on various levels from cell to plant, even supply chain. They are also a set, to be looked at together.

Lead time. Measuring lead-time encourages inventory reduction, one-piece flow, reduction of flow length, and waste reduction. The measure is best done end-to-end from receiving dock to dispatch. Next best is to track only work in process lead-time. Lead-time can be measured on a sample basis, by tagging a small number of components each month at the receiving dock. Build up the distribution – do not just measure average lead-time. You really want to get the shape of the distribution – the narrower the better. Alternately, but not as good, you can use Little's Law – see the section on Factory Physics. A variation on this measure is to track 'Ohno's Time Line' – the time between receiving an order and receiving payment, expressed in $ per hour. This is particularly good since it includes transaction processing time, and puts the emphasis on cash flow.

Customer Satisfaction – Following the first Lean principle, monitoring customers is a basic requirement. If failure is indicated here, this has to be the first priority. Do get this measure from customers, not internally from shipments. An obvious question is – who are your customers? Final or intermediate? Answer: both. Sample them across all relevant dimensions – cost, quality, delivery as basics, but note also soft measures such as the RATER framework: reliability, assurance, tangibles, empathy, responsiveness. See Zeithaml and Bitner, *Services Marketing*, 2006.

Schedule Attainment – An internal measure of consistency. Schedule attainment is the ability to hit the target for quantity and quality on a day-to-day basis line-by-line or cell-by-cell – not weekly for the plant. Again track the distribution. If you have a Heijunka system this is straightforward. Of course, if the schedule is out of line with customer demands, the measure is a waste of time.

Inventory Turns, and 'SWIP to WIP' – Inventory turns is an established measure. An alternative is days of inventory. Arguably if you are measuring lead-time it is not necessary to track inventory turns but do track lead-time dock to dock and WIP inventory turns, rather than overall inventory turns. Why? Because WIP is fully under your own control, raw materials and finished goods are not fully under own control. SWIP is standard work in progress inventory, so measuring the variation between what should be and actual is useful.

17.3.1   QCDMMS

QCDMMS is an acronym for a set of measure categories widely used in Lean organisations and displayed at each line or area.

Quality – Internal scrap, rework, and first time through – expressed in parts per million. 'First time through' percentage is parts entering minus parts scrapped or reworked at each stage. Because rework can happen several times this measure can be negative.

Cost – Typically a productivity measure – units per person per week. Usually not a monetary value. OEE performance may be shown here.

Delivery performance – Inbound from suppliers, outbound to customers. QOTIF (Quality, OnTime in Full). A delivery that is not 100% perfect, on time, and in full scores zero.

Morale – Absenteeism, suggestions or improvements and possibly the result of an attitude audit.

Management – Communications, extent of cross training, attendance at shop floor meetings.

Safety – Accidents, unsafe acts and audit of unsafe conditions.

17.3.2   DTI 7 Measures

The British Department of Trade and Industry (formerly DTI, now BERR) published a set of 7 measures that were developed out of Industry Forum's Improvement Events (Blitz events). These are now widely used in the UK. As a set they are very useful; as individual measures they may have limitations. They are intended as a way of tracking trends within a company, rather than as benchmark measures.

1. **Not Right First Time**   NRFT = (Defective units x 1 million)/ Total units.

2. **Productivity**   P = No of units made / No of direct operator hours.

3. **Stock turns** = Sales turnover per product / (value of RM + WIP + FGI). It is useful to turn this measure into three measures – one for each of raw material, work in process, and finished goods, since only WIP is fully under own control. The bad news is that value of inventory is subject to local accounting conventions and may be subject to manipulation. Best to define each category in terms of purchased material or parts value only, and not to assume any value is added until sold.

4. **Delivery Schedule Attainment** – This is the number of planned deliveries – (number late + number of part deliveries) as a percentage of total planned deliveries. A negative figure is possible. Note that quality does not figure, unless defectives are counted as part deliveries.

5. **OEE** – Overall Equipment Effectiveness. See discussion on OEE in Total Productive Maintenance (TPM) section.

6. **Value Added per Person** – (Output value – Input Value)/Direct Employees. Beware! May encourage inappropriate automation.

7. **Floor Space Utilisation** = Sales turnover in area / Area in square metres.

Further Reading

DTI, *Quality Cost Delivery*, Dept of Trade and Industry, 1998

DTI (now BERR) website: www.berr.gov.uk.

17.3.3   Schonberger's Micro JIT Ratios

Richard Schonberger suggested three quick ratios in 1987 that are still very useful reminders of the real objectives of Lean. They are:

1.   Lead time to work content. Work content is actual work or value adding time. This encourages continuous flow, keep it moving,

synchronised operations. Of course the ideal ratio is 1, but typical ratios run to 100 or even 1000.

2. Process speed to sales rate. This ratio encourages uniform flow to takt. Ideal is 1 but typical is 5 to 1000. It addresses 'Hurry up and wait', and batch and queue. The ratio discourages monuments and encourages a balanced line.

3. Number of pieces to number of workstations. The ideal is one-piece flow with a ratio of one. A good ratio is 2. Typical is 50 or more. This encourages focused cells, and discourages stockrooms and excessive supermarkets.

Further Reading

Richard Schonberger, *World Class Manufacturing Casebook*, Introduction, Free Press, 1987

17.3.4   Supply Chain Measures

Goldratt favours two complementary measures for supply chain effectiveness – Throughput Dollar Days and Inventory Dollar Days.

Throughput Dollar Days (TDD) is a measure that accumulates the cost of inventory only below an agreed 'emergency' level. The idea is that the potential of lost sales is tracked, so this measure focuses attention both on the item and the response time for replenishment. To set the emergency agreed stock level, consultation needs to take place between vendor and buyer. The emergency level is the level below which there is a high probability of losing sales. This is normally well below the safety stock level used in inventory control. If the inventory level falls below this emergency level the measure starts ticking, and accumulates the full throughput value day by day. Throughput is revenue minus direct variable costs. The measure accumulates every day the inventory is below target. So a shortage for 5 days is 5 times greater than the same shortage for one day. High throughput items, of course, attract higher penalties. There is therefore a motivation on the part of the vendor to focus on minimising delays

for valuable items, and on installing the appropriate capacity. Where an item is a component of an assembly, throughput is defined as the revenue of the full end item minus the direct variable costs.

Inventory Dollar Days (IDD) is a measure that tracks flow through the supply chain. It measures both the value of the item and the length of time that it remains in the supply chain. This measure is taken periodically. Where the measure is used in a chain, the time is measured from manufacture to selling. Internally, the measure should be used for raw material, work in process, and finished goods, separately. The claim is that this measure is far more effective than the conventional inventory turn measure. The problems with the inventory turn measure is that it is an overall average figure (some items may have very poor turns, other fairly good), and that it ignores the value of the inventory items.

The measures are complementary. TDD encourages the right minimum inventory to be held. IDD limits overproduction. Both have a time dimension. In a supply chain the measures would be communicated along the chain. The thought is that these measures send out the right messages for supply chain cooperation.

17.3.5   Checking out Measures

Andy Neely and Mike Bourne of Cranfield University and London Business School suggest a framework for checking out each measure, similar to Kipling's Six Honest Serving Men.

Measure: a self-explanatory title

Purpose: why is it being measured? To which business objective does this measure relate?

Target: what is to be achieved, and by when?

Formula: the formula or ratio used.

Frequency: how often should the measure be taken, and reviewed?

Who measures? Who is responsible for collection and reporting?

Source of data: where does it come from?

Who acts? Who is responsible for taking action?

What to do? What action should be taken?

And, you could add:

Limits – What are the control limits within which no action is required?

Further Reading

M Bourne, A Neely, K Platts, J Mills (2002) The success and failure of performance measurement initiatives, *International Journal of Operations & Production Management*, Vol.22 No.11, p.1288-1310

17.4 Target Costing, Kaizen Costing and Cost Down

This final section brings together many of the tools presented in earlier sections. The concept of Target Costing is well established in Lean. The idea is simply that pricing begins with the market.

Target cost = Market price - Target Profit.

So, instead of the price being derived from cost plus profit, the cost is derived from market factors. Target costing is done in anticipation of future demand. In fact, the price may create the demand. Target costing begins with the customer's needs. A customer may in fact want to buy holes not drills, or 'power by the hour', not an aircraft engine.

It is proactive, not reactive. It is a tough system, because there can be no compromise on the target cost. There are variations, for example in the aircraft industry and in Formula 1 there is the target weight.

According to Cooper and Slagmulder, target costing has the cardinal rule, 'The target cost of a product can never be exceeded'. Unless this rule is in place a target costing system will lose its effectiveness and will always be subject to the temptation of adding just a little bit more

functionality at a little higher price. There are three strands to target costing: allowable cost, product level target cost, and component target costs.

Much of the following material on the three strands is derived from Cooper and Slagmulder.

Allowable cost is the maximum cost at which a product must be made in order to earn its target profit margin. The allowable cost is derived from target selling price - target profit margin. Target selling price is determined from three factors: customers, competitive offerings, and strategic objectives. The price customers can be expected to pay depends significantly upon their perception of value. So if a new product or variant is proposed, marketing must determine if and how much customers are prepared to pay for the new features. The position on the product life cycle is important. An innovative lead product may be able to command a higher price.

Customer loyalty and brand name are influential. Then there are the competitive offerings: what functions are being, and are anticipated to be, offered at what prices.

Target Costing and Cost Down

Finally there are strategic considerations about, for example, whether the product is to compete in a new market, and the importance of market share.

Target profit margin is the next factor in determining allowable cost. There are two approaches, according to Cooper and Slagmulder. The first uses the predecessor product and adjusts for market conditions. The second starts with the margin of the whole product line, and makes adjustments according to market conditions.

Product Level Target Costing begins with the Allowable Cost and challenges the designers to design a product with the required functionality at the allowable cost. Sometimes the design team will not know the real allowable cost, but will be set a target which is considered to be a difficult-to-achieve challenge, for motivational reasons. A useful concept is the Waste Free Cost. This concept, also found in value engineering, is the cost assuming that all avoidable waste has been taken out. Another guiding principle is the 'cardinal rule' that cost must not be allowed to creep up: if an extra function is added, there must be a compensating cost reduction elsewhere. The process of moving in increments from the current cost to the target cost is referred to as 'drifting' and is closely monitored. Once the target cost has been achieved, effort stops: there is no virtue in achieving more than is required.

Component target costing aims at setting the costs of each component. This is an important strategic consideration because it involves the question of supplier partnership and trust.

The above figure shows a hierarchy of approaches and tools. There are three routes to addressing component and product target costs. The first is through market-price tradeoffs, involving negotiations between designers and marketers and between OEM and suppliers, on the sensitivity of price, functionality and quality. Core tools here are the Kano model, QFD, increasingly design for Six Sigma, and centrally, value engineering. The design concept of the four objectives and six tradeoffs is also important. All of these are discussed in separate sections.

The second area is inter-organisational development. This involves working with supplier partners to achieve cost down. The various approaches used by Toyota and others were discussed in the Improvement section, and in the Supplier Partnership section - see particularly supplier association and purchasing association. Chained target costing extends this pressure, or cooperation, further upstream along the supply chain.

Each company along the chain is expected (or forced to?) participate by a level-by-level process. Audits play a part here - for example Ford uses its FPS audit tool to assess suppliers and uses activity sampling to identify the extent of cost down opportunity. (See sections on these tools). Toyota uses their supplier support centre.

When waste is identified supplier companies are either helped to remove it or expected to remove it. Ford uses a confidential costing system, called Lean to Cost, to translate the identified waste into money terms.

The third area is Concurrent Cost Management. This can take place both internally or with immediate suppliers. The idea is to address costs at each level, from concept to production. The stages used to be addressed sequentially, but are now increasingly being done concurrently. Concurrent engineering ideas are used. Of particular note is the Toyota 'set based' methodology which gradually homes in on the specifications whilst allowing flexibility and innovation until quite late into the process. Ramp up is an important stage aimed at reducing problems before full scale production of the new product begins. Once the product goes into production, three other actions may follow:

1. Further variants may be launched from the base product or platform at strategic intervals, to maintain competitiveness by adding functionality or passing on advances in technology, or passing on price reductions.

2. Further value engineering (sometimes referred to as value analysis after the initial launch) may take place at regular intervals.

One Japanese company aims to do a value analysis on each of their continuing consumer electronics products once per year. The aim is either to reduce cost or to improve functionality.

3. 'Kaizen costing' is undertaken. Kaizen costing is the post-launch version of target costing, and aims to achieve target cost levels at specific points. Kaizen costing is not really costing in the conventional western sense. The western way is to track variances. How feeble! The kaizen way is to target productivity improvements in people, materials, methods, and machines - as identified by audits, benchmarks, waste analysis, mapping, and activity sampling - in specific periods of time. What the paperwork says about the variances does not matter - what matters is the real tangible improvements in productivity.

Kaizen costing targets three areas: the method or facilities, the product, the overheads. Method is targeted by both the Policy Deployment process focusing on cost down initiatives, level by level, and by local initiatives carried out by the team with or without help from Lean Promotion Office or OEM staff. A typical improvement would be a cell re-balance as explained in the Layout and Cells section of this book. The product is targeted by value engineering. Overheads are targeted using Information value stream mapping and Brown paper mapping.

Further Reading

Robin Cooper and Regine Slagmulder, *Target Costing and Value Engineering*, Institute of Management Accountants / Productivity Press, 1997

Robin Cooper and Regine Slagmulder, *Supply Chain Development for the Lean Enterprise*, Institute of Management Accountants / Productivity Press, 1999

Robert Kaplan and Robin Cooper, *Cost and Effect*, Harvard Business School Press, 1998

Shahid Ansari et al, *Target Costing*, Irwin McGraw Hill, 1997

Brian Maskell and Bruce Baggaley, *Practical Lean Accounting*, Productivity, 2004

Jean Cunningham and Orest Fiume, *Real Numbers*, Managing Times Press, 2003

Joe Stenzel (ed.), *Lean Accounting: Best Practices for Sustainable Integration*, Wiley, 2007

Robin Cooper and Brian Maskell, 'How to Manage Through Worse-before-Better', *MIT Sloan Management Review*, Summer 2008

18 Lean – How it all came about

The Lean Toolbox has traditionally been aimed at practitioners actually moving towards Lean either as change agent or as team members. However, increasingly Lean has attracted students studying the evolution and expansion of Lean into other contexts, such as services and health care operations. The great success of Lean has not just created a great following in practice, but has also sparked a lot of interest in academia, and we will outline in more detail the history of academic research on Lean, as well as provide the main academic references for those wishing to engage in research on Lean.

18.1 Lean before Toyota

Where did Toyota's unique culture come from? Some organisation theorists like Warren Bennis and David Snowden believe that the founder often has huge and sustaining influence. Apparently, Sakichi Toyoda (1867-1930) the founder of Toyota, was a great fan of Samuel Smiles' book *Self-Help*, originally published in 1859. This book is the only book on display at Sakichi Toyoda's birthplace, the shrine of Toyota. Sakichi Toyoda schooled his family, including Kiichiro Toyoda (1894-1952), founder of Toyota Motor.

Smiles' bestselling book is still in print, and was probably the first self-help book written. It tells of the great innovators of the Industrial revolution, such as Watt, Davy, Faraday, Stephenson, Brunel, and Wedgewood, artists such as Reynolds and Hogarth, writers such as Shakespeare, and soldiers such as Wellington and Napoleon. The majority of them beavered away through hard work, often with little technical education but great practical experience, often over considerable periods, with patience and continual experimentation, to realise their goals. And their goals were firmly linked to the needs of customers. They were, in general, good businessmen although their primary motivation was not the accumulation of wealth. Some were

Quakers who believed in a fair deal for their workers and a fair but not excessive profit over the longer term. 'Attention (to detail), application, method, perseverance, punctuality, despatch are the principal qualities required...', 'Accuracy in observation...', (Newton and Darwin were astute observers.) 'Method is like packing things in a box; a good packer will get in half as much again as a bad one....' and 'the shortest way to do many things is to do only one thing at once'. Most worked with 'constant modification and improvement, until eventually it was rendered practical and profitable to an eminent degree.' And 'the highest patriotism and philanthropy consist, not so much in altering laws and modifying institutions as in helping and stimulating men to elevate and improve themselves by their own free and independent individual action.'

Does that sound like the Toyota we often hear about today? See, for example, Liker's *Toyota Way* principles numbers 1, 9, 10, 11, 12, 13, 14. Respect for people, Gemba, Kaizen, Observation – it's all there – but not necessarily in those words.

To reinforce these ideas Terence Kealey has written about how many great innovations in history have come about not through science driving technology, but technology driving advances in science, through hands-on application at the workplace. 'Were the increases in productivity primarily a consequence of the great technical advances such as the spinning jenny or of the myriad small technical advances that innumerable workers and manufacturers made to their machines alongside the big advances? Romantically, we attribute the increases in industrial revolutionary productivity to the great individual innovations such as the jenny, but when the economists do their sums they show that the vast number of small technical improvements overwhelmed the impact of big innovations.' (p169)

References

Samuel Smiles, *Self-Help*, Oxford World's Classics, 2002 (originally published 1859)

Terence Kealey, *Sex, Science and Profits*, Heinemann, 2008

18.2 Toyota: the Birthplace of Lean

The foundation of the Toyota Motor Company dates back to 1918, when the entrepreneur Sakichi Toyoda established his spinning and weaving business based on his advanced automatic loom. He sold the patents to the Platts Brothers in 1929 for £100,000, and it is said that these funds provided the foundation for his son, Kiichiro, to realise his vision of manufacturing automobiles. The tale goes that Sakichi told his son on his deathbed: 'I served our country with the loom. I want you to serve it with the automobile'. At the time the Japanese market was dominated by the local subsidiaries of Ford and General Motors (GM), starting Toyoda's automotive business was fraught with financial difficulties and ownership struggles after Sakichi's death in 1930. Nevertheless, Kiichiro prevailed and began designing his Model AA -- by making considerable use of Ford and GM components! The company was relabelled 'Toyota' to simplify the pronunciation and give it an auspicious meaning in Japanese. Truck and car production started in 1935 and 1936, respectively, and in 1937 the Toyota Motor Company was formally formed. World War II disrupted production, and the post-war economic hardship resulted in growing inventories of unsold cars, leading to financial difficulties at Toyota, leading to the resignation of Kiichiro from the company.

His cousin Eiji Toyoda became managing director – with considerable irony retrospectively – and was sent to the United States in 1950 to study American manufacturing methods. Going abroad to study competitors was not unusual; pre-war a Toyota delegation had visited the Focke-Wulff aircraft works in Germany, where they observed the 'Produktionstakt' concept, which later developed into what we now know as 'takt time'. Eiji Toyoda was determined to implement mass production techniques at Toyota, yet capital constraints and the low volumes in the Japanese market did not justify the large batch sizes common at Ford and GM. Toyota's first plant in Kariya was thus used for both prototype development and production, and had a capacity of 150 units per month.

While the simple and flexible equipment that Kiichiro had purchased in the 1930s would enable many of the concepts essential to TPS, the individual that gave the crucial impulse towards developing the Toyota Production System (TPS) capable of economically producing large variety in small volumes, was Taiichi Ohno (Ōno Taiichi). Ohno had joined Toyoda Spinning and Weaving in 1932 after graduating as mechanical engineer, and only in 1943 joined the automotive business after the weaving and spinning business had been dissolved. Ohno did not have any experience in manufacturing automobiles, but brought a 'common-sense approach' without any preconceptions that has been instrumental in developing the fundamentally different Just-in-Time philosophy. Analysing the Western production systems, he argued that they had two logical flaws. First, he reasoned that producing components in large batches resulted in large inventories, which took up costly capital and warehouse space and resulted in a high number of defects. The second flaw was the inability to accommodate consumer preferences for product diversity. Ohno believed that GM had not abandoned Ford's mass production system, since the objective was still to use standard components enabling large batch sizes, thus minimising changeovers. In his view, the management of Western vehicle manufacturers were (and arguably still are) striving for large scale production and economies of scale.

From 1948 onwards, Ohno gradually extended his concept of small-lot production throughout Toyota from the engine machining shop he was managing. His main focus was to reduce cost by eliminating waste, a notion that developed out of his experience with the automatic loom that stopped once the thread broke, in order not to waste any material or machine time. He referred to the loom as 'a text book in front of my eyes', and this 'jidoka' or 'autonomous machine' concept would become an integral part of the

Toyota Production System. Ohno also visited the U.S. automobile factories in 1956, and incorporated ideas he developed during these visits, most notably the 'kanban supermarket' to control material replenishment. In his book, Ohno describes the two pillars of TPS as autonomation, based on Sakichi's loom, and JIT, which he claims came from Kiichiro who once stated that 'in a comprehensive industry such as automobile manufacturing, the best way to work would be to have all the parts for assembly at the side of the line just in time for their user'. In order for this system to work, it was necessary to produce and receive components and parts in small lot sizes, which was uneconomical according to traditional thinking. Ohno had to modify the machine changeover procedures to produce a growing variety in smaller lot sizes. This was helped by the fact that the much of the machinery Kiichiro had bought was simple, general purpose equipment that was easy to modify and adapt. Change-over reduction was further advanced by Shigeo Shingo, who was hired as external consultant in 1955 and developed the SMED (single-minute exchange of dies) system.

The result was an ability to produce a considerable variety of automobiles in comparatively low volumes at a competitive cost, altering the conventional logic of mass production. In retrospect these changes were revolutionary, yet they were largely necessary adaptations to the economic circumstances at the time, which required low volumes and great variety. By 1950, the entire Japanese auto industry was producing an annual output equivalent to less than three days' of the U.S. car production at the time. Toyota gradually found ways to combine the advantages of small-lot production with economies of scale in manufacturing and procurement. Thus, more than anything, it is this 'dynamic learning capability' that is at the heart of the success of TPS. As Fujimoto in his book on the evolution of TPS concludes:

'Toyota's production organization [..] adopted various elements of the Ford system selectively and in unbundled forms, and hybridized them with their ingenious system and original ideas. It also learnt from experiences with other industries (e.g. textiles). It is thus a myth that the Toyota Production System was a pure invention of genius Japanese automobile practitioners. However, we should not underestimate the entrepreneurial imagination of Toyota's production managers (e.g. Kiichiro Toyoda, Taiichi Ohno, and Eiji Toyoda), who integrated elements of the Ford system in a domestic environment quite different from that of the United States. Thus, the Toyota-style system has been neither purely original not totally imitative. It is essentially a hybrid'.

Astonishingly, TPS went largely unnoticed by the West – albeit not kept as a secret – and according to Ohno only started attracting attention during the first oil crisis in 1973, when Japanese imports threatened Western manufacturers.

References

Fujimoto, T. 1999. *The Evolution of a Manufacturing System at Toyota*, Oxford University Press, 1999

Ohno, T. 1988. *Toyota Production System: beyond large-scale production*, Productivity Press, 1988

18.3 Why do we call it 'Lean'?

The first paper on TPS in English appeared in 1979. It was not published by academics, but by four managers of Toyota's Production Control department – including Fujio Cho, who in 1999 became president of the Toyota Motor Corporation. The Western world took notice. In 1979, the 'Repetitive Manufacturing Group (RMG)' was established to study TPS under the sponsorship of the American Production and Inventory Control Society (APICS). The group held a meeting at Kawasaki's motorcycle plant in Lincoln, Nebraska, in June 1981 and exposed participants to Kawasaki's well-developed JIT system, a clone of Toyota's system. The group included Richard Schonberger and Robert Hall who, based on their experiences, published their books on JIT. In parallel, Yasuhiro Monden of

Tsukuba University published his book on TPS in 1983. Up until this point, the debate in the Western world was largely based around shop-floor techniques, commonly referred to as 'JIT' or 'zero inventory' production.

The next step towards describing the Lean philosophy came with the International Motor Vehicle Program (IMVP) at MIT. The programme was based at MIT, but from the start the idea was to create an international network of faculty at other universities, with Dan Jones as UK team leader, Jim Womack, as research manager, and Dan Roos as programme director.

The programme was geared towards indentifying what drove the Japanese competitive advantage. At the time a range of explanations were given. The most common explanations (and with hindsight, misperceptions) were:

1. **Cost advantage** – Japan was seen to have lower wage rates, a favourable Yen/Dollar exchange rate and lower cost of capital, elements that combine to an 'unfair playing field'.

2. **Luck** - Japan had fuel-efficient cars when the energy crisis came, or it was simply a fortunate effect of the 'business life cycle issue'.

3. **'Japan, Inc.'** – MITI, Japan's Ministry of International Trade and Industry, was suspected of orchestrating a large-scale industrial policy.

4. **Culture** – Cultural differences in Japan allowed for more efficient production, which cannot be replicated in other countries.

5. **Technology** – The use of advanced automation in Japanese factories; 'It was all done with advanced robotics.' Some even suggested that the Japanese were acquiring Western technology, which they then exploited.

6. **Government Policy** – Trade barriers against the U.S., more lenient labour laws in Japan, and a national health care program lowered the overall labour cost.

The IMVP sponsor companies encouraged the research team to look into the issue of why Japan was getting ahead. The research remit, according to Dan Jones, was not only to describe the gap between the western world and Japan, but also 'to measure the size of the gap'. A key challenge was to normalise the labour input that varies greatly by vehicle size and option content, as well as by the degree of vertical integration, i.e. to what extent the manufacturer produced components in house, or buys them in from suppliers. So, while there was a good understanding of the differences in manufacturing practices across regions, the way of executing a valid comparison was far less defined. As Dan Jones remarked, '[..] we had a method, but we did not have a methodology.'

The initial design of the benchmarking methodology was developed by Womack and Jones during 1985/86, and was tested at Renault's Flins plant in 1986. In May that year, John Krafcik went to see Jim Womack to discuss potential research opportunities if he were to enrol at MIT. Krafcik was the first American engineer to be hired by NUMMI, and joined MIT as an MBA student. By summer 1986 Womack and Krafcik formally started the assembly plant study by visiting GM's Framingham assembly plant in Massachusetts.

Another MIT student, John Paul MacDuffie, also became involved in the programme at the time. MacDuffie was working as research assistant to Haruo Shimada from Keio University (a visiting professor at the Sloan School), who was interested in the Japanese transplants in the U.S., trying to understand how well they were able to transfer the Japanese human resource and production systems. Shimada was one of the first researchers allowed to visit and conduct interviews at the new transplants of Honda, Nissan, Mazda, and NUMMI. Shimada used a benchmarking index according to which he classified companies on the spectrum from 'fragile' to 'robust' or 'buffered'. This terminology was initially used by IMVP researchers, but 'fragile' later amended to 'Lean', which was seen to have a more positive connotation. The term

'Lean production' was first used by Krafcik in 1988, and subsequently, Womack et al. of course used the term 'Lean production' to contrast Toyota with the Western 'mass production' system in the 'Machine' book. The name 'Lean' was born!

Further Reading

Holweg, M., 2007. The Genealogy of Lean production. *Journal of Operations Management* 25 (2), 420-437

Krafcik, J. 1988. The Triumph of the Lean Production System. *Sloan Management Review* (Fall), 41-52

Sugimori, Y., K. Kusunoki, K., Cho, F., Uchikawa, S. 1977. Toyota Production System and Kanban System; Materialization of Just-in-Time and Respect-for-Human System. *International Journal of Production Research*, 15 (6), 553–564

Womack, J.P., Jones, D.T., Roos, D. 1990. *The Machine that Changed the World*, HarperCollins, New York

See also 'Bicheno's Top 100 Books on Lean', which is available at www.leanenterprise.org.uk .

19 Further Resources – Where to get help

19.1 Companion Volumes

A range of related books and games is available from PICSIE Books, please see www.picsie.co.uk and www.amazon.co.uk or www.amazon.com.

The Lean Toolbox for Service Systems, by John Bicheno (Paperback – 2008)

Fishbone Flow: Integrating Lean, Six Sigma, TPM and TRIZ, (using fishbone diagrams), by John Bicheno (Spiral-bound - 2006)

Six Sigma and the Quality Toolbox: for Service and Manufacturing, by John Bicheno and Philip Catherwood (Paperback - 2005)

In addition, several class-room games are available:

1. **The Buckingham Lean Game.** An ideal game to learn Lean manufacturing. Unlike other games this game includes several products, changeover, quality and right size machines. It can be adapted to multiple environments.
2. **The Buckingham Supply Chain Game and LEAP Game.** These two bundled games are excellent simulations of both the distribution chain and the supply chain. The LEAP Game was developed at Cardiff University.
3. **The Buckingham Heijunka Game.** This more advanced game illustrates how to transform a current state into a future state using Heijunka principles. An extensive set of Powerpoint slides on Lean scheduling and mapping concepts relating to the game (including batch sizing, mixed model, pitch, supermarkets, types of kanban, etc) is included.

4. **The Buckingham Service Operations Game.** Built around an actual situation, this game several Lean service concepts including service mapping, failure demand, 5S in service, and customer interactions.

19.2 Research Centres, Research Programmes and Web Resources

- Association for Manufacturing Excellence, AME http://www.ame.org/
- Lean Enterprise Research Centre, Cardiff Business School, UK (where there are free downloads) at www. leanenterprise.org.uk
- APICS – The Association for Operations Management, www.apics.org
- Centre for Process Excellence and Innovation, University of Cambridge: http://www-innovation.jbs.cam.ac.uk
- Manufacturing Management Research Center, Tokyo University http://www.ut-mmrc.jp/e_index.html
- Lean Aerospace Initiative at MIT: http://web.mit.edu/lean/
- Lean Blog: www.leanblog.org
- Lean Enterprise Academy, UK (LEA): www.leanuk.org,
- Lean Enterprise Institute (based in Boston, USA): LEI: www.Lean.org,
- University of Kentucky's Lean Center: http://www.mfg.uky.edu/
- The International Motor Vehicle Program (IMVP) website features many working papers for download – http://imvp.mit.edu

19.3 Articles, Books and Videos

The papers we mention in this book can be purchased and downloaded individually.

- *Harvard Business Review* articles and case studies are available at http://harvardbusinessonline.hbsp.harvard.edu

- For papers from the *MIT Sloan Management Review* see http://sloanreview.mit.edu/smr/
- Other academic papers can be purchased individually from Internet libraries. For papers from the *Journal of Operations Management* see www.sciencedirect.com, or for the *International Journal of Operations and Production Management*: www.emeraldinsight.com. Also check http://scholar.google.com, as many papers are available online for free (in a working paper format)
- The Association for Manufacturing Excellence (AME) has a range of videos / DVDs. See www.ame.org. Likewise the Society for Manufacturing Engineers (SME) has an ever-expanding set of DVD's on Lean topics. See www.sme.org.

19.4 Certification

AME and SME each have a Lean Certification program at Bronze, Silver and Gold levels. An examination must be taken, and a diary of relevant work submitted. The Gold level requires a professional interview. At the time of writing this is in North America only, but is coming to the UK, Australia and elsewhere.

The Lean Enterprise Research Centre, Cardiff Business School, UK began the first Masters degree in Lean Operations in the world in 1998. In 2006 a Masters degree in 'Lean in Service' began. These are part-time executive programs taught largely on-site at participating companies.

The Lean Enterprise Research Centre has also established a 'Lean Learning Ladder' to give recognition to organisations having their own internal development programs.

Index

Topics

People

Lightning Source UK Ltd.
Milton Keynes UK
21 January 2011
166070UK00001B/2/P